D1485366

ANDERSONIAN LIBRARY
WITHDRAWN
FROM
LIBRARY
STOCK
UNIVERSITY OF STRATHCLYDE

Methods in RECEPTOR RESEARCH

Part I

METHODS IN MOLECULAR BIOLOGY

Edited by

ALLEN I. LASKIN
ESSO Research and Engineering Company
Linden, New Jersey

JEROLD A. LAST
Harvard University
Cambridge, Massachusetts

VOLUME 1: Protein Biosynthesis in Bacterial Systems, *edited by Jerold A. Last and Allen I. Laskin*

VOLUME 2: Protein Biosynthesis in Nonbacterial Systems, *edited by Jerold A. Last and Allen I. Laskin*

VOLUME 3: Methods in Cyclic Nucleotide Research, *edited by Mark Chasin*

VOLUME 4: Nucleic Acid Biosynthesis, *edited by Allen I. Laskin and Jerold A. Last*

VOLUME 5: Subcellular Particles, Structures, and Organelles, *edited by Allen I. Laskin and Jerold A. Last*

VOLUME 6: Immune Response at the Cellular Level, *edited by Theodore P. Zacharia*

VOLUME 7: DNA Replication *edited by Reed B. Wickner*

VOLUME 8: Eukaryotes at the Subcellular Level, *edited by Jerold A. Last*

VOLUME 9: Methods in Receptor Research (in two parts) *edited by Melvin Blecher*

Methods in RECEPTOR RESEARCH

(in two parts)

Part I

Edited by Melvin Blecher

Department of Biochemistry
Georgetown University Medical Center
Washington, D.C.

MARCEL DEKKER, INC. New York and Basel

MARCEL DEKKER, INC.

270 Madison Avenue, New York, New York 10016

LIBRARY OF CONGRESS CATALOG CARD NUMBER: 75-40846

ISBN: 0-8247-6414-5

Current printing (last digit):
10 9 8 7 6 5 4 3 2 1

PRINTED IN THE UNITED STATES OF AMERICA

To Jocelyn

PREFACE

The Editor need not remind the reader of the currency of receptor research; the burgeoning literature is an ample reminder. The reader will also have frequently encountered the frustration arising from sketchy methods often provided in journal articles; hence, the popularity of volumes dealing exclusively with methods. This volume, therefore, is an effort to collect in laboratory manual form a broad spectrum of the detailed methods employed currently by many of the world's foremost receptor research laboratories.

We have elected to exclude from this volume methods involving receptors for steroid and thyroid hormones since we feel that such methods, having much in common with each other but little in common with those used with other hormones, warrant separate treatment in another volume. Therefore we will deal here exclusively with receptors for polypeptide hormones, prostaglandins, catecholamines, opiates, acetylcholine, lectins, and cholera toxins.

The reader will note a certain amount of seeming redundancy and overlap among the chapter titles and methods employed. Thus, more than one chapter is devoted to several of the hormones (insulin, glucagon, prostaglandins and gonadotropins), and many authors use the chloramine-T method for preparing radioiodinated polypeptides. The Editor feels, however, that the redundancy and overlap are more apparent than real. In these instances, different laboratories will have different experimental approaches and will utilize methods and variations unique to their system, even though studying the same ligand.

Finally, while the experimental approaches, laboratory methods, and data-handling techniques described herein are given in terms of specific ligands and receptor macromolecules, many of the chapters provide information which has general applicability. To cite two examples, Catt et al. in Chapter 9 describe general methods for the determination of binding constants and concentrations of receptors, while De Meyts in Chapter 11 deals extensively with general methods for determining the physicochemical parameters of steady-state binding and for studying cooperative interactions among receptor sites.

Melvin Blecher

CONTRIBUTORS TO PART I

VANN BENNETT,* Department of Pharmacology and Experimental Therapeutics, The Johns Hopkins University School of Medicine, Baltimore, Maryland

MELVIN BLECHER, Department of Biochemistry, Georgetown University Medical Center, Washington, D. C.

KEVIN J. CATT, Section on Hormonal Regulation, Reproduction Research Branch, National Institute of Child Health and Human Development, National Institutes of Health, Bethesda, Maryland

PEDRO CUATRECASAS,† Department of Pharmacology and Experimental Therapeutics, The Johns Hopkins University School of Medicine, Baltimore, Maryland

PIERRE De MEYTS, Diabetes Branch, National Institutes of Arthritis, Metabolism and Digestive Diseases, National Institutes of Health, Bethesda, Maryland

MARIA L. DUFAU, Section on Hormonal Regulation, Reproduction Research Branch, National Institute of Child Health and Human Development, National Institutes of Health, Bethesda, Maryland

FRANCES M. FINN, Protein Research Laboratory, University of Pittsburgh School of Medicine, Pittsburgh, Pennsylvania

MARY A. FLETCHER, Divisions of Endocrinology, Metabolism, and Immunology, Department of Medicine, University of Miami School of Medicine, Miami, Florida

STEVEN GOLDSTEIN, Department of Biochemistry, Georgetown University Medical Center, Washington, D. C.

---

*Current affiliation: The Biological Laboratories, Harvard University, Cambridge, Massachusetts

†Current affiliation: The Wellcome Research Laboratories, Burroughs Wellcome Company, Research Triangle Park, North Carolina

KLAUS HOFMANN, Protein Research Laboratory, University of Pittsburgh School of Medicine, Pittsburgh, Pennsylvania

ARTHUR KARLIN, Department of Neurology, College of Physicians and Surgeons, Columbia University, New York, New York

JEAN-MARIE KETELSLEGERS, Section on Hormonal Regulation, Reproduction Research Branch, National Institute of Child Health and Human Development, National Institutes of Health, Bethesda, Maryland

ROBERT J. LEFKOWITZ, Departments of Medicine and Biochemistry, Duke University Medical Center, Durham, North Carolina

GERALD S. LEVEY, Divisions of Endocrinology, Metabolism, and Immunology, Department of Medicine, University of Miami School of Medicine, Miami, Florida

MARK G. McNAMEE, Department of Neurology, College of Physicians and Surgeons, Columbia University, New York, New York

STEPHEN L. POHL,* Department of Internal Medicine, Division of Endocrinology, New England Medical Center Hospital, Boston, Massachusetts

S. RAMACHANDRAN, Applied Science Laboratories, Inc., Department of Production, Research, and New Products, State College, Pennsylvania

LEO E. REICHERT, Jr., Department of Biochemistry, Division of Basic Health Sciences, Emory University, Atlanta, Georgia

BRIJ B. SAXENA, Cornell University Medical College, New York, New York

RAMON VALDERRAMA, Department of Neurology, College of Physicians and Surgeons, Columbia University, New York, New York

CHERYL L. WEILL, Department of Neurology, College of Physicians and Surgeons, Columbia University, New York, New York

---

*Current affiliation: Endocrinology Division, Department of Internal Medicine, University of Virginia School of Medicine, Charlottesville, Virginia

CONTENTS

Cumulative Indexes appear in Part II

CONTENTS OF PART II

Methods in RECEPTOR RESEARCH

Part I

Chapter 1

METHODS OF ISOLATION AND CHARACTERIZATION OF THE ACETYLCHOLINE RECEPTOR

Arthur Karlin
Mark G. McNamee
Cheryl L. Weill
Ramon Valderrama

Department of Neurology
College of Physicians and Surgeons
Columbia University
New York, New York

I. INTRODUCTION

The acetylcholine receptor transduces the binding of acetylcholine into a change in the ionic permeability of the synaptic membrane. A complete molecular description of this phenomenon is the goal of much current research. Such a description would include the chemical and physical structure of the receptor and its disposition in the membrane; and the nature of its interactions with specific ligands such as acetylcholine, with permeating cations such as Na^+ and K^+, and with other membrane components. Methods for assay and purification of the receptor have been developed, and some facets of the structure and interactions of the receptor have been brought to light (see Refs. 1 and 2 for reviews). We will review the methodology underlying the recent progress in acetylcholine receptor research, stressing procedures with which we have had direct experience.

II. PROBING RECEPTORS IN INTACT CELLS

A. General Approaches

It is possible to order receptor activators (acetylcholine and congeners) in terms of the concentration eliciting a half-maximal response (the apparent dissociation constant of the activator-receptor complex), and to order competitive inhibitors [(+)-tubocurarine and congeners] in terms of their apparent dissociation constants. Much work of this kind has been carried out with frog skeletal muscle and with cells (electroplax) from the electric tissues of fish [1-3]. Activation of the receptors in postsynaptic membranes of these cells has been shown to lead to nearly equal increases in the permeability of the postsynaptic membrane to Na^+ and K^+. These are primary characteristics that one seeks to correlate with properties of the purified receptor, namely, the specific ordering of the binding of activators and inhibitors and the specific control of cation permeability by receptor in reconstituted membrane.

The characterization of receptor in cells has been refined by the introduction of the analysis of endplate "noise." This analysis leads to estimates of the conductance and the duration of individual open receptor channels and of the dependence of these parameters on the nature of the activator [4-8]. Of great importance has been the introduction of new agents, not among the classical agonists and antagonists, which specifically react with the receptor, namely, the α-neurotoxins of snake venoms [9] and certain affinity labels [10]. The use of these agents has led not only to basic new information about the receptor in situ, but has made possible much of the progress in the isolation and characterization of the receptor. Much of the work with these agents directly relevant to receptor isolation was done on the intact electroplax of the electric eel (*Electrophorus electricus*) [e.g., 10, 11]; and a brief description of the techniques employed follows.

B. Electroplax

1. Physiology

The electroplax of the electric tissue of *Electrophorus* is a large, flat cell, whose physiological function, the generation of transcellular current, is a consequence of its asymmetry [12, 13]. The membrane on one side of the cell is innervated and electrically and chemically excitable, while that on the other side receives no innervation and is not excitable. Following neural triggering by way of a depolarizing postsynaptic potential, the innervated membrane generates an action potential of about 150 mV. The potential difference across the noninnervated membrane does not change, and, therefore, the potential change across the whole cell follows the action potential. A series of several thousand cells firing simultaneously can thereby generate a considerable electromotive force.

The electroplax in the organ of Sachs, the most posterior electric organ, are particularly large and widely spaced. These are dissected following the procedure of Schoffeniels [14]. A transverse slice (~3 cm wide) of the tail through the organ of Sachs is taken, and a row of electroplax is dissected out and mounted in a dish of eel ringers solution [165 mM NaCl, 2.3 mM KCl, 1.2 mM K_2HPO_4, 0.3 mM KH_2PO_4, 2 mM $CaCl_2$, 2 mM $MgCl_2$, 10 mM glucose (pH 7.1)]. The dissection is performed under a dissecting microscope (focal length 200 mm) at an overall magnification in the range of 16 to 40X. Each cell is contained within a compartment composed of sheets of connective tissue. With practice, these can be largely removed without damaging the cell. The cell, freed of its sheath, is mounted in a holder such that it separates two bathing solutions (see Ref. 10 for details). The cell is impaled from the non-innervated side with a glass microelectrode (5-10 MΩ); and the potential differences across the cell and across either of the two membranes can be recorded simultaneously. In the simplest case, which suffices for most applications, the response of the cell is taken as the decrease in the potential difference across the innervated membrane after the application in the solution bathing this membrane of an activator for a fixed period of time, say 80 sec. On the time scale of these measurements, there is no apparent desensitization of the depolarizing response. Replacement of the solution of activator with eel Ringers solution leads to a rapid repolarization of the membrane potential.

The graded depolarizing response to acetylcholine and congeners reflects the state of the receptor. It is also possible to monitor the state of two other systems, the action potential-generating system and the active Na^+- and K^+-transport system. An action potential may be elicited by a brief depolarizing pulse across the cell, picked up across the bath electrodes and displayed on an oscilloscope. The state of the Na^+-pump is indicated by the rate of recovery from a depolarization and by an overshoot in the recovery of a few millivolts above the resting potential [15]. Agents which affect only the activator-elicited depolarization—but not the resting potential, the action potential, or the recovery phase—may be judged to have physiological specificity. These simple techniques proved most convenient for the investigation of the effects of covalent reactions on the receptor response. The solution bathing the innervated membrane flows at 1 to 2 cm/sec. This maintains the concentration of all reactants in the solution constant and helps reduce the effects of unstirred layers close to the membrane.

2. Covalent Modification of the Response

It was found that the disulfide reducing agent, dithiothreitol (0.2-1 mM in eel Ringers solution, pH 8.0), applied for about 10 min to the innervated membrane, has profound effects on the specificity of the receptor: The responses to monoquaternary activators are diminished [16], the responses to some bisquaternary activators (e.g., decamethonium) are enhanced [10]; and the bisquaternary compound, hexamethonium, normally an inhibitor, acts as an activator of the reduced receptor [17]. Furthermore, all the changes due to reduction by dithiothreitol are reversed following a brief application of a mild oxidizing agent [e.g., 1 mM 5,5'-dithiobis-(2-nitrobenzoic acid) at pH 8.0] to the innervated membrane [10, 16, 18].

Neither dithiothreitol nor 5,5'-dithio(2-nitrobenzoic acid) have any effect on the resting potential, action potential, or repolarization.

A number of reagents which alkylate sulfhydryl groups were found to prevent the reversal by oxidizing agents of the effects of reduction [10, 16, 17]. Most importantly, maleimide derivatives bearing quaternary ammonium moieties with affinity for the acetylcholine binding site were found to prevent the reversal by oxidizing agents at concentrations three orders of magnitude lower than that of N-ethylmaleimide, which has no such affinity [10, 17]. Moreover, the rate of reaction of the quaternary ammonium maleimides is decreased in the presence of reversible receptor ligands such as hexamethonium. These quaternary ammonium maleimides conform to the criteria of affinity labels [19] of the <u>reduced</u> receptor. None of them appear to have any irreversible effects on the <u>unreduced</u> receptor. The interpretation of these results is that there is an easily reducible disulfide in the close vicinity of the acetylcholine-binding site on the receptor and that, following its reduction, one of the sulfhydryl groups formed is well-positioned to react with these affinity labels when bound to the site [10].

Two types of affinity labels were in fact found. One type reacts with the reduced receptor, preventing both reoxidization of the disulfide and activation of the receptor by reversible agonists. The most highly specific compound of this type tested is 4-(N-maleimido)benzyltrimethylammonium iodide (MBTA) [10]. A second type also reacts with the reduced receptor at the acetylcholine-binding site and prevents reoxidation of the disulfide; but also causes <u>activation</u> of the receptor, and membrane depolarization [10, 20]. This type is exemplified by bromoacetylcholine bromide (BAC) and by the <u>p</u>-nitrophenyl ester of <u>p</u>-carboxyphenyltrimethylammonium iodide (NPTMB).

III. ASSAY OF RECEPTORS BY AFFINITY LABELING

A. Synthesis of a Radioactive Affinity Label [21]

1. N'-(4-N,N-dimethylaminobenzyl)maleamic acid (I)

A freshly prepared ether solution of <u>p</u>-amino-N,N-dimethylbenzylamine [22], prepared by tin- and HCl-reduction of <u>p</u>-nitro-N,N-dimethylbenzylamine [23], was added dropwise to a vigorously stirred solution of 1.2 equivalents of maleic anhydride in ether. After 1 hr at room temperature, the solution was cooled in ice, filtered, and the precipitate washed with ether to remove excess maleic anhydride. The combined yield for the reduction and acylation was 93%, giving white plates of m.p. 208 to 209°C after recrystallization twice from 80% aqueous ethanol.

Analysis: Molecular ion found at m/e 248; calculated, 248. Infrared maxima at 1680 cm^{-1} due to carboxyl, at 1596 cm^{-1} due to amide carbonyl, and at 1540 cm^{-1} due to amide II band. Calculated for $C_{13}H_{16}O_3N_2$: C, 62.89; H, 6.50; N, 11.28. Found: C, 62.60; H, 6.68; N, 11.13.

2. 4-(N'-maleimido)-N,N-dimethylbenzylamine (II)

Cyclization of I to II was by the method of Cava et al. [24]. A pasty mixture of 0.5 g of I, 88 mg sodium acetate (anhydrous), and 1 ml of acetic anhydride was stirred over a steam bath for ~ 5 min until a yellow solution was formed. This was poured into ice water; the mixture extracted twice with methylene chloride, adjusted to neutrality with sodium bicarbonate; and extracted twice more with methylene chloride. The latter two extracts were combined and dried over magnesium sulfate. The solvent was removed under reduced pressure, and 380 mg of bright-yellow crystals (m.p. 86 to 87°C) were obtained from benzene-hexane solution.

Analysis: Molecular ion found at m/e 230; calculated 230. Infrared maximum (KBr) at 1710 cm^{-1} due to the carbonyls. Nuclear magnetic resonance peaks (CCl_4) δ 2.26 (singlet, 6H, $(CH_3)_2$-); 3.48 (singlet, 2H, -CH_2 N=); 6.86 (singlet, 2H, vinyl); 7.48 (AB quartet, 4H aromatic). II polymerizes on standing. It is conveniently removed from polymer by dissolving in carbon tetrachloride and filtering.

3. MBTA

One-half gram of II was dissolved in 5 ml methyl iodide and stirred overnight. MBTA was precipitated with ether and recrystallized from hot acetonitrile and benzene to yield 0.8 g of yellow crystals, m.p. 204 to 205°C.

Analysis: Molecular ion, minus methyl iodide, found at m/e 230; calculated, 230. Infrared maximum at 1710 cm^{-1} due to carbonyls. Nuclear magnetic resonance peak (D_2O) δ 3.63 (singlet, 9H, $(CH_3)_3$-); 4.75 (singlet, 2H, -CH_2 N-); 6.88 (singlet, 2H, vinyl); 7.68 (quartet, J = 9 Hz $\Delta\nu$ 18 Hz, 4H aromatic).

4. 4-(N-maleimido)benzyltri-[^{3}H]methylammonium iodide ([^{3}H]MBTA)

Into a vial containing 131 mg of II in 0.5 ml of methylene chloride was transferred 5.7 μl of tritiated methyl iodide (2.2 Ci/mmol, 13 mg) in 0.5 ml of ether on a vacuum line. The vial was sealed and stored at room temperature in the dark for six days. The vial was opened, 5 ml of ether added, and the sample was spun on a centrifuge and decanted. The precipitate was washed with ether four times. The resulting bright-yellow powder was dissolved in 0.8 ml of acetonitrile and 0.2 ml of methylene chloride. The nuclear magnetic resonance spectrum was identical to that of unlabeled MBTA. The solution was added dropwise to 10 ml of ether, with stirring. The precipitate was separated and dried in vacuo to yield 39 mg of [^{3}H] MBTA. This was dissolved in 100 ml of acetonitrile to yield a 90-μM solution, samples of which were sealed into glass vials and placed in liquid nitrogen for storage.

Solutions of [^{3}H] MBTA for labeling experiments were prepared as follows: A vial containing [^{3}H] MBTA in acetonitrile was removed from the liquid nitrogen and opened. The acetonitrile was removed under vacuum, and the residue dissolved in 0.1 mM HCl to make ~ 0.25 mM [^{3}H] MBTA. A sample of this was taken for determination of concentration, and the remainder was divided into screwtop vials, frozen in liquid nitrogen, stored at -20°C, and used over a period of two weeks. MBTA hydrolyzes to the maleamic acid derivate 4-(N-maleamido)benzyltrimethylammonium iodide (III). The concentration of MBTA and of III in the HCl solution was calculated from the absorbance at 224, 236.7, 260, and 290 nm; and the respective molar extinction coefficients for MBTA: 32,500, 15,900, 942, 451; and for III: 20,700, 15,900, 8990, 4930. The absorbance at the isosbestic point, 236.7 nm, yields the sum of MBTA and III. The absorbance at the remaining three wavelengths were used pairwise to calculate the individual concentrations of MBTA and III; and the results, in all cases agreeing closely, were averaged. The concentration of MBTA in dilutions of the stock solution was determined taking into account the following rates of hydrolysis of MBTA: at pH 4, 0.004/day at -20°C, 0.01/hr at 0°C, and 0.08/hr at 23°C; and at pH 7.1, 0.016/min at 25°C.

B. Labeling Receptor with [^{3}H] MBTA

1. General Considerations

MBTA was inferred from physiological experiments to react with a sulfhydryl group formed by reduction near the acetylcholine-binding site 1,000-fold faster than with other sulfhydryl groups not near the binding site [10, 21]. On the surface of the intact electroplax following reduction, there are of the order of 10^4-fold more nonreceptor than receptor sulfhydryl groups; therefore, despite the 1,000-fold greater rate of reaction of receptor, about 10-fold as many nonreceptor as receptor sulfhydryls would be expected to react with MBTA. The specificity of the reaction improves markedly with purification (all of the reaction is specific with purified receptor); but, in general, one must have a means to distinguish specific (i.e., site-directed) from nonspecific reaction. As stated before, the reaction of MBTA with the receptor, as inferred physiologically, is retarded in the presence of reversible ligands of the acetylcholine-binding site; thus, in the presence of a high concentration of such a ligand, MBTA

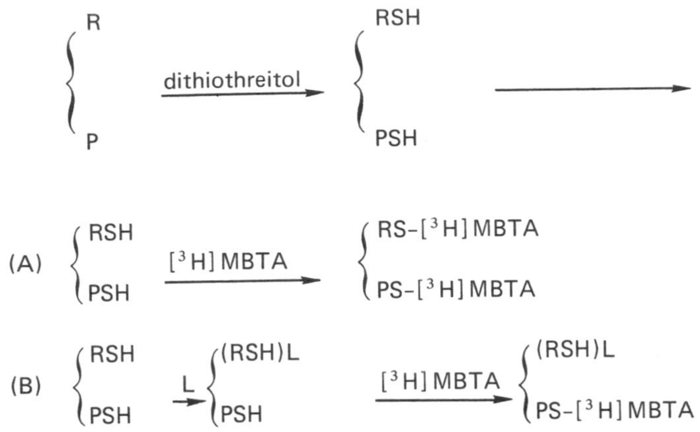

$$\left\{\begin{matrix} R \\ P \end{matrix}\right. \xrightarrow{\text{dithiothreitol}} \left\{\begin{matrix} RSH \\ PSH \end{matrix}\right. \longrightarrow$$

$$(A)\ \left\{\begin{matrix} RSH \\ PSH \end{matrix}\right. \xrightarrow{[^3H]MBTA} \left\{\begin{matrix} RS\text{-}[^3H]MBTA \\ PS\text{-}[^3H]MBTA \end{matrix}\right. \quad \text{or}$$

$$(B)\ \left\{\begin{matrix} RSH \\ PSH \end{matrix}\right. \xrightarrow{L} \left\{\begin{matrix} (RSH)L \\ PSH \end{matrix}\right. \xrightarrow{[^3H]MBTA} \left\{\begin{matrix} (RSH)L \\ PS\text{-}[^3H]MBTA \end{matrix}\right.$$

SCHEME 1

should react only with nonspecific sulfhydryls, and not with the specific protectible sulfhydryls. Scheme 1 illustrates a general mechanism applicable to all labeling by [^{3}H] MBTA. In this scheme, R is receptor, P is all other proteins, and L is a ligand binding to the acetylcholine-binding site. The extent of labeling in reaction (A) minus that in reaction (B) should equal the extent of specific labeling, RS-[^{3}H] MBTA. For L, 1 mM carbamylcholine, 1 mM hexamethonium, and 125 nM α-neurotoxin of Naja naja siamensis have been used with similar results. In addition, it is possible to reoxidize the receptor specifically using 0.5 μM dithiobischoline, which acts as an affinity-reoxidizing agent of the reduced receptor [18]; and this agent has been used in reaction (B) as the receptor-protecting agent, L, with results indistinguishable from those obtained with noncovalently binding ligands. It should be noted that, under the usual conditions of reduction by dithiothreitol, few disulfide bonds other than the one at the acetylcholine-binding site of the receptor are reduced. In cases where there are proteins other than receptor present, as in cells and membrane fractions, the sulfhydryl groups which are nonspecifically labeled by [^{3}H] MBTA are, for the most part, present before reduction.

2. Labeling Receptor in Intact Electroplax [21, 25]

Electroplax are dissected as discussed in Sec. II. B. It is important that the connective tissue sheath covering the innervated membrane be carefully and almost completely removed, so that access of reagents to the cell surface is a free as possible. The cell membrane itself, however, should remain undamaged, thereby minimizing access of MBTA to the interior of the cell. Typically, electroplax are labeled as follows [25]:

Two groups of about 15 cells each are treated in parallel. One group is labeled without protection (A) and the other group is labeled with protection (B, C, or D). Initially, both groups are treated with 0.2 mM dithiothreitol in a tris-Ringers solution [165 mM NaCl, 5 mM KCl, 2 mM $CaCl_2$, 2 mM $MgCl_2$, 2 mM tris (pH 8.0)] for 10 min; then washed for 10 min in eel Ringers solution [same as tris-Ringers, except 1.5 mM phosphate (pH 7.1) replaces tris] containing 10 mM glucose (RG). Thereafter, one group is treated according to procedure A and one according to B, C, or D:

A: RG (20 min); 12 nM [^{3}H]MBTA (10 min)

B: RG (5 min); 0.5 μM dithiobischoline (5 min); RG (10 min); 12 nM [^{3}H]MBTA (10 min)

C: RG (15 min); 1 mM hexamethonium (5 min); 12 nM [^{3}H]MBTA in 1 mM hexamethonium (10 min)

D: 0.28 nM *N. n. siamensis* α-neurotoxin (10 min); RG (10 min); 12 nM [^{3}H]MBTA (10 min)

All solutions other than dithiothreitol are made up in eel ringers solution (no glucose), pH 7.1. The cells are transferred from solution to solution in small baskets made of 25-ml beakers cut off at the bottom and covered with fine nylon net held by a rubber band. These are suspended in rapidly stirring solutions in larger beakers. Finally, both groups are washed for 25 min with five changes of RG, blotted, and weighed. The cells are dissolved in NCS solubilizer (Amersham-Searle) and counted in a toluene-base scintillant; or they are dissolved in 2% sodium dodecyl sulfate at 50°C for 2 hr preparatory to gel electrophoresis (Sec. VI. B), and an aliquot counted in Scintisol-Complete (Isolab). The specific labeling saturates at 10 pmol/g wet weight of cell, or ~2 pmol/mg of protein [21].

In a modification of this labeling procedure, the reduced cells are labeled with a mixture of 12 nM [^{3}H]MBTA and 9.6 μM N-ethyl-[2,3-^{14}C]maleimide (2 Ci/mole). Since MBTA reacts with the reduced receptor about 5,000-fold faster than does N-ethylmaleimide [10, 26], there is much less reaction of the latter with the reduced receptor than of the former, despite the 800-fold higher concentration. The [^{14}C]-N-ethylmaleimide reacts nondiscriminately with all available sulfhydryls and is used as a general tag for protein. The ratio of ^{3}H-activity to ^{14}C-activity is thus a measure of the ratio of the quantity of receptor to the quantity of total protein.

3. Labeling Membrane Fractions

Membrane fractions may be labeled using the same scheme used with cells. The separation of dithiothreitol, the protecting agent, and [^{3}H]MBTA from membrane fragments and vesicles can be achieved by either repeated centrifugation and resuspension, or by gel filtration. A membrane fraction of Electrophorus main organ electric tissue is labeled by the centrifugal method according to Scheme 1, with some modifications. The eel Ringers solution contains 4 mM EDTA and no glucose. The membrane protein concentration is 1 to 2 mg/ml. The membranes are washed by sedimenting at 4°C in a 65 rotor (Beckman), at settings of 65,000 rpm and 15 min, and by resuspension in ringers solution three times. The first two times, the pellets are resuspended in the centrifuge tubes (vol 8.5 ml). The third time, they are removed from the tubes and resuspended in an all-glass, handhomogenizer. The reaction with $[^3H]$MBTA is stopped after 2 min with an excess of β-mercaptoethanol. As with intact cells, the specific labeling (A-B) of the membranes was found to saturate as a function of $[^3H]$MBTA concentration [27]. The separations can also be conveniently obtained by filtration through Bio-Gel P-6 (Bio-Rad).

4. Labeling in Solution

The procedure used for labeling 300 μg of receptor purified from Torpedo californica electric tissue will exemplify the method [26]. To 300 μl of a solution containing 1 mg of receptor protein/ml of TNP50 [0.2% Triton X-100, 50 mM NaCl, 10 mM Na-PO_3, 1 mM EDTA, 3 mM NaN_3 (pH 7.0)] is added 100 μl of TNP 50 and 50 μl of 2 mM dithiothreitol in 200 mM tris-Cl (pH 8.3). (The tris brings the final pH to 8.0.) The mixture is incubated for 20 min at 25°C, and then 75 μl 0.56 M Na-PO_4 (pH 6.7) is added, bringing the pH to 7.0. The mixture is cooled to 4°C and layered on a 24 × 0.9-cm column of Bio-Gel P-6, 100 to 200 mesh (Bio-Rad Laboratories); pre-equilibrated with TNP150 (same as TNP50 except containing 150 mM NaCl); and eluted with TNP150 at 0.4 ml/min at 4°C. About 90% of the initial protein is collected at the void volume in 2.5 ml, completely separated from the dithiothreitol. It is not necessary to actually determine the protein concentration at this stage, since the elution properties of the column are nearly constant and may be checked before use with a readily detected protein such as cytochrome-C. To the 2.5 ml of reduced receptor at 25°C is added 50 μl of 25 μM $[^3H]$MBTA. After 2 min, the reaction is quenched with 25 μl of 0.1 M β-mercaptoethanol. The mixture is cooled and again chromatographed on the P-6 column as before. The eluted fractions are sampled for radioactivity (10 μl into 5 ml of scintillant), and fractions containing about 90% of the radioactive protein are pooled. There is complete separation from small labeled products. The pooled fractions of labeled receptor contain about 200 μg protein in 2.5 ml. In this procedure, at least 95% of the available sites are labeled.

This procedure can be modified as long as an excess of dithiothreitol (at least 10-fold) and of [^{3}H] MBTA (at least twofold) over receptor sites is maintained. The concentration of [^{3}H] MBTA should not be raised above 1 μM, however, to avoid nonspecific reaction. With receptor of a specific activity of 2 nmol sites per mg protein or above, there is insignificant nonspecific reaction of [^{3}H] MBTA under these conditions. With less purified receptor preparations (for example, a detergent extract of membrane), the extent of nonspecific reaction of [^{3}H] MBTA can be determined by adding α-neurotoxin at a final concentration of 1 μg/ml to an aliquot of the reduced receptor and incubating for ~15 min at 25°C before adding [^{3}H] MBTA.

5. [^{3}H] MBTA Assay

In the preceding labeling procedure, the concentration of dithiothreitol is lowered to well below that of the [^{3}H] MBTA to be added. This is most practical for the saturation labeling of fairly large quantities of receptor. For the purpose of assaying a large number of samples, however, it is cumbersome. Conditions have been determined under which it is not necessary to remove the dithiothreitol before adding [^{3}H] MBTA and which result in complete reaction of the specific sites with [^{3}H] MBTA [26, 28]. This is possible because MBTA reacts with the reduced receptor about 80-fold faster than with dithiothreitol, and it reacts with dithiothreitol about 25-fold faster than with "ordinary" protein sulfhydryl groups. (Ratios of this order can be inferred from Table I of Ref. 10; a detailed analysis of these reactions will appear elsewhere [28].) Since the concentration of [^{3}H] MBTA seen by the three sulfhydryl-bearing classes is the same

$$\frac{d[\text{receptor}]}{k_R[\text{receptor}]} = \frac{d[\text{dithiothreitol}]}{k_D[\text{dithiothreitol}]} = \frac{d[\text{protein SH}]}{k_P[\text{protein SH}]}$$

or

$$\begin{aligned}(1/k_R)\ln([\text{receptor}]_0/[\text{receptor}]_t)\\ = (1/k_D)\ln([\text{dithiothreitol}]_0/[\text{dithiothreitol}]_t)\\ = (1/k_P)\ln([\text{protein SH}]_0/[\text{protein SH}]_t)\end{aligned}$$

where the subscript 0 indicates t=0, and the k_X are the second-order rate constants. When 95% of the receptor, for example, has reacted with [^{3}H] MBTA, the extent of reaction of dithiothreitol with [^{3}H] MBTA is

$1-(20)^{-80} = 0.038$; thus, if 4% of the total number of moles of dithiothreitol is less than the total of [^{3}H] MBTA, then the alkylation of the reduced receptor is nearly complete before [^{3}H] MBTA is consumed by reaction with dithiothreitol.

Initially in the development of this assay, N-ethylmaleimide was included to react with dithiothreitol, thereby to spare [^{3}H] MBTA [26]. In fact, the rate of reaction of reduced receptor with [^{3}H] MBTA had been underestimated, and the inclusion of N-ethylmaleimide was found to be unnecessary [28]. The complete procedure presently used follows.

In 15-ml, screwtop, plastic tubes (Becton-Dickinson), quadruplicate 50-μl samples containing 1 to 50 μg of protein in TNP50 are mixed with 50 μl of 0.4 mM dithiothreitol in 0.2% Triton X-100, 150 mM NaCl, 20 mM tris-Cl, 1 mM EDTA, 3 mM NaN_3 (pH 8.3); to give a final buffer composition of 0.2% Triton X-100, 100 mM NaCl, 5 mM Na-PO_4, 10 mM tris-Cl, 1 mM EDTA, 3 mM NaN_3 (pH 8.0). (The concentrations of all the components of the medium may be varied somewhat, but the final pH should be 8.0.) All incubations are at 25°C. After 20 min, 10 μl of 0.53 M Na-PO_4 (pH 6.7) is added to bring the pH to 7.0. To two of the samples (set A), 400 μl of TNP150 is added; and to the other two samples (set B), 400 μl of TNP150 containing 3 μg N. n. siamensis α-neurotoxin per ml. After 15 to 20 min, 500 μl of 2 μM [^{3}H] MBTA in 0.2% Triton X-100-150 mM NaCl (unbuffered) is added to each sample, and the reaction is quenched after 3 min with 25 μl of 0.1 M β-mercaptoethanol. Finally, about 100 μg of reduced, carboxamidomethylated, succinylated lysozyme [29] is added as carrier, and the tubes are capped and put in ice. The tubes may be put in the freezer for later processing, or the procedure may be carried forward immediately.

To each tube is added 14 ml of saturated ammonium sulfate solution, the top is replaced, and the tube inverted. The contents are filtered through a 24-mm glass fiber filter (Whatman GF/A) held in a polypropylene filter holder (Gelman). The tubes, including the top, are washed twice with 15 ml of 50% saturated ammonium sulfate solution, and the washings filtered. Each filter is washed three more times with ~25 ml of 50% saturated ammonium sulfate solution, twice with 25 ml of ice-cold 1 N HCl, and twice with 15 ml of acetone. (This wash procedure reduces the background retention to about 0.05% of the applied ^{3}H activity.) The dry filters are placed in counting vials and digested in 0.1 ml of H_2O and 0.6 ml of NCS solubilizer for 1 hr at 50°C. Finally, 10 ml of a toluene-base scintillant is added. The extent of site-specific labeling is the difference between the means of set A and set B. With purified receptor, all labeling is blocked by N. n. siamensis toxin. The factor converting cpm to moles of label is obtained by preparing duplicate standards containing an aliquot of the 2-μM [^{3}H] MBTA solution spotted and dried on a glass filter, digested with NCS, and counted like the other samples. The factor is the quotient of the total quantity in moles of [^{3}H] MBTA plus hydrolyzed [^{3}H] MBTA (Sec. III. A) divided by the cpm and, depending on the efficiency of counting, has had in our hands a value of 1 to 1.5×10^{-15} moles/cpm. Specific activities obtained by this assay agree closely with those obtained by the previous labeling procedure (Sec. III. B. 4).

IV. ASSAY OF RECEPTORS BY THE BINDING OF α-NEUROTOXINS

A. Background

The venoms of elapid and hydrophid snakes contain highly toxic polypeptides [9, 30]. A class of these, the α-neurotoxins (60 to 74 amino acid residues), act specifically on the receptor in the postsynaptic membrane of vertebrate muscle and electroplax. They produce a nondepolarizing block similar to that produced by (+)-tubocurarine, and this block is reversed only very slowly by washing. Lee and his associates [9] demonstrated that ^{131}I-labeled α-neurotoxins bind specifically to neuromuscular junctions and that this binding is inhibited by (+)-tubocurarine. The fruitful application of this approach to electric tissue began with the demonstration by Changeux et al. [11] that α-bungarotoxin acts on the electroplax of *Electrophorus* as it does on vertebrate muscle and that, in addition, it displaces the binding of [^{14}C] decamethonium to subcellular membrane vesicles prepared from *Electrophorus* electric tissue. Miledi et al. [31], also proceeding on the basis of the work of Lee, demonstrated the binding of [^{131}I] α-bungarotoxin to a crude membrane fraction of *Torpedo marmorata*. They showed that the toxin-binding component is solubilized from the membranes by a solution of Triton X-100. The rate of binding of α-bungarotoxin to the membranes was shown to be decreased in the presence of high concentrations of carbamylcholine and of (+)-tubocurarine. Raftery et al. [32], Meunier et al. [33], and Fulpius et al. [34] described the binding of various radioactively labeled α-toxins to membrane fragments from *Electrophorus* electric tissue and to detergent extracts of the membrane fractions. It was confirmed that toxin binding is retarded in the presence of reversible receptor ligands.

Various methods are currently employed to determine the extent of binding of α-toxins. All take advantage of the high rate of association and the low rate of dissociation of the receptor-α-toxin complex. The equilibrium dissociation constants are of the order of 10^{-10} to 10^{-9} M [e.g., 34]. It is thus possible to separate the complex from free toxin without appreciable dissociation of the complex within intervals of the order of an hour. In addition, the α-toxins are small (mol wt 7000-8000) and positively charged, while the complexes are large (mol wt $\sim 3 \times 10^5$) and negatively charged.

Separation has been effected by gel filtration [31, 32, 35], by precipitation and collection of the complex on Millipore filters [36, 37], by adsorption of the complex to DEAE-cellulose disks [38, 39, 40], and by sedimentation in a sucrose density gradient [41]. In our hands, gel filtration has proved most reliable and also fast enough for a small number of samples.

B. Toxin Purification and Tritiation

The principal α-neurotoxin of Naja naja siamensis was purified from lyophilized venom (Miami Serpentarium) following the procedure of Karlsson et al. [42] exactly. The toxin is radioactively labeled by the reductive methylation of lysine residues with formaldehyde and $[^3H]NaBH_4$ [43, 44, 45; cf. 39]. The procedure is carried out in a well-ventilated hood as follows: To 4 mg of toxin in 660 μl of 50 mM NaCl-10 mM Na-PO_4 (pH 6.7) is added 475 μl of 0.2 M $NaHCO_3$ (pH 9.0) and 145 μl of 40 mM formaldehyde (a 10-fold molar excess of formaldehyde). After 1 hr at room temperature, 6 μmol of $[^3H]NaBH_4$ (8.2 Ci/mmol; Amersham-Searle) in 500 μl of dry dimethylformamide is added, and, after an additional 2 hr at room temperature, the reaction is terminated by the addition of 10 μl of glacial acetic acid. The well-mixed solution is allowed to sit for at least 15 min and then applied to a 0.9 $\times$ 20-cm column of Bio-Gel P-6 (100-200 mesh) and eluted with 90 mM ammonium acetate (pH 6.5) at 820 μl/min. Ten fractions of 500 μl each around the void volume, containing about 80% of the applied protein, are pooled and concentrated by lyophilization to 500-μl volume, applied to a 0.9 $\times$ 50-cm Bio-Gel P-30 column, and eluted with 50 mM Na-PO_4 (pH 7.0) at 150 μl/min. Nine 0.5-ml fractions eluting around 1.7 void volumes and containing about 90% of the applied protein (by absorbance at 279 nm) are combined.

The product is stored at a concentration of ~0.5 mg/ml at -196°C. The concentration is determined both by A_{279}, using a molar extinction coefficient at 279 nm of 8300 [42]; and by Lowry determination, using purified unlabeled toxin as the standard. Specific activities are based on the latter. The toxin is labeled to the extent of ~0.5 mole of methyl groups per mole of toxin, and under typical counting conditions yields about 1,600 cpm/pmol. The biological activity of $[^3H]$toxin is indistinguishable from unlabeled toxin by a bioassay using mice [46]; but this would not reveal small differences. All of the $[^3H]$toxin is bound by an excess of receptor.

C. Binding Assay

For precise determination of $[^3H]$toxin binding, we used the following procedure. To ~50 μg of purified receptor is added ~4 μg of $[^3H]$toxin in a final volume of ~150 μl and final medium of 0.2% Triton X-100, 100 mM NaCl, 10 mM Na-PO_4, 1 mM EDTA (pH 7.0). The mixture is incubated at 25°C for 1 hr and then applied to a 20 × 0.375-cm column (a modified 2-ml Mohr pipet) of Bio-Gel P-30 and eluted at 4°C with TNP150 at 0.04 ml/min (controlled by a syringe drive). Fractions of 100 μl are collected and 10-μl aliquots are counted. There is a peak of ^3H activity at the void volume containing receptor-toxin complex and a second peak centered at two void volumes containing free toxin. The three fractions of the first peak with the greatest ^3H activity (containing about 90% of the ^3H activity of the peak) are pooled, and the contained ^3H activity and protein (Lowry) are determined. The specific labeling is calculated on the basis of these determinations on the isolated complex. Routinely, ~80% of the applied receptor protein is recovered in the pooled fractions, so that if the loss in accuracy is tolerable, protein need not be determined and much smaller quantities of receptor could be used. Less than 0.5% of the free [^3H]toxin peak overlaps the pooled peak fractions.

Applying the procedure to receptor purified from *Electrophorus* and and from *Torpedo*, we have found that in both cases the specific activity determined by [^3H]toxin binding is about twice that determined by the [^3H]MBTA assay (Sec. III. B. 5) [47, 48]. In the latter assay, all labeling by [^3H]MBTA is blocked by toxin; by contrast, less than 50% of the [^3H]toxin binding is blocked by prelabeling receptor with MBTA. As previously inferred [48b], a site of toxin binding overlaps the site of specific reaction of [^3H]MBTA; however, there may be an additional site of toxin binding.

V. ISOLATION AND PURIFICATION OF RECEPTORS

A. Membrane Fractions

The receptor resides in the cell membrane, principally at the synapse. The isolation of a subcellular membrane fraction enriched in synaptic membrane results in a purification of the receptor in the range of 5- to 10-fold. Membrane fractions are potentially useful for the investigation of the interactions of the receptor with other components in the membrane. Furthermore, in some preparations of closed vesicles, the permeability control properties of the receptor are retained and can be investigated in relative isolation from other cellular functions [49, 50].

A simple, large-scale procedure results in the isolation of a membrane fraction from 600 g of *Electrophorus* main organ electric tissue, containing nearly all of the receptor (about 20 nmol), 15% of the protein (about 2 g), ~40% of the acetylcholinesterase, and 40% of the Na^+- K^+-ATPase. The procedure is as follows [26]: To 600 g of fresh main organ electric tissue, 2,400 ml of 0.89 M sucrose is added and the mixture homogenized in a 4-liter commercial blendor (Waring) on "low" for 10 sec. This is sufficient to disperse the tissue, leaving large, multinucleated cell fragments with connected nerve nets. The homogenate is centrifuged in an L2 65B centrifuge (Beckman) set at 19,000 rpm and 35 min (4°C). A 19 rotor is used throughout, and the speed and time settings are given. The pellets, plus 1,200 ml of 0.89 M sucrose, are homogenized on "low" for 5 sec and centrifuged as before. The pellets, plus 1,200 ml of 0.3 M sucrose, are homogenized on "high" for ~30 sec and checked by phase microscopy to see that no nucleated cell fragments remain and that the nerve nets are detached from the membrane fragments. This homogenate is centrifuged at 6,000 rpm for 20 min to bring down nuclei, nerve nets, and connective tissue. The supernatant is centrifuged at 19,000 rpm overnight (8-16 hr). The pellets, plus 100 ml of 1 mM NaEDTA (pH 7.0), are homogenized in a VirTis "45" homogenizer at 80% full-speed for 30 sec, and the final suspension is stored at -20°C. This procedure is also applicable to *Torpedo* electric tissue.

More discriminating, small-scale fractionations of *Electrophorus* [51] and of *Torpedo* [52, 53] electric tissues have been described, which separate membrane rich in acetylcholinesterase from membrane rich in receptor.

B. Receptor Solubilization

Receptor is an intrinsic membrane protein. Surfactants are required for its solubilization in aqueous solution. It is not, however, soluble in chloroform-methanol [1]. Receptor can be extracted with Triton X-100 or with other nonionic detergents from the large-scale membrane fraction described in Sec. V. A. The extract contains all of the receptor and ~20% of the protein of the membrane fraction [26]. For the purposes of receptor purification, however, it is not necessary to prepare a membrane fraction first; rather, a crude, particulate fraction can be extracted. Although this latter extract contains more extraneous protein than the former, affinity chromatography of the two types of extract results in similar specific activities of purified receptor.

A procedure applicable to *Electrophorus* and to *Torpedo* electric tissue is as follows [47, 48]: About 600 g of tissue, either freshly dissected or thawed after storage in liquid N_2, is minced with scissors and homogenized in 2200 ml of 1 mM EDTA (pH 7.4) using a 4-liter-capacity Waring blendor (20 sec at "low"). All operations are at 0 to 4°C. The homogenate is filtered through four layers of gauze and centrifuged in a Beckman 19 rotor at settings of 19,000 rpm and 45 min, giving a total particulate fraction. The resulting sediment is rehomogenized (20 sec at "high") in 1,200 ml of 1 M NaCl-2 mM $NaPO_4$-1 mM EDTA (pH 7.0), in order to solubilize acetylcholinesterase [54]; and is resedimented at settings of 19,000 rpm and 55 min. [Much more of the acetylcholinesterase of Electrophorus (80%) is solubilized than of Torpedo (20%).] The pellet is suspended (10 sec at "low") in 1,200 ml of 1 mM EDTA (pH 7.4) and centrifuged as before. The pellet is resuspended (20 sec at "medium") in 210 ml (final volume) of 3% Triton X-100-50 mM NaCl-10 mM $NaPO_4$-1 mM EDTA-3 mM NaN_3 (pH 8.0) and stirred for 1 hr. The suspension is centrifuged in a 60 rotor at 60,000 rpm for 30 min and the supernatant collected. The pH of the supernatant is adjusted to 7.0 with 0.6 M NaH_2PO_4. Typically, the extract of 600 g of Torpedo tissue contains about 1 g of protein, 25 nmol of acetylcholinesterase catalytic sites, and 400 nmol of receptor-MBTA-reactive sites; and that of Electrophorus tissue, about 1 g of protein, 2 nmol of acetylcholinesterase catalytic sites, and 40 nmol of receptor-MBTA-reactive sites. The extract can be stored at 0°C for a few days or at -196°C for longer periods of time.

C. Affinity Chromatography

1. Background

The most effective procedure for further purification of receptor in a detergent extract is by affinity chromatography. A variety of receptor ligands have been attached covalently to agarose beads to produce affinity gels (Table 1).

Extract is applied to the affinity gel, which adsorbs the receptor. After weakly binding proteins are washed from the gel, the receptor is selectively eluted with a high concentration of a cholinergic ligand such as carbamylcholine, hexamethonium, (+)-tubocurarine, or benzoquinonium. Each laboratory has used unique conditions for purification and assay of receptor, and the results have not been uniform (see discussion in Sec. IV. A).

2. Procedure

a. Affinity Gel

The affinity gel contains $[-NH(CH_2)_3NH(CH_2)_3-NHCOCH(NHCOCH_3)CH_2CH_2S]COC_6H_4\overset{+}{N}(CH_3)_3$ linked to 4%-agarose beads. This is made as follows: To ~25 ml of packed Affi-Gel 401 (a homocysteinyl derivative of agarose indicated by the brackets above, available from Bio-Rad Laboratories, and which arrives partially oxidized) is added 25 ml of 20 mM dithiothreitol in 0.2 M tris-Cl (pH 8.0), and this mixture is stirred with a magnetic stirrer for ~1 hr at room temperature, thereby bringing all the sulfhydryl groups on the gel into the reduced state. The mixture is then

TABLE 1. Purification of Receptor

Gel ligand[a]	Specific activity[b] (μmol sites/g protein)	Molecular weight SDS-PGE[c] (10^3 daltons)	Ref.
	Electrophorus		
N. naja toxin	7	42, 54	[41]
N. n. siam. toxin	11	160[d]	[39]
$-HN\phi(-OCH_2CH\overset{+}{N}Et_3)_2$	6	43, 48	[36]
$-HN\phi\overset{+}{N}Me_3$	4	44, 50	[44]
$-OC\phi\overset{+}{N}Me_3$	4[e]	40, 47, 53	[26, 47]
$-NH(pyridinium)^+CH_3$	6	37, 42, 49	[40]
	Torpedo		
$-HN(CH_2)_3\overset{+}{N}Me_3$	10	40, 50, 65	[55]
N. n. siam. toxin	2	45, 50	[56]
N. n. siam. toxin	3, 8[f]	46, -, -[g]	[57, 58]
$-OC\phi\overset{+}{N}Me_3$	4[e]	39, 48, 58, 64	[48]
N. n. siam. toxin	12	34, 36, 38, 44	[45]

[a] Toxin is linked directly to agarose. The small ligands are linked to agarose by an arm 7 to 22 atoms long.

[b] Specific activity after final step of purification. Assay is by toxin binding, except (e) and (f).

[c] These are the apparent molecular weights of major staining bands on polyacrylamide gel electrophoresis of reduced receptor preparation in SDS, except that (d) is unreduced.

[e] Assay is by affinity labeling with $[^3H]$MBTA. Toxin binding gives about twice the specific activity.

[f] This value is the sum of two classes of ACH-binding sites which are only 50% blocked by high concentrations of (+)-tubocurarine.

[g] Molecular weights of additional components are not given.

diluted fourfold with ice-cold distilled water and centrifuged in a clinical centrifuge for a few minutes to sediment the gel. All operations are conveniently carried out in 50-ml capacity, conical, screwtop, plastic centrifuge tubes (Falcon). The gel is washed by resuspension and sedimentation in 10 volumes of ice-cold water three times, suspended in 200 ml of 0.1 M NaCl-50 mM Na-PO_4 (pH 7), and sedimented again. Samples (~ 0.2 ml) of the final supernatant and of the gel are taken for sulfhydryl determination. The gel is transferred into one 50-ml tube with this phosphate buffer, again sedimented, and the supernatant removed. To the gel is added 25 ml of 20 mM NPTMB (see following paragraph) in the phosphate buffer, and the mixture is stirred for ~40 min at room temperature. After sedimentation, the gel is distributed into four tubes and washed with about 10 volumes of cold water three times, and duplicate samples of the gel are taken for sulfhydryl determination. About 25 ml of 20 mM iodoacetamide in phosphate buffer is then added to the gel to block any residual sulfhydryl groups, and the mixture is stirred for 45 min at room temperature. The gel is washed three times, and duplicate samples are taken for sulfhydryl determinations. The gel is finally suspended in 0.02% NaN_3 and stored at 4°C.

Sulfhydryl determinations are performed following the Ellman procedure [59]. Gel samples are pipetted into tared centrifuge tubes and the gel packed by a 2-min centrifugation in a clinical centrifuge. The supernatant is removed, and the weight of the packed gel determined (usually ~ 100 mg). To the packed gel is added 1 ml of 0.33 mM 5,5'-dithiobis-(2-nitrobenzoic acid) in 200 mM tris-Cl (pH 8.0). The mixture is agitated briefly and centrifuged. The absorbance of the supernatant at 412 nm is divided by 13,600 to obtain the molar concentration of sulfhydryl, from which can be calculated the moles of sulfhydryl per milligram of gel. The difference between the sulfhydryl content before and after reaction with NPTMB is one estimate of the content of p-carboxyphenyltrimethylammonium groups in the derivatized gel (usually about 1 μmol/g). (Recently, samples of Affi-Gel 401 from Bio-Rad have titers close to 3 μmol/g and are less satisfactory for receptor purification.)

A second assay for the content of p-carboxyphenyltrimethylammonium groups is by hydrolysis of a weighed sample of packed gel (100 mg) in 1 ml of 10 mM NaOH for 30 min at room temperature. The absorbance of the supernatant at 220 nm divided by 9.15 gives the quantity of p-carboxyphenyltrimethylammonium ion in μmol/ml of supernatant, and since there is 1 ml of supernatant, in μmol per weight of packed gel. The blank for this assay is Affi-Gel 401 treated similarly. Azide should be washed from gel samples before this assay because of its absorbance at 220 nm.

b. Synthesis of NPTMB [60]

<u>p-Dimethylaminobenzoic acid, ester with p-nitrophenol.</u> A solution of 10 g (0.06 moles) of p-nitrophenol, 6 g of p-dimethylaminobenzoyl

chloride [61], and 20 g of pyridine is refluxed for 1 hr. The cooled semi-crystalline mass is poured onto cracked ice, the mixture filtered, and the solid washed with water and air-dried to yield 10.5 g of crude product, melting at 165-190°C (dec.). After recrystallization from about 250 ml of acetonitrile, the product weighs 6.5 g and melts at 195-197°C (dec.). Further recrystallization does not alter the melting point.

Analysis: Calculated for $C_{15}H_{14}N_2O_4$: C, 63.92; H, 4.92; N, 9.77. Found: C, 64.11; H, 5.17; N, 9.73.

(p-Carboxyphenyl)trimethylammonium iodide, ester with p-nitrophenol (NPTMB). A mixture of 3 g (0.01 moles) of *p*-dimethylaminobenzoic acid, ester with *p*-nitrophenol, and 15 ml of methyl iodide in 75 ml of acetonitrile is heated in a sealed tube at 115 to 120°C for 5 hr. The unreacted solid (1.2 g) is removed by filtration and the filtrate concentrated to dryness to yield 2.3 g of residue. The residue is extracted with 200 ml of hot water, and the filtered solution is lyophilized to yield 1.2 g of light-yellow solid, melting at 175 to 180°C. This material is crystallized from 50 ml of acetonitrile and 50 ml of anhydrous ether. The product, which weighs 890 mg, melts at 185 to 187°C. Recrystallization does not change the melting point.

Analysis: Calculated for $C_{16}H_{17}N_2O_4I$: C, 44.87; H, 4.00; N, 6.54; I, 29.65. Found: C, 44.47; H, 4.06; N, 6.66; I, 29.84.

The absence of unquaternized ester was shown by thin layer chromatography using 1:1 $CHCl_3$-CH_3OH as the solvent. In this system the unquaternized ester moves rapidly, whereas the quaternized ester remains at the origin.

c. Application and Elution of Receptor

Two procedures for adsorbing receptor to the affinity gel have been used. In the first, the extract is applied to the affinity gel packed in a column; and in the second, the extract is stirred with the affinity gel, which is then separated and washed by centrifugation and packed in a column. There are also two procedures for eluting the affinity gel, one optimal for *Electrophorus* receptor and the other optimal for *Torpedo* receptor. The first requires higher salt and carbamylcholine concentrations than the second.

The procedures for *Electrophorus* are as follows: Approximately 200 ml of extract (pH 7.0) is applied to 2 ml of affinity gel (bed dimensions, 0.5 × 10 cm) at a rate of about 0.9 ml/min. All solutions are applied to the gel from glass syringes (50 or 20 ml) driven by an infusion pump (Harvard Apparatus Co., Model No. 975). This flow rate is possible only if the protein concentration of the extract is not appreciably greater than 4 mg/ml. The extract can be diluted if necessary with TNP50. All operations are at 4°C. Alternately, 200 ml of extract can be stirred with ~4 ml of affinity gel for 1 hr. The gel is sedimented and washed with 100 ml of TNP50 in 50-ml centrifuge tubes in a clinical centrifuge, and is then packed into a 0.5-cm diam column.

The column is successively eluted at a rate of 0.5 ml/min with 25 ml of TNP50, ~10 column volumes of TNP150, and ~5 column volumes of 50 mM carbamylcholine chloride in TNP100. A typical summary of the elution of protein, acetylcholinesterase, and acetylcholine receptor is shown in Table 2.

Nearly all of the receptor eluted from the column with carbamylcholine is eluted after less than one column volume of the carbamylcholine solution is applied. It is found convenient and conservative of receptor to assay the protein in the individual fractions before pooling, using fluorescamine [64]. This permits the determination of as little as 0.5 μg of protein. After the pooling of fractions, proteins are determined by the Lowry method, using bovine serum albumin as the standard. This requires about 10 μg of protein for an accurate determination.

Torpedo extract is much richer in receptor than Electrophorus extract, and more affinity gel is used. The extract (~ 200 ml) is either passed through a 0.9 × 25-cm column of affinity gel (16 ml) at about 0.8 ml/min, or alternately, it is stirred with 16 ml of gel for 1 hr. In the latter case, the gel is sedimented and washed with ~200 ml of TNP50 and then packed in the column. The elution sequence and the recoveries of protein, acetylcholinesterase, and receptor are shown in Table 3.

Torpedo receptor is purified only about 10-fold by affinity chromatography, compared with the 200-fold purification of Electrophorus receptor. However, the specific activity of the receptor in the former extract is about 20-fold that in the latter, and the receptors are purified by affinity chromatography to about the same specific activities: ~3 μmol of $[^3H]$-MBTA sites per g of protein, or ~6 μmol of $[^3H]$toxin sites per g of protein. The recoveries of receptor are not satisfactory (~20%), but no means of improving the recovery from this affinity gel has been found. It is important to note, however, that the protein not eluted with carbamylcholine from the affinity gel has been eluted in sodium dodecyl sulfate, and has the same polypeptide components by gel electrophoresis as does the receptor eluted in carbamylcholine (unpublished results).

Purified receptor can be stored at 0°C for a few days, or in liquid nitrogen for longer periods, with no loss in specific activity [48]. Storage at -20°C, however, does result in a large decrease in specific activity.

D. Sucrose Density Gradient Sedimentation

An approximately 30% increase in the specific activities of both Electrophorus [47] and Torpedo [48] receptors purified by affinity chromatography

TABLE 2. Affinity Chromatographic Purification of Receptor from *Electrophorus electricus*

	Protein[b]		ACHE[c]		Receptor[d]	
Elution sequence	mg	%	nmol	%	nmol	%
Extract[a]						
Input	1145	100.0	2.14	100.0	15.3	100.0
Not bound	1136	99.2	1.04	48.3	0	0
TNP50[e]	10.8	0.9	0.08	3.9	0	0
TNP150	1.1	0.1	0.47	21.8	1.96	12.8
50 mM CARB in TNP100[f]	1.2	0.1	0.04	1.9	3.81	25.0
Total recovery	1149	100.3	1.63	75.9	5.77	37.8

[a]Extract (186 ml) from 660 g of tissue in 3% Triton X-100-50 mM NaCl-10 mM $NaPO_4$-1 mM EDTA-3 mM NaN_3 (pH 7.0).

[b]Protein was determined by the Lowry method with bovine serum albumin as the standard [62].

[c]Acetylcholinesterase activity was measured by Ellman's procedure [63] and is expressed as nanomoles of catalytic sites, assuming 80,000 daltons per catalytic site and 10 moles acetylthiocholine hydrolyzed per minute per gram enzyme.

[d]Receptor activity was determined by specific [^{3}H] MBTA labeling (see text).

[e]TNPxx contains 0.2% Triton X-100, xx mM NaCl, 10 mM Na-PO_4 1 mM EDTA, 3 mM NaN_3 (pH 7.0).

[f]Carbamylcholine was removed from fractions by dialysis before assay for ACHE and receptor.

can be achieved by sedimentation in a sucrose density gradient. About 1 mg of receptor in 0.5 ml TNP150 is layered over 12 ml of a 5 to 20% sucrose gradient also containing TNP150. [The receptor from the affinity column can be concentrated before this step if necessary in a collodion bag ultrafiltration apparatus (Schleicher and Schuell).] The receptor is spun at 40,000 rpm for 14 hr at 4°C in an SW 41 rotor (Beckman). Fractions of 0.4 ml are collected and analyzed for sucrose concentration (by refractometry), protein, acetylcholinesterase, and receptor. Acetylcholinesterase is separated from receptor in the gradient, and the ratio

TABLE 3. Affinity Chromatographic Purification of Receptor from Torpedo californica

Elution sequence	Protein		ACHE		Receptor	
	mg	%	nmol	%	nmol	%
Extract						
Input	1133	100	19.2	100	371	100
Not bound	900	79.4	16.3	85.0	0	0
TNP50[a]	6.9	0.6	0.39	2.0	0	0
TNP100	8.5	0.8	0.21	1.1	1.5	0.6
10 mM CARB in TNP90	26.3	2.3	1.27	6.6	73.5	19.8
Total recovery	942	83	18.2	95	75	20

[a]TNPxx contains 0.2% Triton X-100, xx mM NaCl, 100 mM Na-PO_4, 1 mM EDTA, 3 mM NaN_3 (pH 7.0).

of catalytic sites to receptor sites is reduced from about 1:100 initially to about 1:2,000 in the receptor peak. The peak fractions have uniform specific activity of receptor and contain the same polypeptide components by dodecyl sulfate-acrylamide gel electrophoresis [47]. By this procedure, at least, the receptor is homogeneous. Torpedo receptor tends to aggregate, and two peaks of receptor have been obtained; these have the same specific activities and the same polypeptide components. The average specific activity of both Electrophorus and Torpedo receptor after this procedure is about 4 μmol/g of protein by the [^{3}H]MBTA assay. Sucrose density gradient centrifugation is also a convenient means for changing the concentration or kind of detergent (for example, cholate for Triton X-100) in the solution of the receptor.

VI. CHARACTERIZATION OF PURIFIED RECEPTORS

A. Initial Characteristics

1. Specific Activity

The specific activity of purified receptor is expressed in terms of the number of sites per weight of protein. The value for the number of sites is dependent on the method of its determination and is subject to conceptual, systematic, and experimental errors. Three types of agents have been used to quantitate binding sites: rapidly equilibrating receptor ligands such as acetylcholine, the slowly dissociating α-neurotoxins, and the covalently reacting affinity label [^{3}H] MBTA. As noted in (Sec. IV. C), the ratio of toxin sites to [^{3}H] MBTA sites is about two [47, 48]; in addition, others have found that the ratio of toxin sites to acetylcholine sites is also about two [36, 40, 65a]. The basis for these ratios is not known. The hypothesis that there are two distinct classes of toxin-binding sites explains the results simply; however, it is not the only possibility. In any case, the extent of binding of toxin at saturation may not be an accurate measure of the number of "functional" acetylcholine-binding sites. The binding of acetylcholine, as determined by equilibrium dialysis, has been reported by some laboratories [36, 65a] to be due to the progressive saturation of a single class of noninteracting sites; but, by contrast, another laboratory reports that the binding appears to be complex, with two or more classes of interacting sites [57]. In the latter case especially, assumptions have to be made about the binding characteristics of acetylcholine to discriminate "specific" from "nonspecific" binding. The criteria that all the specific acetylcholine binding should be blocked both by toxin and by (+)-tubocurarine are helpful in defining the specific sites, but are not always applied. In the case of [^{3}H] MBTA, all the labeling of purified receptor is blocked by toxin and depends on the prior reduction of a disulfide located near the acetylcholine-binding site; nevertheless, it is still an assumption that, at saturation, one and only one [^{3}H] MBTA attaches per functional acetylcholine-binding site.

On the experimental level, accuracy in the determination of the number of sites depends on, among other things, the conversion of cpm to moles of bound label. This requires that the quantity of the labeling agent in a counting standard be known accurately, and that the specific radioactivity of the labeling agent attached to the receptor be the same as the specific activity of the agent in the counting standard. The stock solution of radioactively tagged toxin, for example, usually contains both labeled and unlabeled species. The preferential binding by the receptor of one of these species would give misleading results, since the conversion factor would be based on the stock solution. The specific activity of receptor depends on at least four determinations: the radioactivity and the quantity of protein in the receptor sample; and the radioactivity and the quantity of ligand, toxin, or affinity label in a standard. In many cases, however, laboratories report only the highest specific activities obtained, and not the average value and its variation. From the trend of specific activities reported (Table 1), it probably can be safely concluded that there are 1 to 2×10^5 g of protein per mole of sites binding toxin and 2 to 3×10^5 g of protein per mol of sites binding acetylcholine and susceptible to labeling with [^{3}H] MBTA.

2. Hydrodynamics

The hydrodynamic properties of receptor in detergent have not yet been determined unambiguously [65b]. By gel filtration on agarose, the receptor has a Stokes radius close to that of β-galactosidase of mol wt 540,000; the sedimentation velocity of the receptor in a sucrose density gradient, however, is closer to that of catalase of mol wt 240,000. It was inferred that the molecular weight of the receptor is about 3×10^5 and that it has considerable detergent bound to it [66].

An important characteristic of the receptor is its polypeptide composition. The apparent molecular weights of the polypeptide components have been determined by gel electrophoresis in sodium dodecyl sulfate (SDS) (Table 1). Despite considerable variation in results, most laboratories find that the smallest component has a molecular weight of ~40,000, and that there is at least one more component in Electrophorus and at least two more in Torpedo receptor. It is unclear at this time whether the differences in reported compositions lie in the receptor preparations, or the gel techniques, or both. At present, only one component has been shown definitely to be a receptor subunit: By gel electrophoresis of both Electrophorus [26] and Torpedo [48] receptor specifically labeled with $[^3H]$MBTA, it has been demonstrated that all the radioactivity is associated with the ~40,000-dalton component, which thus bears all or part of the acetylcholine-binding site. This was in fact first demonstrated by labeling and extracting intact electroplax [25]. The techniques employed in the electrophoresis of purified receptor are given in Sec. VI. B.

B. Gel Electrophoresis

1. System

The gel system used [26] is a modification of that described by Fairbanks et al. [67]. The stock solution A is made by dissolving 40 g of acrylamide (Bio-Rad) and 1.5 g of N,N'-methylenebisacrylamide (Bio-Rad) in water to a final volume of 100 ml. Stock solution B contains 1 M tris-acetate and 20 mM EDTA (pH 8.0). Stock solution C contains 10% (wt/vol) of sodium dodecyl sulfate (Sequanal grade; Pierce Chemical Co.). All stock solutions are filtered through Whatman No. 1 filter paper; A and B are kept at 4°C, and C at room temperature. For 5.6% gels, the following additions are made in rapid succession: 7 ml of A, 5 ml of B, 5 ml of C, 30 ml of water, 1 ml of 7.5% ammonium persulfate (wt/vol), 1 ml of 1.25% N,N,N'N'-tetramethylethylenediamine (TEMED) (vol/vol), and additional water to 50 ml total volume. This mixture is rapidly added to glass tubes (14×0.6-cm, i.d.) to a height of 12 cm and carefully overlayed with a solution containing 0.1% sodium dodecyl sulfate, 0.15% ammonium persulfate, and

0.05% TEMED. The gels form in ~30 min. After 2 hr, the overlay solution is poured off and the gels washed and overlayed with "running buffer" [1% sodium dodecyl sulfate, 100 mM tris-acetate, 2 mM EDTA (pH 8.0)]. The gels are aged at least 12 hr before use. We find that this system resolves close receptor bands better than the original system of Fairbanks et al. [67].

2. Sample Preparation

The aim for analytical gels is to apply to the gel ~15 μg of protein or ~3 to 5 μg per band in a volume of 15 to 25 μl. The protein must be saturated with SDS, and disulfides completely reduced. The former requires an ionic strength below ~200 mM and preferably much lower [68]; dialysis of the samples is usually required to achieve this, since receptor is in a Triton X-100 solution containing appreciable salt. The procedure used to prepare such samples for electrophoresis is as follows:

The sample solution (usually 1 ml or less) is dialyzed against 2% SDS-10 mM tris-acetate, 2 mM EDTA (pH 8.0) for 2 to 4 hr at room temperature. This is conveniently done in a microdialyzer (Hoefer Instrument Co.). The sample is transferred to a conical centrifuge tube and 9 volumes of dry acetone is added to precipitate the protein, which is sedimented in a clinical centrifuge [29]. The supernatant is carefully removed with a pulled-out Pasteur pipet. If the initial concentration of protein in the sample is < 50 μg/ml, the sample can be concentrated by lyophilization, or carrier [29] can be added to 50 μg/ml before precipitation. The protein pellet is washed with about half as much acetone as used in the first step, to remove residual Triton X-100, and after centrifuging again, dried in a vacuum dessicator on a water aspirator for ~30 min. To the precipitated protein is added a volume of sample buffer [2% SDS-10 mM tris-acetate-2 mM EDTA-20 mM dithiothreitol (pH 8.0)], which will give a final protein concentration of about 1 mg/ml. A few crystals of sucrose, to give a final concentration of about 10%, and 25 μl (per ml of sample) of a 0.05% aqueous solution of Pyronin Y are also added. The sample is incubated at 50°C for 0.5 to 2 hr. Small amounts of undissolved material, appearing as a slight turbidity, are removed by sedimentation.

A somewhat more complicated procedure for sample preparation guards better against possible proteolysis and should be used in initial runs at least [26]. Receptor in Triton X-100 is precipitated with 9 volumes of acetone and washed with 5 volumes of acetone. The precipitate is dried as described earlier, redissolved in 1% SDS-2 mM tris-acetate (pH 8.0) at 50°C, and dialyzed for at least 2 hr against the same SDS solution, to lower the salt concentration. The receptor, the concentration of which should be > 50 μg/ml, is again precipitated and washed with acetone, dried, and redissolved in sample buffer as in the previous paragraph.

Samples are layered on top of the gels under the overlay of running buffer, using a 20- or 25-μl microcap (Drummond) with a syringe control (Hamilton). The tubes are carefully filled with more running buffer and then the level of buffer in the upper chamber is brought above the top of the gel tubes. A mixture of protein molecular-weight standards (e.g.: lysozyme, 14,000; light chain of γ-globulin, 23,500; aldolase, 40,000; heavy chain of γ-globulin, 50,000; catalase, 60,000; bovine serum albumin, 68,000; phosphorylase, 94,000; β-galactosidase 135,000; myosin, 210,000) containing 2 to 4 μg of each protein per 20 to 25 μl is prepared and applied in an identical manner. Samples and standards can be stored at -20°C, and following the addition of fresh dithiothreitol and heating at 50°C for 30 min, yield the same patterns as freshly prepared samples. Samples are run into the gels at 2 mA per gel, and then the gels are run at 8 mA per gel for ~2.5 hr, after which time the Pyronin Y has migrated about 8.5 cm. Immediately after electrophoresis, the gels are removed from the tubes by reaming, the fronts marked with india ink, and the gels fixed and stained with Coomassie Brilliant Blue according to Fairbanks et al. [67], or, if not fixed and stained, frozen and stored at -20°C.

In order to determine the distribution of radioactively tagged polypeptides, gels are frozen and sliced into 1-mm slices using a gel slicer (MRA Corp.). In this case the gels, the running buffer, and (if the gels are stained) the staining and destaining solutions all contain 10% glycerol. The slices of unfixed, unstained gels are collected in counting vials, shaken at room temperature with 1 ml of 1% SDS-2 mM tris-acetate (pH 8.0) for 1 to 2 days, and counted in 10 ml of Scintisol Complete (Isolab). Slices of fixed gels are dried in vials at 50°C for ~4 hr, 0.1 ml of water and 0.5 ml of NCS (Amersham-Searle) is added, and the vials are tightly capped and incubated at 45 to 50°C for 24 hr. A toluene-base scintillant (10 ml) is added for counting.

The results of the application of these techniques to the receptors purified from *Electrophorus* [26, 47] and from *Torpedo* [48] and labeled with [^{3}H]MBTA are shown in Figures 1 and 2. The similarities and differences between the two receptors are evident. Both receptors have a subunit of ~40,000 daltons which is specifically and exclusively labeled by [^{3}H]MBTA. A second putative subunit of ~48,000 daltons is also common to both preparations. The other components are dissimilar in molecular weight. The possible functions of any but the ~40,000-dalton subunits are unknown. It is worth noting that all amino acid analyses carried out so far on purified receptors are for a mixture of the polypeptide components. Present techniques, however, allow the amino acid analysis of polypeptides isolated in gel slices [69].

The techniques described can be profitably applied to cells [25] and membrane fragments [27] which have been covalently labeled with a site-directed reagent. The receptor in cells and membrane fractions from

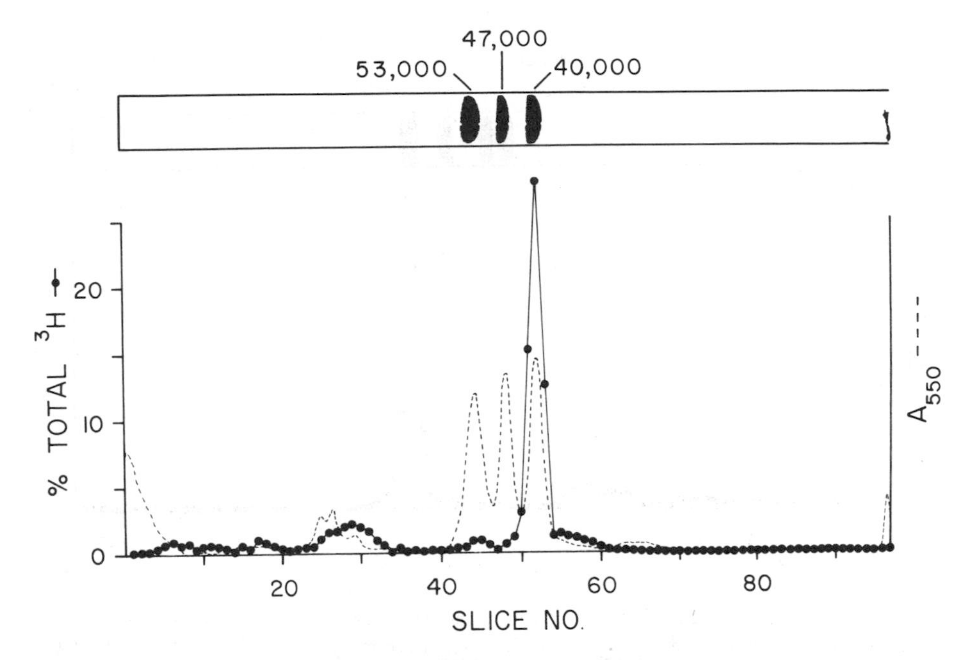

FIG. 1. Acrylamide gel electrophoresis in sodium dodecyl sulfate of purified receptor from Electrophorus electricus affinity labeled with [^{3}H] MBTA. Top: Photograph of gel stained with Coomassie Brilliant Blue. Bottom: ^{3}H activity (—•—) (45,000 cpm total) in 1-mm slices of the same gel superimposed on the densitometer trace, A_{550}, (----) obtained before slicing.

Electrophorus tissue constitutes a small fraction of the total protein, and the receptor subunits cannot be seen on SDS-gels by staining; nevertheless, a specifically labeled component can be identified by comparing the distributions of radioactivity on gels of material labeled with and without protection (Sec. III. B). In this manner, the existence and the molecular weight of a specifically labeled component can be determined before it is purified.

C. Immunological Methods

Antisera against receptor purified from Electrophorus have been raised in rabbits [70, 71]. These antisera precipitate receptor-toxin complexes and, in preliminary experiments, have been found to block the response of the isolated electroplax to carbamylcholine. The immunized rabbits develop a flaccid paralysis which resembles human myasthenia gravis. Immunology coupled with electrophysiology is a potent approach to the

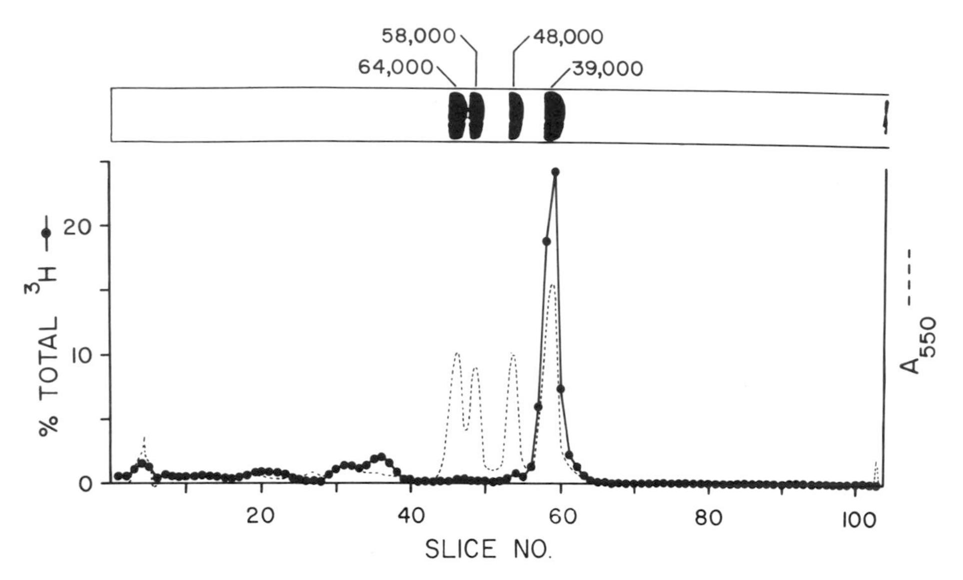

FIG. 2. Acrylamide gel electrophoresis in sodium dodecyl sulfate of purified receptor from Torpedo californica affinity labeled with [^{3}H] MBTA. Details as in Fig. 1: ^{3}H (—●—), A_{550}, (----).

questions of function and accessibility of receptor and of individual subunits, in situ.

In practice, New Zealand rabbits (~5 kg) are innoculated in the foot pads with 25 to 50 μg of purified receptor in 1 ml of TNP150, emulsified in 1 ml of Freunds Complete Adjuvant (Colorado Serum Co.). A similar innoculation is given three weeks later. A few days after the second injection, the rabbits begin to show signs of paralysis, which progresses rapidly. The serum is collected when the paralysis is severe.

The antisera raised against purified receptor from Electrophorus precipitates both Electrophorus receptor and receptor from Torpedo, and vice versa (unpublished results). This can be seen by double diffusion in agar plates and also by a quantitative precipitation assay using [^{3}H] MBTA-labeled receptor. This assay is an adaptation of that of Patrick et al. [70]; they measured the precipitation of the complex of purified Electrophorus receptor and ^{125}I-labeled α-neurotoxin from Naja naja naja. The use of a small covalent label such as [^{3}H] MBTA offers the advantage that there is no possibility of dissociation from the receptor during the lengthy incubations and also that less of the receptor surface is occluded by the tagging molecule. The reactions are carried out in a 0.4-ml capacity, capped, polypropylene microcentrifuge tube (A.H. Thomas Co.). To 0.5 to 1 μg of labeled receptor (~2000 cpm/μg) in TNP50 (volume 150 μl) is added 20 μl of a mixture of normal and immune rabbit sera, and the whole is incubated for 24 hr at 4°C. Goat antirabbit IgG is then added in an amount determined to give maximum precipitation (~40 μl), and the incubation is continued for 20 hr at 4°C. The tubes are centrifuged for 2 min in a microfuge (Beckman) and the supernatants collected with pulled-out Pasteur pipets and saved for counting. The pellets are washed twice with 200 μl of TNP50 and then dissolved in 200 μl of 1% acetic acid. The dissolved precipitate is counted in 10 ml Scintisol Complete (Isolab). The quantity of labeled receptor in the precipitates is plotted against the quantity of antireceptor added. The initial slope is a measure of serum titer.

D. Reconstitution of Membrane

Strong proof that a purified preparation of receptor is both complete and functional would be provided by its incorporation into a characterized lipid membrane with reconstitution of the control of Na^+ and K^+ permeability characteristic of the receptor in intact cells. Such a result has not yet been convincingly achieved, but some promising preliminary results have been reported. Hazelbauer and Changeux [50] were able to dissolve membrane vesicles from _Torpedo_ electric tissue in sodium cholate solution and, by dialyzing away the cholate, to reconstitute vesicles, which in some cases showed increased Na^+ permeability in the presence of carbamylcholine. This suggested that solubilization per se does not necessarily inactivate the permeability control function of the receptor. Kemp et al. [72] added a detergent extract of rat muscle to an artificial thin lipid film, which spontaneously showed increasing permeability with time. The rate of increase was greater in the presence of carbamylcholine.

Purified receptor from _Electrophorus_ and from _Torpedo_ have been incorporated into phospholipid vesicles by two procedures [47], both adaptations of methods used by Racker [73, 74] to reconstitute a variety of membrane systems. The first consists of the slow dialysis of a cholate suspension of purified receptor (cholate is exchanged for Triton X-100 by sucrose density gradient centrifugation) and phospholipid against detergent-free buffer, and the second involves sonication of detergent-depleted receptor with phospholipids. Presently, we use a weight ratio of 10:1 of phospholipid to receptor. In either case, vesicles incorporating receptor are obtained which are separable from phospholipid vesicles not incorporating receptor by flotation in a sucrose density gradient. The receptor incorporated by the first method is fully active by [^{3}H] MBTA assay. Some activity is lost in the second method. The vesicles have a reasonably low permeability to $^{22}Na^+$, which can be increased by the addition of antibiotic ionophores such as gramicidin A, but the addition of receptor activators such as carbamylcholine has had no reproducible effect on the permeability. Others [55], however, have reported that occasionally vesicles reconstituted from receptor and lipids, both purified from Torpedo, show increased permeability to Na^+ in the presence of carbamylcholine and that this effect is blocked by α-toxin. Conditions for the consistent reconstitution of membrane containing functional receptor have yet to be described.

ACKNOWLEDGMENTS

The support provided by research grants NS-07065 from the National Institute of Neurological and Communicative Disorders and Stroke and BMS75-03026 from the National Science Foundation and by a grant-in-aid from the New York Heart Association, Inc. are gratefully acknowledged. Ramon Valderrama is a Fellow of the Muscular Dystrophy Associations of America.

REFERENCES

1. A. Karlin, Life Sci., 14:1385 (1974).
2. H. P. Rang, Quart. Rev. Biophys., 7:283 (1974).
3. M. V. L. Bennett, in Fish Physiology (W. S. Hoar and D. J. Randall, eds.), Academic Press, New York and London, 1971, p. 347.
4. B. Katz and R. Miledi, J. Physiol., 224:665 (1972).
5. B. Katz and R. Miledi, J. Physiol., 230:707 (1973).
6. C. R. Anderson and C. F. Stevens, J. Physiol., 235:655 (1973).
7. E. M. Landau and D. Ben Haim, Science, 185:944 (1974).
8. D. Colquhoun, V. E. Dionne, J. H. Steinbach, and C. F. Stevens, Nature, 253:204 (1975).
9. C. Y. Lee, Ann. Rev. Pharmacol., 12:265 (1972).
10. A. Karlin, J. Gen. Physiol., 54;245s (1969).
11. J.-P. Changeux, M. Kasai, and C. Y. Lee, Proc. Natl. Acad. Sci. U.S.A., 67:1241 (1970).
12. M. Altamirano, C. W. Coates, H. Grundfest, and D. Nachmansohn, J. Gen. Physiol., 37:91 (1953).
13. R. D. Keynes and H. Martins-Ferreira, J. Physiol., 119:315 (1953).
14. E. Schoffeniels, Biochim. Biophys. Acta, 26:585 (1957).
15. A. Karlin, Proc. Natl. Acad. Sci. U.S.A., 58:1162 (1967).

16. A. Karlin and E. Bartels, Biochim. Biophys. Acta, 126:525 (1966).

17. A. Karlin and M. Winnik, Proc. Natl. Acad. Sci. U.S.A., 60:668 (1968).

18. E. Bartels, W. Deal, A. Karlin, and H. Mautner, Biochim. Biophys. Acta, 203:568 (1970).

19. S. J. Singer, in Molecular Properties of Drug Receptors (R. Porter and M. O'Conner, eds.), Churchill, London, 1970, p. 229.

20. I. Silman and A. Karlin, Science, 164:1420 (1969).

21. A. Karlin, J. Prives, W. Deal, and M. Winnik, J. Mol. Biol., 61: 175 (1971).

22. G. M. Bennett and G. H. Willis, J. Chem. Soc., 1929:264.

23. E. Stedman, J. Chem. Soc., 1927:1905.

24. M. P. Cava, A. A. Deana, K. Muth, and M. J. Mitchell, in Organic Synthesis, Vol. 41, Wiley, New York, 1961, p. 93.

25. M. J. Reiter, D. A. Cowburn, J. M. Prives, and A. Karlin, Proc. Natl. Acad. Sci. U.S.A., 69:1168 (1972).

26. A. Karlin and D. A. Cowburn, Proc. Natl. Acad. Sci. U.S.A., 70: 3636 (1973).

27. A. Karlin and D. A. Cowburn, in Neurochemistry of Cholinergic Receptors (E. DeRobertis and J. Schacht, eds.), Raven Press, New York, 1974, p. 37.

28. A. Karlin, M. G. McNamee, and D. A. Cowburn, Anal. Biochem., in press.

29. A. M. Wiener, T. Platt, and K. Weber, J. Biol. Chem., 247:3242 (1972).

30. E. Karlsson, Experientia, 29:1319 (1973).

31. R. Miledi, P. Molinoff, and L. T. Potter, Nature, 229:554 (1971).

32. M. A. Raftery, J. Schmidt, D. G. Clark, and R. G. Wolcott, Biochem. Biophys. Res. Commun., 45:1622 (1971).

33. J.-C. Meunier, R. W. Olsen, A. Menez, P. Fromageot, P. Boquet, and J.-P. Changeux, Biochemistry, 11:1200 (1972).

34. B. Fulpius, S. Cha, R. Klett, and E. Reich, FEBS Lett., 24:323 (1972).

35. G. Biesecker, Biochemistry, 12:4403 (1973).

36. J.-C. Meunier, R. Sealock, R. Olsen, and J.-P. Changeux, Eur. J. Biochem., 45:371 (1974).

37. G. I. Franklin and L. T. Potter, FEBS Lett., 28:101 (1972).

38. J. Schmidt and M. A. Raftery, Anal. Biochem., 52:349 (1973).

39. R. P. Klett, B. W. Fulpuis, D. Cooper, M. Smith, E. Reich, and L. D. Possani, J. Biol. Chem., 248:6841 (1973).

40. H. W. Chang, Proc. Natl. Acad. Sci. U.S.A., 71:2113 (1974).

41. J. Lindstrom and J. Patrick, in Synaptic Transmission and Neuronal Interaction (M. V. L. Bennett, ed.), Raven Press, New York, 1974, p. 191.

42. E. Karlsson, H. Arnberg, and D. Eaker, Eur. J. Biochem., 21:1 (1971).

43. R. H. Rice and G. E. Means, J. Biol. Chem., 246:831 (1971).

44. G. Biesecker, Biochemistry, 12:4403 (1973).

45. D. E. Ong and R. N. Brady, Biochemistry, 13:2822 (1974).

46. D. A. Boroff and U. Fleck, J. Bacteriol., 92:1580 (1966).

47. M. G. McNamee, C. L. Weill, and A. Karlin, in Protein Ligand Interactions (H. Sund and G. Blauer, eds.), Verlag Walter de Gruyter, Berlin, 1975, p. 316.

48a. C. L. Weill, M. G. McNamee, and A. Karlin, Biochem. Biophys. Res. Commun., 61:997 (1974).

48b. J. M. Prives, M. J. Reiter, D. A. Cowburn, and A. Karlin, Mol. Pharmacol., 8:786 (1972).

49. M. Kasai and J. -P. Changeux, J. Memb. Biol., 6:58 (1971).

50. G. L. Hazelbauer and J. -P. Changeux, Proc. Natl. Acad. Sci. U.S.A., 71:1479 (1974).

51. J. R. Duguid and M. A. Raftery, Arch. Biochem. Biophys., 159:512 (1973).

52. J. B. Cohen, M. Weber, M. Huchet, and J. -P. Changeux, FEBS Lett., 26:43 (1972).

53. J. R. Duguid and M. A. Raftery, Biochemistry, 12:3593 (1973).

54. H. I. Silman and A. Karlin, Proc. Natl. Acad. Sci. U.S.A., 58:1664 (1967).

55. D. Michaelson, R. Vandlen, J. Bode, T. Moody, J. Schmidt, and M. A. Raftery, Arch. Biochem. Biophys., 165:796 (1974).

56. E. Heilbronn and C. Mattson, J. Neurochem., 22:315 (1974).

57. M. E. Eldefrawi and A. T. Eldefrawi, Arch. Biochem. Biophys., 159:362 (1973).

58. R. C. Carroll, M. E. Eldefrawi, and S. J. Edelstein, Biochem. Biophys. Res. Commun., 55:864 (1973).

59. G. L. Ellman, Arch. Biochem. Biophys., 82:70 (1959).

60. K. Losee and J. Bernstein, private communication.

61. A. Graf and B. Langer, J. Prakt. Chem., 148:161 (1937).

62. O. H. Lowry, N. J. Rosebrough, N. J. Farr, and A. L. Randall, J. Biol. Chem., 193:265 (1951).

63. G. L. Ellman, K. D. Courtney, G. V. Andres, and R. M. Featherstone, Biochem. Pharmacol., 7:88 (1961).

64. S. Udenfriend, S. Stein, P. Bohlen, W. Dairman, W. Leimgruber, and M. Weigele, Science, 178:871 (1972).

65a. M. A. Raftery, J. Schmidt, R. Vandlen, and T. Moody, in Neurochemistry of Cholinergic Receptors (E. DeRobertis and J. Schact, eds.), Raven Press, New York, 1974, p. 5.

65b. C. Tanford, Y. Nazaki, J. A. Reynolds, and S. Makino, Biochemistry, 13:2369 (1974).

66. J.-C. Meunier, R. W. Olsen, and J.-P. Changeux, FEBS Lett., 24:63 (1972).

67. G. Fairbanks, T. L. Steck, and D. F. H. Wallach, Biochemistry, 10:2606 (1971).

68. J. A. Reynolds and C. Tanford, Proc. Natl. Acad. Sci. U.S.A., 66:1002 (1970).

69. T.-H. Liao, G. W. Robinson, and J. Salnikow, Anal. Chem., 45:2286 (1973).

70. J. Patrick, J. Lindstrom, B. Culp, and J. McMillan, Proc. Natl. Acad. Sci. U.S.A., 70:3334 (1973).

71. H. Sugiyama, P. Benda, J.-C. Meunier, and J.-P. Changeux, FEBS Lett., 35:124 (1973).

72. G. Kemp, J. O. Dolly, E. A. Barnard, and C. E. Warner, Biochem. Biophys. Res. Commun., 54:607 (1973).

73. E. Racker, J. Biol. Chem., 247:8198 (1972).

74. E. Racker, Biochem. Biophys. Res. Commun., 55:224 (1973).

Chapter 2

PREPARATION OF ACTH-SENSITIVE ADRENOCORTICAL MEMBRANES*

Frances M. Finn and Klaus Hofmann

Protein Research Laboratory
University of Pittsburgh School of Medicine
Pittsburgh, Pennsylvania

*The experimental work reported in this paper was generously supported by Grant AM-01128 from the U.S. Public Health Service.

I. INTRODUCTION

A discussion of methodology for preparation of adrenocorticotropic hormone (ACTH) receptor(s) is somewhat premature, since such a molecule has not as yet been isolated and characterized. ACTH receptors have been located on the surface of adrenocortical cells by experiments showing that ACTH attached to carriers, presumably too large to enter the cell; for example, cellulose [1], agarose [2], or polyacrylamide [3] elicit steroidogenesis in adrenal cell preparations.

Crude particulates that bind ACTH and contain a hormone-sensitive adenylate cyclase have been described [4-6], and an attempt to solubilize such a preparation yielded a material that did not sediment at 105,000 g. In view of its behavior on gel filtration using Bio-Gel A-15, the bulk of this material was considered to be of very high molecular weight (3×10^6 daltons) [7].

A first consideration when attempting receptor isolation must be the availability of abundant starting material, since the concentration of receptor in target tissue is likely to be very low. In view of their size and ready availibility, bovine adrenals were selected as a tissue source for preparation of adrenocortical membranes. The present paper provides a detailed description of the preparation and certain properties of bovine adrenocortical plasma membranes.

II. PREPARATION OF MEMBRANES

The size and shape of bovine adrenals facilitate rapid dissection of cortical from medullary tissue, and, in addition, the zonation of the cortex itself is somewhat simpler than that in other animals and humans. The zona fasciculata constitutes the major zone in cattle adrenals; the reticularis is not well-defined, and frequently zona fasciculata cells extend all the way to the medulla [8]. On the average, 4.4 g (wet weight) of cortical tissue can be obtained from a single beef adrenal. Adrenals from young steers are preferable to glands from older animals, since the latter contain more connective tissue and, as a result, the cortical material is more difficult to homogenize.

In a typical preparation, 20 to 30 steer adrenals are collected in an ice-cold "medium" (prepared by adjusting the pH of distilled water to 7.5 by addition of $NaHCO_3$) and are transported to the laboratory in an ice bath. The need for cooling at this point is not critical, since we have frequently observed that the adenylate cyclase system of plasma membranes prepared from adrenals stored at ambient temperature for as long as 1 hr after collection is still highly responsive to ACTH stimulation.

The remainder of the operations involved in plasma membrane preparation are, however, performed at ice-bath temperature.

The glands are defatted and bisected longitudinally. A convenient starting point for bisection is at the insertion of the adrenal artery. Once the cortical tissue is exposed to air, care must be exercised to prevent tissue dehydration. Membranes prepared from partially dehydrated tissue appeared as myelin figures when examined in the electron microscope. For this reason, all subsequent dissection is performed under 0.25 M sucrose.* Petri dishes (100 × 15 mm) lined with 90-mm filter paper circles are half-filled with the sucrose solution and placed inside ice-filled crystallizing dishes (100 × 50 mm). For demedulation, each adrenal half is cut along the border between medulla and cortex, and the medullary tissue is removed by a combination of teasing and cutting. Single cortices are blotted, quickly transferred to a petri dish containing a measured amount of ice-cold 0.25 M sucrose, and the cortical tissue is scraped from the capsule with a spatula. The suspension is then transferred to a tared beaker. The adrenals can be subdivided into batches most conveniently at the stage where cortex and capsule are separated from one another. Usually each batch contains tissue from ten adrenals. The suspension is weighed and diluted with 0.25 M sucrose to a final tissue concentration of 1 g/9 ml. The diluted suspension is homogenized in a Dounce homogenizer (Blaessig Glass, Rochester, N. Y.) using the loosely fitting pestle (10 strokes) and rehomogenized by hand in a glass-glass homogenizer with a clearance of 1 mm (1 stroke). The homogenate is diluted to 1,200 ml with medium, the suspension is stirred for 5 min with a magnetic stirrer, and filtered through four layers of 20/12 standard surgical gauze.

The homogenization procedure is a compromise between opposing goals: namely, to homogenize thoroughly enough to ensure a good yield of membranes, and to preserve the integrity of the plasmalemma sufficiently so that it will sediment at a reasonably low centrifugal force.

The filtrate is centrifuged (Sorvall RC2 B centrifuge with GSA rotor) at 10,400 g for 20 min, and the supernatant is decanted. The pellet is resuspended in medium with the Dounce homogenizer (3 strokes) and diluted to 400 ml with medium. This suspension is centrifuged at 10,400 g for 20 min, and the supernatant is aspirated and discarded. The pellets from the various batches are now combined and resuspended in medium; the suspension is diluted with medium to 100 ml per batch and an equal volume of 63% sucrose wt/wt is added with stirring. Centrifuge tubes

---

*The pH of all sucrose solutions was adjusted to 7.5 with $NaHCO_3$.

(Sorvall 290-ml polycarbonate bottles), each containing 80 ml of 45% sucrose wt/wt, are layered with 100 ml of the above suspension and centrifuged for 20 min at 10,400 g. After centrifugation, a top layer—a gelatinous interphase adhering to the wall—and erythrocytes at the bottom of the tube are visible. The top layer (rich in membranes) is collected by aspiration, care being taken not to remove the gelatinous interphase. The pooled supernatants are mixed with an equal volume of medium to lower the density of the suspension, to facilitate concentration of the membranes by centrifugation. The resulting suspension is centrifuged at 23,000 g for 30 min (Beckman Type 21 rotor), the supernatant is aspirated, and the pooled pellets are resuspended in 0.25 M sucrose with the Dounce homogenizer and diluted to a final volume of 50 ml. This membrane suspension is used for isopycnic centrifugation.

Loading the zonal rotor (Ti-14, Beckman-Spinco) with the sucrose gradient for isopycnic centrifugation should begin during the period when the membrane preparation is centrifuging in the two-phase discontinuous gradient, so that the zonal rotor will be filled when the final suspension is ready for injection. The gradient is established using a Beckman high-capacity gradient pump, Model 141. In earlier research [9] we employed a linear gradient composed of 32% wt/wt and 45% wt/wt sucrose with a "cushion" of 45% sucrose (150 ml). We have subsequently found that the yield of membranes can be markedly improved by increasing the volume of the lighter-density solution at the expense of the cushion. Thus, 100 ml of the lighter solution is first pumped to the rotor, then a linear gradient is begun. Under these conditions, membranes concentrate (as judged by protein concentration) in the 32% sucrose solution (Fig. 1).

The membrane suspension is injected into the core of the filled rotor, spinning at 3,000 rpm, and is followed by an overlay (15 ml) of 0.25 M sucrose. Centrifugation is carried out at 47,000 g at 4°C over a period of 16 hr to allow enough time for particles to equilibrate with the sucrose concentration corresponding to their buoyant density. The centrifuge is then decelerated to 3,000 rpm, and the rotor contents are displaced by 45% sucrose pumped to the periphery of the rotor at a rate of ~7.5 ml/min. Tygon tubing (1 mm i.d.) is used on the outlet to reduce mixing. Fractions (25 ml each) collected in ice-cooled plastic bottles are used immediately for protein and enzyme determinations. The yield of membrane protein averages ~0.3 mg/g wet weight of cortical tissue. Since protein is only 38.6% of the total membrane weight [10], the yield on a weight basis is, therefore, ~0.8 mg/g starting material.

The efficiency of the isopycnic separation was checked by adding ^{14}C-labeled membranes [11] to a fresh membrane suspension and subjecting this mixture to isopycnic centrifugation. Over 65% of the labeled membranes were recovered in the fractions, corresponding to 32% wt/wt sucrose.

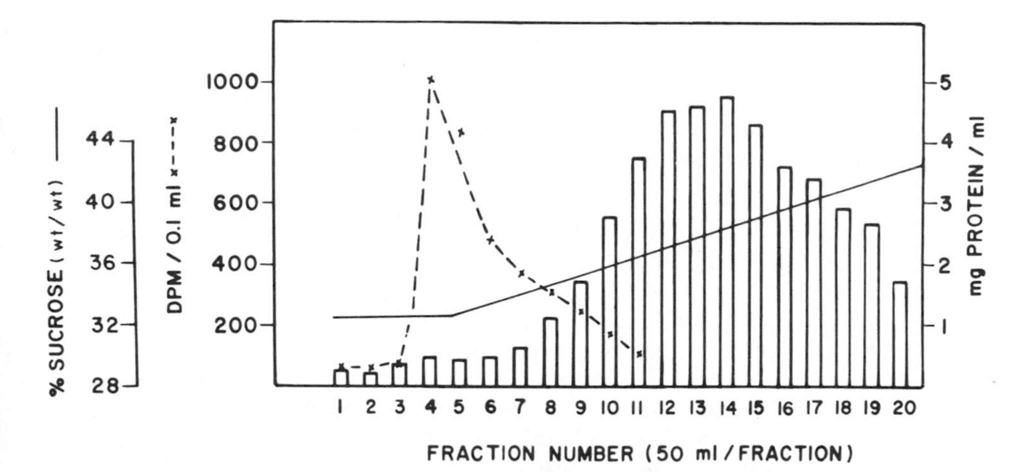

FIG. 1. Distribution of protein and radioactively labeled membranes in various fractions derived from zonal centrifugation. Labeled membranes (see text) were added to a crude membrane preparation from the discontinuous gradient, and the mixture was subjected to isopycnic centrifugation. Fifty adrenals were processed and zonal centrifugation was performed using a Ti-15 rotor.

Adrenal cortical plasma membranes are apparently much more sensitive to fragmentation during homogenization than membranes from liver tissue. Long membrane sheets (described for liver membrane preparations) that readily sediment in low centrifugal fields are not obtained from cortical tissue. Electron microscopic examination of adrenal plasma membranes in intact cells reveals the presence of very few desmosomes and tight junctions [9]. Consistent with this observation are data on the chemical composition of isolated membranes which demonstrate a high lipid:protein ratio [10].

When a mixture of labeled membranes and fresh, crude membranes from the discontinuous gradient was centrifuged at 23,000 g for 30 min, over 30% of the label remained in the "clear" supernatant. We have noted a negative correlation between the number of velocity centrifugations and membrane yield.

In an attempt to obtain a greater enrichment of membranes, purified membranes were subjected to a second isopycnic centrifugation using a shallower sucrose gradient. This procedure necessitated collecting and resuspending the membrane fraction. A teflon-glass homogenizer (A. H. Thomas, type C) was used for this purpose. Electron micrographs of the resultant membranes clearly indicated that substantial fragmentation had occurred during this process.

The higher centrifugal forces necessary to sediment the more fragmented cortical membranes result in significant contamination with mitochondria. However, the mitochondria have a buoyant density different from that of plasma membranes, and the two are readily separated by isopycnic centrifugation.

The protein concentration of membrane-rich fractions from zonal centrifugation (usually fractions 1 through 5) is determined by a modification of the Lowry method [12]. Suitable aliquots (usually 1 ml) of the zonal material are diluted to 6 ml with medium and collected by centrifugation at 23,000 g for 20 min (Beckman type 30.2 rotor). This step is included to dilute the sucrose, since impurities in the sucrose tend to give falsely elevated protein values. The supernatant is decanted and, at this point, the appearance of the pellet should be examined. Plasma membrane pellets are clear and virtually colorless. A tan or brown coloration usually indicates the presence of mitochondria. The pellet material is solubilized in 4 ml of 0.1 N sodium hydroxide containing sodium dodecyl sulfate (SDS) (0.75 mg/ml). Detergent is essential for solubilization and its presence has no effect on subsequent color development. The suspensions are stirred for 30 min at room temperature; aliquots (0.2, 0.5, and 1.0 ml) are diluted to 2 ml with the SDS containing sodium hydroxide; and the Lowry color reaction is performed in the usual manner.

III. CRITERIA OF HOMOGENEITY

A. Marker Enzymes

An accepted method for assessing the degree of purification of subcellular organelles involves evaluation of the change in specific activity of characteristic marker enzymes. However, as concerns plasma membranes, this approach has some drawbacks. This is not to say that marker enzyme analysis should be ignored, but rather that the limitations of this method should be recognized.

The enzyme 5'-nucleotidase is considered primarily a plasma membrane enzyme for, although its presence has been demonstrated in other organelles [13], the highest specific activity is observed in plasma membrane preparations. Determination of 5'-nucleotidase depends on measuring the liberation of inorganic phosphate from 5'-AMP. Using this substrate, we observed little difference in specific activity between homogenate and purified membranes. The reason for this became clear when a pH-activity profile was executed using 5'-AMP, 2'-AMP, and β-glycerophosphate as substrates. Most of the activity ascribed to 5'-nucleotidase in the homogenate was apparently due to alkaline phosphatases. Specific

5'-nucleotidase activity in the membrane was actually enriched at least 30-fold over that present in the homogenate [10].

A second problem, particularly troublesome in the case of membrane-bound marker enzymes, is that of enzyme instability. The preparation of bovine adrenocortical plasma membranes requires ~20 hr from dissection to harvesting from the zonal rotor. During this period, many of the enzymes one might use to judge homogeneity are progressively inactivated. Furthermore, the rates of inactivation are dependent on environmental factors, i.e., the presence of proteolytic and lipolytic enzymes, tonicity and pH of the medium, and others. Thus, a valid comparison of the specific enzyme activities of various fractions cannot be made by collecting samples during the preparative steps.

Conversely, the purity of membranes can be inferred by evaluating the activities of marker enzymes for cellular constituents likely to be contaminents. Assaying mitochondrial enzymes, for example, cytochrome oxidase [14], is particularly helpful in the case of adrenocortical membranes since mitochondria are the most likely source of contamination. A comparison of the cytochrome oxidase activity of various fractions from the zonal centrifugation was in agreement with the results of electron microscopic findings (see Sec. III. B). The specific activity of cytochrome oxidase in the mitochondrial fraction was approximately 48 units, the low-density fractions containing mainly membranes exhibited an activity of only 1 to 4 units.

The criticisms discussed for plasma membrane marker enzymes apply to these measurements as well.

B. Electron Microscopy

Electron microscopic visualization of suitably stained preparations affords another method for assessing homogeneity. However, even examination of a truly representative field may be misleading, since adrenocortical plasma membranes have few features that distinguish them from other subcellular membranes. The ability to form vesicles is not unique for plasma membranes; however, if the vesicles are sufficiently large, contamination by membranes derived from small organelles can be excluded. The presence of free ends among the vesicles is also diagnostic.

The preparation described here contains mainly vesicles with a circumference >2 μm, together with occasional membranes which show free ends. In addition, myelin fragments and partially disorganized lipofuchsin granules were observed occasionally. Very few mitochondria or mitochondrial fragments were present [9].

A more specific electron microscopic test for purity involves the cytochemical localization of 5'-nucleotidase by the in situ precipitation of lead phosphate as a result of the hydrolysis of 5'-AMP [13]. Provided that controls using 2'-AMP (to exclude nonspecific phosphatase activity) are included, and that other subcellular fractions are examined, this technique appears to be reasonably specific for plasma membranes. When the adrenocortical membranes were examined by this technique, precipitates of lead phosphate were concentrated both in large vesicles and in membranes with free ends.

IV. HORMONE-RECEPTOR INTERACTION

A. Labeling the Hormone

In order to carry out meaningful binding studies, radioactively labeled ACTH exhibiting full biological activity must be available. Iodination, whether in the presence of ICl, chloramine-T-KI, I_2, I_3^-, or peroxidase-catalyzed iodination, leads to oxidation or oxidative cleavage of tryptophyl peptide bonds [15]. In addition, histidine residues can undergo iodination, and oxidative changes may occur at serine and methionine. The oxidation of methionine using the chloramine-T method has been observed with gastrin [16]. Oxidation of tryptophan and methionine residues destroys the biological activity of ACTH. Dedman et al. [17] clearly demonstrated that peroxide oxidation, presumably of methionine to methionine sulfoxide, drastically lowers the biological activity of ACTH [I] (Table 1). Furthermore, the sensitivity of biological activity to replacement of tryptophan by phenylalanine [VI] [18], N-methyl tryptophan [III] [19], or o-nitrophenylsulfenyl tryptophan [IV] [20, 21] has been amply documented. Attempts to purify iodinated ACTH have been only partially successful. Lefkowitz et al. [4] separated the iodinated molecule from natural porcine ACTH by gradient elution on carboxymethyl cellulose; however, the purified material had only 32 units/mg when assayed by adenylate cyclase activation.

In view of all the drawbacks of this method of labeling, one wonders why its use has been so widespread. The argument that only iodination can provide the level of specific radioactivity required to measure hormone-receptor binding can be questioned, at least for ACTH interaction with its receptor. Results from early studies [22] with a mouse adrenal tumor suggested that ACTH affinity for this preparation could be characterized by two separate binding constants, 10^{12} M and 10^{7} M. Curiously, though, the apparent K_M for adenylate cyclase activation that can be calculated from the results presented corresponds only to the lower constant, $\sim 10^{7}$ M. Using a biologically fully active ^{14}C-labeled derivative of ACTH, (Gln^5) $ACTH_{1-20}$ amide [V], of low specific radio-

TABLE 1. Amino Acid Sequence of ACTH and Analogs[a]

	1 2 3 4 5 6 7 8 9 10 11 12 13 14 15 16 17 18 19 20 21 22 23 24 39
I	H-Ser-Tyr-Ser-Met-Glu-His-Phe-Arg-COHN-Trp-Gly-Lys-Pro-Val-Gly-Lys-Lys-Arg-Arg-Pro-Val-Lys-Val-Tyr-Pro----------Phe-OH
II	H-Ser-- Pro-OH
III	H-Ser-------------------------------CO-N(CH_3)-Trp -- Pro-OH
IV	H-Ser---------------------------------------Trp(NPS)-- Pro-OH
V	H-Ser--------------Gln ----Phe*--Val-$CONH_2$
VI	H-Ser--------------Gln --------------------Phe--Val-$CONH_2$
VII	Boc-Ser-- Gly-OH
VIII	H-Lys(Boc) ------------- Lys(Boc)-Lys(Boc) ----------------- Lys(Boc)---------- Pro-O-t-Bu

[a]Boc = *tert*-butoxycarbonyl; O-t-Bu = *tert*-Butylester; Phe* = [^{14}C] Phe; NPS = orthonitro phenylsulfenyl.

activity, we were able to confirm the existence of the lower binding constant as well as the enzyme K_M [9] with bovine adrenocortical plasma membranes. Saez et al. [23], using iodinated $ACTH_{1-24}$ [II] with a much higher specific radioactivity, have been unable to confirm the existence of the high affinity sites (10^{12} M) in isolated plasma membranes prepared from human, sheep, and rat adrenals according to the method described in this review. Thus, it would seem unnecessary to resort to the use of ^{131}I or ^{125}I in order to detect ACTH-receptor interaction.

More recently a technique that circumvents the undesirable features of the usual iodination has been described by Brundish and Wade [24]. In this procedure the carboxy terminal-protected tetradecapeptide $ACTH_{11-24}$ [VIII of $ACTH_{1-24}$ [II] is iodinated with iodine monochloride. This segment of the molecule contains a tyrosine in position 23, and none of the other residues present are likely to undergo iodination or oxidation. The iodinated peptide is then coupled to the protected amino terminal portion $ACTH_{1-10}$ [VII], the iodine is catalytically exchanged for tritium, and the product deprotected. After column chromatography, the specific radioactivity was 46 Ci/mmol, and biological assay on isolated adrenal cells indicated that the labeled peptide possessed full steroidogenic activity.

We have found that $ACTH_{1-24}$ [II] can be conveniently labeled by methylation with formaldehyde followed by reduction with sodium borohydride in a manner similar to that described by Rice and Means [11] for labeling proteins. $ACTH_{1-24}$ [II] (Synacthen, Ciba) (0.496 mg; 0.169 μmol) is dissolved in 1 ml of 0.01 N HCl, and the solution is lyophilized immediately. The residue is dissolved in 0.25 ml of water; and the solution is added to a 50-ml round-bottom flask containing 0.5 ml of 0.2 M sodium borate buffer (pH 9.0), 0.5 ml of an aqueous solution of [^{14}C]formaldehyde (0.025 mCi; 0.42 μmol), and 10 μl of an aqueous solution of formaldehyde (1.25 μmol). The volumes given in this procedure should be adhered to precisely to avoid precipitation of the peptide. After 10 min at room temperature, four 10-μl portions of a freshly prepared aqueous solution of sodium borohydride (0.423 μmol/10 μl) are added over a period of 5 min. The solution is then lyophilized, the residue is redissolved in 0.5 ml of deaerated 10% acetic acid, and the solution is applied to a Sephadex G-25 column (0.9 × 140 cm) equilibrated with the same solvent. The column is eluted with deaerated 10% acetic acid, fractions (2 ml each) are collected, and aliquots (10 μl) are examined for radioactivity. Radioactivity is eluted in two peaks corresponding approximately to tubes 15 through 20 and 30 through 35. The contents of the tubes comprising the first fraction are pooled and lyophilized. The specific activity of the sample was 10.7 mCi/mmol. This material exhibited approximately 140% the activity of the underivatized molecule when assayed using the rat adrenal cell assay [25].* Radioactivity can either be incorporated via [^{14}C]formaldehyde or, if higher specific radioactivity is desired, [^{3}H] sodium borohydride may be used for reduction.

B. Binding Studies

The hormone binding capabilities of adrenocortical plasma membranes can be measured using either fresh or frozen preparations. Binding measurements have been performed with preparations stored for as long as 18 months at -15°C without noticeable change in hormone affinity.

The binding studies, as originally described, were performed with (Gln^5) $ACTH_{1-20}$ amide [V]. The label was introduced synthetically as [^{14}C]phenylalanine, and the resulting peptide had a specific activity of 0.117 mCi/mmol. Many of the binding specificity studies have since been repeated using the methylated [^{14}C]$ACTH_{1-24}$ preparation (specific activity 10.7 mCi/mmol) described, with the same results; however the use of a hormone with higher specific radioactivity (hence more dilute solutions) necessitates the incorporation of measures to prevent nonspecific binding. At high dilution, ACTH, in neutral or basic solutions, binds rapidly to paper, glass [26], Sephadex, talc, and many other seemingly inert materials. Indeed, the binding of ACTH to microfine precipitated silica (QUSO G_{32}; Philadelphia Quartz Co.) serves as a method of separation of ACTH from so called "damaged" molecules during the iodination reaction [27]. We have noted that neither siliconization nor the inclusion of albumin effectively prevents this nonspecific adsorption. The exclusive use of nonwettable plastics seems to offer the best solution to this problem. When more concentrated hormone solutions are used, the nonspecific adsorption becomes negligible.

For binding studies, the membrane fractions were diluted to 200 μg protein per ml with a 1:1 vol/vol mixture of medium and 37% sucrose wt/wt and suspended with a Dounce homogenizer. Aliquots of these suspensions were added to polyallomer centrifuge tubes (Beckman, 0.5 × 3.5 in.) containing [^{14}C](Gln^5) $ACTH_{1-20}$ amide [V] (final concentration, $0.5 - 5.0 \times 10^{-5}$ mmol/ml) or the labeled hormone, together with varying amounts of structurally related unlabeled peptides dissolved in water (1 ml). All solutions are at ice-bath temperature. The tubes are shaken and kept at 0°C for 10 min. Saez at al. [23] have shown that binding is very rapid even at this temperature. After centrifugation for 30 min at

---

*We thank Dr. G. Sayers of the Physiology Department, Case Western Reserve University, School of Medicine, Cleveland, Ohio, for this assay.

23,000 g, samples (0.5-1.0 ml) are withdrawn and counted in a scintillation spectrometer. A tube containing radioactive hormone and 1 ml of the 1:1 mixture of sucrose and medium in place of membranes serves as a control. Disappearance of radioactivity from the supernatant is taken as a measure of binding. Alternatively, the pellet may be counted directly.

V. ADENYLATE CYCLASE ASSAY

Occupation of the receptor by a physiologically active peptide results in activation of a plasma membrane enzyme adenylate cyclase. Hence, the demonstration that the isolated subcellular fraction contains an adenylate cyclase specifically responsive to ACTH provides another criterion of purity of the preparation. Hormone-sensitive adenylate cyclase is found in the zonal fractions enriched in plasma membranes and coincides with 5'-nucleotidase activity. This adenylate cyclase can also be activated by fluoride ion. Fluoride activation precludes hormone stimulation, and the resulting activity is usually greater.

Unlike hormone binding, adrenocortical adenylate cyclase activity is rather unstable. Large losses (50% or more) in activity may occur within 24 hr when the membranes are stored even at 4°C. For this reason, it is essential to use fresh membranes for enzyme assays.

The substrate [^{3}H]ATP contains, in addition to the expected hydrolysis products, radioactivity detectable on thin layer chromatography in a position corresponding to the R_f for 3'5'-cyclic AMP (cAMP). Approximately 1% of the total radioactivity is found here. Since only ~0.1% of substrate ATP is converted to cAMP during the course of a normal enzyme assay, it was considered advisable to remove this contaminant before using the substrate. The purification of ATP is performed as follows:

A solution of [^{3}H]ATP (1.0 mCi) and ATP (20 mg) in water (1 ml) is applied to a water-washed AG1-X2 (200-400 mesh, Bio-Rad Laboratories) column (0.4 × 3.5 cm in a Pasteur pipet) in the acetate cycle. The column is eluted with 0.01 N HCl-0.01 M NaCl until the level of radioactivity in the effluent is negligible (~200 ml of solvent). This serves to remove AMP, cAMP, ADP, and any neutral or basic impurities. ATP is then eluted with 0.1 N HCl-0.05 M NaCl (~70 ml). The pH of the ATP solution is adjusted to 7.0 with concentrated ammonium hydroxide, and the solution is desalted on a Norite A-filter cel column. To prepare this column, Norite A is washed by repeated centrifugation at 164 g for 10 min and resuspension in 10 vol of a solution of 0.15 N ammonium hydroxide:ethanol 1:1. Six or seven washings should produce a clear supernatant and afford charcoal particles of sufficient size to be retained on the column. The supernatant is aspirated after the final wash and the pellet is dried in vacuo over P_2O_5. The column is prepared from a dry mixture of 2.5 g of filter cel and 5 g of washed charcoal poured into a 1-cm diam glass tube having one end constricted and plugged with cotton. The sample is applied to the dry column and followed by a 10-ml water wash. Then the ATP is eluted using water aspirator vacuum with the ammonia-ethanol mixture, the effluent collected in a 500-ml round bottom flask, concentrated in vacuo at 35°C, and lyophilized. The residue is dissolved in 12 ml of 0.12 M tris-HCl buffer (pH 7.3), containing 0.03 M theophylline and 13.2 mM $MgSO_4$.

For determination of adenylate cyclase, the following reagents were mixed in polyallomer tubes at ice-bath temperature:

1. ATP regenerating system.
   a. 0.1 ml of an aqueous solution (50 mg/ml) of P-enolpyruvate (trisodium salt hydrate; Sigma Chemical Company).
   b. 0.01 ml of pyruvate kinase (rabbit skeletal muscle, type II, 380 units/mg; Sigma Chemical Company).
2. 0.01 ml of 0.8 M dithiothreitol or dithioerythritol.
3. 0.1 ml of a 0.1% solution of bovine serum albumin (pH 7.5) containing appropriate concentrations of the various peptides.
4. 0.2 ml of the ATP solution. Sodium fluoride, when used, is dissolved in the ATP solution at a concentration of 10^{-2} M.
5. 0.4 ml of plasma membranes (1 mg protein per ml) suspended in medium. This suspension is prepared by diluting the desired zonal fractions with 2 vol of medium, collecting the membranes by centrifugation at 23,000 g for 30 min, and resuspending the pellet in sufficient medium to obtain the desired protein concentration.

The samples are transferred to a Dubnoff shaker and incubated for 15 min at 30°C. The tubes are transferred to an ice bath, cAMP (0.1 ml of an aqueous solution, 1.52 mmol) is added, and the tubes are placed in a boiling-water bath for 2 min. At this point, cAMP determinations may be performed, or the samples may be stored frozen at -15°C.

cAMP is determined by a modification of the method of Krishna et al. [28]. The incubation mixtures are centrifuged at 23,000 g for 20 min and the supernatants applied to water-washed AG1-X2 (acetate cycle) columns (3.5 × 0.4 cm) which are eluted with 0.1 N H_3PO_4. Fractions (2 ml each) are collected, and absorbance at 260 nm serves to locate the desired material. cAMP and AMP elute as a single peak (usually fraction 7), while ATP and ADP are retained on the column. This chromatographic step was introduced when it was found that the barium sulfate precipitation of ATP was not quantitative. The fraction exhibiting maximum absorbance is adjusted to pH 7 with N NaOH (0.1 ml), and 0.4 ml each of 0.25 M barium hydroxide and 0.25 M zinc sulfate are added. The suspensions are centrifuged at 1,280 g for 10 min, and the clear supernatants are withdrawn and perculated through water-washed Dowex 50W-X4 (H^+ cycle) columns (3.5 × 0.4 cm) until the miniscus reaches the top of the resin bed. The columns are then eluted with 0.1 N H_3PO_4. Fractions (2 ml then 3 ml are collected separately and their absorbance at 260 nm is measured to determine recovery of cAMP. Aliquots (0.5-1.0 ml) of the 3-ml eluate are counted in Scintisol Complete (Isolab, Inc. Akron, Ohio) with a liquid scintillation spectrometer.

Adrenocortical membranes contain ATPases and phosphodiesterase that influence the levels of substrate and product. The conditions described for assessing adenylate cyclase activity have been shown to provide an assay that is linear with respect to time and proportional to membrane protein concentration. The ATP-regenerating system maintains a constant level of substrate and the inclusion of theophylline inhibits the phosphodiesterase [28].

The degree of hormone stimulation of adenylate cyclase varies from one preparation to another (Fig. 2). The reason for this variation is unclear; however, we have observed that during the hot summer months,

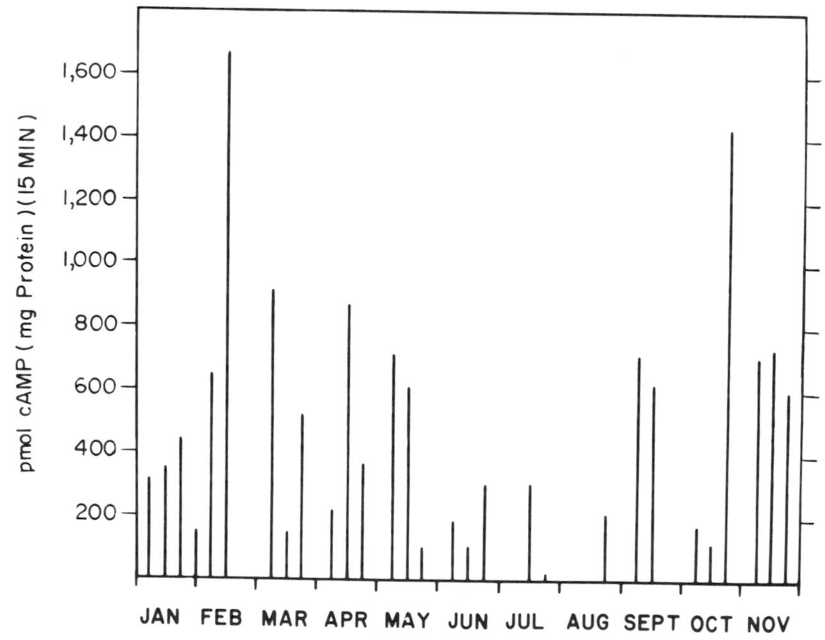

FIG. 2. Variability of response to stimulation by $ACTH_{1-24}$ of adrenocortical membrane adenylate cyclase. Each bar represents a different membrane preparation. Basal (unstimulated) enzyme activity has been subtracted.

hormone stimulation is frequently poor and basal adenylate cyclase levels are increased two- to threefold above normal. During the same period, the level of NaF stimulation may be normal. We have tentatively attributed this result to the presence of a higher level of circulating ACTH caused by the added stress of hot weather on the animals.

Lipid composition, the results of binding studies and adenylate cyclase activation, and a discussion of properties of the membranes considered to be beyond the scope of the present communication may be found in Refs. 9, 10, and 29-31.

REFERENCES

1. B. P. Schimmer, K. Ueda, and G. H. Sato, Biochem. Biophys. Res. Commun., 32:806 (1968).
2. R. C. L. Selinger and M. Civen, Biochem. Biophys. Res. Commun., 43:793 (1971).
3. M. C. Richardson and D. Schulster, J. Endocrinol., 55:127 (1972).
4. R. J. Lefkowitz, J. Roth, W. Pricer, and I. Pastan, Proc. Natl. Acad. Sci. U.S.A., 65:745 (1970).
5. L. A. Kelly and S. B. Koritz, Biochim. Biophys. Acta, 237:141 (1971).
6. M. Ide, A. Tanaka, M. Nakamura, and T. Okabayashi, Arch. Biochem. Biophys., 149:189 (1972).
7. I. Pastan, W. Pricer, and J. Blanchett-Mackie, Metabolism, 19:809 (1970).
8. L. Nicander, Acta Anat., 14, Suppl. 16 (1952).
9. F. M. Finn, C. C. Widnell, and K. Hofmann, J. Biol. Chem., 247:5695 (1972).
10. T. P. Seltzman, F. M. Finn, C. C. Widnell, and K. Hofmann, J. Biol. Chem., 250:1193 (1975).
11. R. H. Rice and G. E. Means, J. Biol. Chem., 246:831 (1971).
12. O. H. Lowry, N. J. Rosebrough, A. L. Farr, and R. J. Randall, J. Biol. Chem., 193:265 (1951).
13. C. C. Widnell, J. Cell Biol., 52:542 (1972).
14. S. J. Cooperstein and A. Lazarow, J. Biol. Chem., 189:665 (1951).
15. N. M. Alexander, J. Biol. Chem., 249:1946 (1974).

16. B. H. Stagg, J. M. Temperley, H. Rochman, and J. S. Morley, Nature, 228:58 (1970).

17. M. L. Dedman, T. H. Farmer, and C. J. O. R. Morris, Biochem. J., 78:348 (1961).

18. K. Hofmann, R. Andreatta, H. Bohn, and L. Moroder, J. Med. Chem., 13:339 (1970).

19. M. Fujino, C. Hatanaka, O. Nishimura, R. Nakayama, M. Shitaka, and R. Fujiwara, Proc. 8th Symp. Pept. Chem., Osaka, 1971, p. 101.

20. Y. C. Kong, W. R. Moyle, and J. Ramachandran, Proc. Soc. Exptl. Biol. Med., 141:350 (1972).

21. S. Seelig, S. Kumar, and G. Sayers, Proc. Soc. Exptl. Biol. Med., 139:1217 (1972).

22. R. J. Lefkowitz, J. Roth, and I. Pastan, Ann. N.Y. Acad. Sci., 185:195 (1971).

23. J. M. Saez, A. M. Morera, A. Dazord, and P. Bataille, J. Steroid Biochem., 5:9251 (1974).

24. D. E. Brundish and R. Wade, J. Chem. Soc., Perkin Trans. I, 1973, p. 2875.

25. G. Sayers, R. L. Swallow, and N. D. Giordano, Endocrinology, 88: 1063 (1971).

26. S. A. Berson and R. S. Yalow, J. Clin. Invest., 47:2725 (1968).

27. G. Rosselin, R. Assan, R. S. Yalow, and S. A. Berson, Nature, 212:355 (1966).

28. G. Krishna, B. Weiss, and B. B. Brodie, J. Pharmacol. Exptl. Ther., 163:379 (1968).

29. F. M. Finn, J. A. Montibeller, Y. Ushijima, and K. Hofmann, J. Biol. Chem., 250:1186 (1975).

30. K. Hofmann, W. Wingender, and F. M. Finn, Proc. Natl. Acad. Sci. U.S.A., 67:829 (1970).

31. K. Hofmann, J. A. Montibeller, and F. M. Finn, Proc. Natl. Acad. Sci. U.S.A., 71:80 (1974).

Chapter 3

$[^3H]$ (-)-ALPRENOLOL: A NEW TOOL FOR THE STUDY OF BETA-ADRENERGIC RECEPTORS

Robert J. Lefkowitz

Departments of Medicine and Biochemistry
Duke University Medical Center
Durham, North Carolina

I. INTRODUCTION

The problem of direct identification of adrenergic receptors by binding studies has proved rather difficult due, primarily, to the large number of competing uptake processes for catecholamines present in tissues. These uptake processes include at least two well-defined uptake processes [1], one neuronal and one nonneuronal, uptake into nerve storage vesicles [2], possible binding to several known metabolizing enzymes for catecholamines [3], as well as possible uptake onto other nonspecific sites as yet not defined. Nonetheless, despite these rather imposing problems, identification of beta-adrenergic receptors can be made on the basis of direct binding studies by strict application of rigorous criteria for receptor identification [4-7].

Ahlquist clearly delineated, more than 25 years ago, the fact that there are two distinct classes of adrenergic receptors for catecholamines, which he termed alpha and beta [8]. The receptors were identified on the basis of differing orders of potencies of agonist drugs. Alpha-adrenergic receptors are those which interact most strongly with epinephrine and norepinephrine and only weakly, if at all, with isoproterenol. Beta-adrenergic receptors are those which interact most strongly with isoproterenol and only very weakly with norepinephrine [8]. Subsequent work from many laboratories has confirmed and extended this classification. In addition, specific antagonist compounds, not available to Ahlquist, have also been used to validate the classification. Thus, there are specific alpha-adrenergic antagonists, such as phentolamine and phenoxybenamine, which block only the alpha-adrenergic class of receptors. Specific antagonists such as propranolol or dichlorisoproterenol block only the beta class of receptors and are referred to as beta-adrenergic antagonists [9]. The beta-adrenergic receptors have been further subclassified into beta$_1$- and beta$_2$-adrenergic receptors [10]. The beta$_1$ receptors interact strongly not only with isoproterenol and epinephrine but also with norepinephrine, which is generally from one-third to one-tenth as potent as isoproterenol and generally almost equipotent with epinephrine. Beta$_1$-adrenergic receptors are found in the heart and adipose tissue [10]. Beta$_2$-adrenergic receptors are characterized by a strong interaction with isoproterenol, but norepinephrine is much weaker, being generally anywhere from one hundreth to one thousandth as potent as isoproterenol. Beta$_2$-adrenergic receptors appear to mediate the metabolic effects of catecholamines in skeletal muscle and liver, as well as the beta-adrenergic effects of catecholamines on the smooth muscle of the trachea, uterus, and intestinal tract [10, 11]. Beta$_1$- and beta$_2$-adrenergic receptors can also be distinguished by the use of specific antagonist compounds. Thus, practolol appears to be relatively more active at beta$_1$-adrenergic receptors, whereas butoxamine is more active at beta$_2$-adrenergic receptors [12-14].

At a biochemical level, relatively little work appears to have been directed at the molecular identification of alpha-adrenergic receptors. Several studies with relatively low specific activity, [^{3}H]- or [^{14}C]alpha-adrenergic antagonists and whole tissue preparations appeared to label largely nonspecific sites which were unrelated to the receptors themselves [15, 16]. The major problem with attempts to identify alpha-adrenergic receptors in subcellular fractions has been the lack of a well-characterized biochemical consequence of alpha-adrenergic receptor interaction with which to correlate binding studies. Considerably more effort has been directed toward the identification and study of the beta-adrenergic receptors. In large part, this is attributable to the fact that the consequence of beta-adrenergic receptor interaction in many tissues has been clearly identified as adenylate cyclase activation [17]. In membrane fractions in which beta-adrenergic receptors are being sought, a well-characterized enzymatic activation (adenylate cyclase) can be studied in parallel with binding experiments, thus providing a convenient tool for correlation and validation of binding data. Thus, in attempting to identify beta-adrenergic receptors in membrane fractions derived from catecholamine-sensitive tissues, one can deduce the characteristics to be expected of the beta-adrenergic receptor-binding sites from the observed characteristics of the interaction of adenylate cyclase with catecholamines. These characteristics are typical of beta-adrenergic receptors when compared with experiments performed in more intact tissue preparations. The binding characteristics to be expected of beta-adrenergic binding sites are as follows: (1) Kinetic characteristics: Interaction of beta-adrenergic agonist catecholamines with adenylate cyclase is a rapid process leading to the establishment of fully stimulated rates of enzyme activity within one minute [5]. Similarly, interaction of antagonists with the system is also rapid, since addition of such antagonists to catecholamine-stimulated enzyme preparations leads to a return to basal enzyme activity, which is fully established within one minute after the addition of the antagonist [5, 18]. Thus, the binding interaction, when studied directly, should exhibit a very rapid rate of association and dissociation. (2) Specificity characteristics: beta-Adrenergic specificity should be demonstrated for binding affinity. The order of potency of classical agonists should be isoproterenol > epinephrine > norepinephrine; and beta-blockers such as propranolol should interact strongly, whereas alpha-blockers such as phentolamine should be weak or inert. Further, on the basis of recently developed data on beta-adrenergic subtypes and adenylate cyclase [19], it is to be expected that binding specificity should display either a beta$_1$-

or $beta_2$ -adrenergic specificity, depending on the characteristics of the response patterns apparent with the intact tissue preparations as well as with the membrane preparations. In addition, one would expect stereospecificity for the interaction of both agonists and antagonists inasmuch as intact tissue preparations and subcellular membrane preparations containing adenylate cyclase both display marked stereospecificity in their response to agonists and antagonists [5]. The (-)-isomers of compounds are generally considerably more active than the (+)-isomers. It should be pointed out that, as compared with most of the polypeptide hormone systems, where relatively few analogs are available for study, the adrenergic system is characterized by the very wide array of agonist and antagonist compounds which are readily available for structure activity studies. (3) Affinity: Not only should the order of potencies of agonists and antagonists be appropriate to the beta-adrenergic receptor, but the affinities of both agonists and antagonists for the binding sites should be in at least reasonable accord with those deduced from physiological studies.

II. [^{3}H] CATECHOLAMINE BINDING

Over the past few years, a great deal of data has accumulated dealing with attempts to use ^{3}H-labeled catecholamine agonists such as isoproterenol [20, 21], epinephrine [22, 23], and norepinephrine [24-27] to perform binding studies with membrane fractions from a wide variety of tissues. The binding characteristics of sites identified with these agents have been reasonably uniform from tissue to tissue. In certain respects these characteristics have resembled what would be expected of beta-adrenergic receptors. In particular, all potent beta-agonist compounds tested inhibit [^{3}H] catecholamine binding to these sites, whereas alpha-adrenergic compounds are generally much weaker or inert. The major structural requirement for binding to the sites appears to be an intact catechol moeity, which also typifies most of the potent beta-adrenergic agonist compounds. However, in certain respects the binding specificity of these sites is different from what might be expected of beta-adrenergic receptor-binding sites. Most notably, binding to these sites is nonstereospecific, is inhibited by a variety of physiologically inert "catechol" compounds, and does not display particularly high affinity for beta-adrenergic antagonist compounds. (Dissociation constants (K_d) for potent beta-adrenergic antagonists are on the order of 10^{-4} M, compared to those for adrenergic receptors, which are generally in the range of 10^{-8} to 10^{-9} M.) In addition, the time course of binding of [^{3}H] catecholamine agonists, although quite rapid in some systems, such as turkey erythrocytes [20, 23], is considerably slower in membranes from a variety of tissues and has been even more slowly reversible [22, 24-27]. A variety of explanations have been put forward in an attempt to explain the nature of these catecholamine-binding sites. In view of the discrepancies noted, it is difficult to reconcile the view that all of the observed binding is in fact occurring at the level of beta-adrenergic receptors. On the other hand, some of the explanations put forth, such as that binding is occurring at the level of catechol-*o*-methyl transferase [28, 29], are clearly erroneous and have been refuted in the literature [30, 31]. Others have thought that this binding is merely a reflection of catecholamine oxidation [31, 32]. A component of the binding may, in fact, be due to such oxidation, especially at 37°C. However, the general interpretation of catecholamine-binding data as being due solely or even primarily to oxidative reactions is clearly an over-simplification and not consistent with much of the published data. The exact nature of catecholamine binding to adenylate cyclase-containing membrane fractions is not entirely clear at this time, and it is undoubtedly a complex phenomenon. It is clear, however, in the view of this author, that [^{3}H] catecholamine binding is not the method of choice for delineating the molecular characteristics of beta-adrenergic receptors in subcellular membrane fractions. Rather, newly developed methodology, to be detailed in Sec. III, provides a much more confident and secure identification of beta-adrenergic receptors by a direct binding approach in membranes from a variety of tissues. These methods will now be presented in detail.

III. [^{3}H] (-)-ALPRENOLOL BINDING ASSAY

A. [^{3}H] (-)-Alprenolol

Alprenolol is a potent, competitive beta-adrenergic antagonist [33]. Its chemical structure is shown in Figure 1. Alprenolol, like all potent beta-adrenergic antagonists, possesses an asymmetric carbon and hence exists in two optical isomeric forms, (-) and (+). The (-)- and (+)-isomers of the compound are available from Hassle Pharmaceuticals, FACK, Sweden as the hydrochloride salts. There are several features about this molecule which make it particularly suited for performing binding studies.

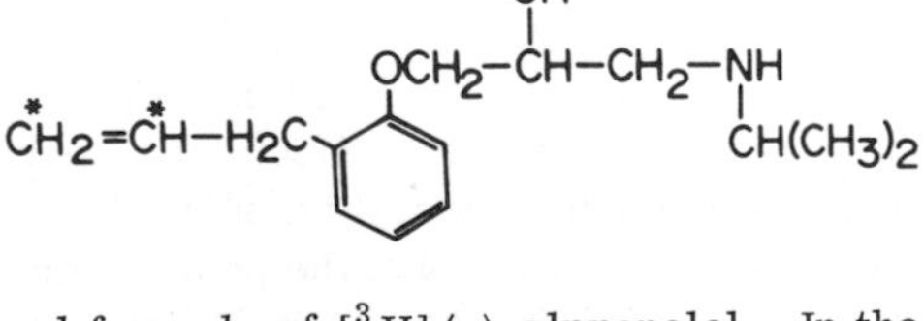

FIG. 1. Structural formula of [^{3}H] (-)-alprenolol. In the labeled compound, the double bond has been saturated with tritium. Mass spectroscopy reveals that the structure of this compound is in fact that of [^{3}H] dihydroalprenolol.

The first is its very high affinity for the beta-adrenergic receptors. The affinity of (-)-alprenolol for the beta-adrenergic receptors, as determined in a variety of intact and subcellular tissue preparations, is apparently equal to or greater than that of (-)-propranolol [4, 5, 34]. The high affinity of the compound for the beta-adrenergic receptors makes it particularly useful for identifying beta-adrenergic receptors directly by binding studies. The second useful feature of the molecule is the presence of an unsubstituted double bond in the aliphatic side chain. This permits tritiation to high specific activities by catalytic reduction. Catalytic reduction permits the insertion of two tritium atoms per alprenolol molecule. Thus, the theoretical maximum specific activity of the tritiated product is 60 Ci/mmol. The specific activity achieved has generally been in the range of 17 to 33 Ci/mmol. The labeled material is prepared by New England Nuclear Co., Boston, Mass., on request, and upon submission of a sample of (-)-alprenolol hydrochloride. The material is quite stable when stored in absolute ethanol at -20°C. Aqueous solutions, however, appear to degrade rapidly to a biologically inactive product.

We have found $[^3H]$(-)-alprenolol as received from New England Nuclear, after tritiation contains less than 5% contamination with chemically degraded products. This small contaminant can be removed by thin layer chromatography on silica gel plates (IB-F Baker) in a solvent system of acetone:benzene:acetic acid, 70:25:5. Native and $[^3H]$(-)-alprenolol cochromatograph with R_f's of ~0.2. $[^3H]$(-)-Alprenolol can be virtually quantitatively eluted from the plates with absolute ethanol. Prior to use in binding assays, the ethanol may be removed by evaporation under a stream of nitrogen, and the labeled material then dissolved in water.

The biological activity of the tritiated material is essentially identical to that of unlabeled (-)-alprenolol when this is evaluated as ability to competitively inhibit isoproterenol activation of frog erythrocyte membrane adenylate cyclase [5]. The K_D of the labeled and unlabeled antagonists for inhibition of the catecholamine-sensitive adenylate cyclase is in the range of 3 to 10 nmol [5].

B. Adenylate Cyclase-Containing Membrane Fractions

The ideal material for performing binding studies to identify beta-adrenergic receptors is a membrane fraction containing catecholamine-responsive adenylate cyclase. Since the effects of catecholamines in stimulating adenylate cyclase are generally mediated by beta-adrenergic receptors, such membranes must contain the beta-adrenergic receptor structures. Purified membrane preparations are preferable to crude homogenates but are by no means absolutely necessary. Our most extensive studies have utilized frog erythrocyte membranes [4-6]. We have been able to perform very detailed studies utilizing a crude membrane preparation from frog erythrocytes which is essentially a washed 30,000 g pellet containing all the nuclear and plasma membrane fragments [4-6]. A simple centrifugation over a 50% sucrose cushion results, however, in a fraction which, although containing only about one-fifth to one-tenth the protein, contains virtually all the [^{3}H] (-)-alprenolol-binding activity and adenylate cyclase activity [35]. We have also used canine cardiac membranes [7], rat adipose membranes, rat skeletal muscle membranes, and several others [36].

C. Incubation Conditions

Optimal incubation conditions for binding assays must be determined individually in each system and, accordingly, only guidelines can be provided here. As noted, effects of adrenergic agonists and antagonists on adenylate cyclase, as well as on physiological processes, are quite rapid. The same is true of [^{3}H] (-)-alprenolol binding to sites in membrane fractions [4, 5]. Equilibrium binding is generally achieved within two to five minutes, and most of the binding (~80%) is established as rapidly as can be conveniently measured (within ~30 sec). Similarly, the dissociation rate of membrane-bound [^{3}H] (-)-alprenolol is also very rapid (ranging from 10 to 30 sec for 50% dissociation) in the systems that we have evaluated. Accordingly, incubations should be performed for relatively short periods of time, as equilibrium is rapidly established, and this provides minimal time for degradative processes to operate. In order to compare the characteristics of [^{3}H] (-)-alprenolol-binding sites to those of the adenylate cyclase-coupled beta-adrenergic receptors, incubation conditions for the binding assays should be comparable to those under which adenylate cyclase experiments are performed.

Due to the scarcity of the [^{3}H] (-)-alprenolol-binding sites (~0.3 pmol/mg membrane protein), it is important to use very concentrated membranes in the incubations. If dilute membrane fractions are used, the number of beta-adrenergic receptor-binding sites present may be too low to accurately quantitate by the binding assay. With crude membrane fractions, membrane protein concentrations during incubation have been in the range of 2 to 3 mg of membrane protein per 150 μl assay. With more purified membrane preparations, the protein concentration can be 5 to 10 times lower [35].

Assays are conveniently performed in plastic, disposable 12 × 75-mm test tubes and are generally incubated for 10 min at 37 °C. Incubations may be performed in a total volume of 150 μl, although the volume of the incubation does not appear to be critical. Incubation mixtures contain [^{3}H] (-)-alprenolol, 2 to 10 nmol (10,000-50,000 cpm); 50 mM tris-HCl, pH 8.1; 10 to 20 mM $MgCl_2$; and unlabeled adrenergic drugs or water. Reactions are generally started by addition of membranes.

The fraction of $[^3H]$ (-)-alprenolol binding that occurs to beta-adrenergic receptors will undoubtedly vary from one membrane fraction to another. In certain systems, such as the frog erythrocyte membrane, a very high percentage of the binding, generally 75% or greater, is to these receptors as determined by criteria to be detailed below. In crude, dog heart membranes, a much lower fraction of the binding is to the beta-adrenergic receptors. Binding which occurs to sites other than the beta-adrenergic receptors is termed "nonspecific" binding. Nonspecific binding is composed of several features. These include $[^3H]$ (-)-alprenolol trapped in the membrane pellet obtained during centrifugation (see Sec. III. D), as well as irreversible binding to nonreceptor sites of unknown character or physiological significance. We have found that this nonspecific binding is generally not displaced by adrenergic drugs at any concentration. Accordingly, we have operationally defined "specific" binding of $[^3H]$ (-)-alprenolol to the membrane sites as that binding which can be displaced by (-)-alprenolol or (-)-propranolol at a concentration of 10 μM. Accordingly, in all experiments, incubation tubes which contain $[^3H]$ (-)-alprenolol and membranes without any added adrenergic agents represent total binding; whereas tubes containing $[^3H]$ (-)-alprenolol and membranes, plus 10 μM (-)-alprenolol or propranolol, represent the nonspecific binding. The difference in the binding observed between these two sets of incubation tubes is the specific binding.

The affinity of a variety of adrenergic agents, both agonists and antagonists, for the $[^3H]$ (-)-alprenolol-binding sites is conveniently assessed by performing dose-response curves with each agent. Adrenergic agents are added to the incubations over a range of concentrations corresponding to the range of concentrations over which they have effects on the catecholamine-sensitive adenylate cyclase, either as agonists or antagonists. The ability of the agents to compete with $[^3H]$ (-)-alprenolol for the receptor-binding sites is assessed by measuring their ability to lower specific binding. Displacement equivalent to that observed with 10 μM alprenolol or propranolol is taken as complete or 100% inhibition of specific binding. It must, of course, be experimentally verified in each individual system that the maximum displacement which can be observed is, in fact, achieved by 10 μM alprenolol. In erythrocyte and cardiac membranes, addition of higher concentrations results in no further lowering of the bound $[^3H]$ (-)-alprenolol.

D. Separation of Receptor-bound from Free [^{3}H] (-)-Alprenolol

An effective way of rapidly separating membrane-bound from free [^{3}H] (-)-alprenolol is by centrifugation in a Beckman microfuge B-152. The small microfuge tubes used for the assay contain a total volume of 400 μl. They can be obtained from Beckman, Brinkman, or Sarstedt Co. Prior to beginning an experiment, an appropriate number of such tubes is set up and numbered to correspond to the incubation tubes. At the completion of the 10-min incubations described, duplicate 50-μl aliquots of incubation mixture are placed into small microcentrifuge tubes, containing 300 μl of incubation buffer. At this stage of the procedure, the microfuge tubes should be in a horizontal position, so that the membranes do not settle into the buffer layer. Once the membranes have been added to the microfuge tubes, centrifugation (for 1 min) should be started immediately. Membranes are generally sedimented within a few seconds. At the end of one minute, there is a tightly packed pellet at the bottom of the tubes. The great rapidity with which the microfuge reaches maximum speed and the attendant rapidity with which the membranes are sedimented minimizes the time during which dissociation of membrane-bound [^{3}H] (-)-alprenolol can occur. After all membrane samples have been centrifuged, the supernatant fluid is aspirated, generally using a No. 18 spinal needle attached to a vacuum line. An additional 300 μl of incubation buffer is carefully layered over the tightly packed pellet and then aspirated. This serves to wash only the uppermost layer of the pellet and the contiguous surface of the inner part of the microfuge tube. The tips of the microfuge tubes are then cut off into scintillation-counting vials and 500 μl of 10% SDS containing 10 mM EDTA is added. Each individual pellet is removed from the bottom of the plastic microfuge tubes with a long needle and then allowed to float freely in the SDS. Samples are shaken overnight, during which period of time the membranes are completely solubilized. Fifteen milliliters of a Triton-toluene-based scintillation fluid are then added, and the samples are counted in a liquid scintillation spectrometer.

An alternate method of separating membrane-bound from free [^{3}H] (-)-alprenolol is by vacuum filtration using glass fiber filters (Whatman GF/C). After incubations are completed, aliquots are placed into 2 ml of ice-cold buffer and immediately filtered through the GF/C glass fiber filters. The filters are then washed with 10 ml of cold buffer. The wash with <u>cold</u> buffer does not lead to dissociation of any receptor-bound [^{3}H] (-)-alprenolol. Filters are dried, added to 10 ml of Triton-toluene-based scintillation fluid and then counted in a liquid scintillation spectrometer. Nonspecific binding is determined by filtering aliquots of membranes which have been incubated with $[^3H]$(-)-alprenolol in the presence of 10^{-5} M alprenolol or propranolol. Adsorption of $[^3H]$(-)-alprenolol to the glass fiber filters is almost nil (< 0.2%) and is not further reduced by unlabeled propranolol. This is in contrast to the situation with several types of Millipore filters, where adsorption of $[^3H]$(-)-alprenolol to the filter is extensive.

IV. INTERPRETATION OF RESULTS

Table 1 presents typical data derived from experiments performed in this laboratory. The cpm of $[^3H]$(-)-alprenolol bound represent the actual counts in each of the membrane pellets determined by liquid scintillation spectrometry. In this experiment, the difference between the cpm bound in the control tubes ("total binding") and in those containing 10 μM alprenolol represents the specific binding of $[^3H]$(-)-alprenolol to receptor-binding sites in the frog erythrocyte membranes. As can be seen, this is 1,062 cpm. (-)-Alprenolol at increasing concentration causes progressive inhibition of this specific binding. (+)-Alprenolol has a similar effect at approximately 100-fold higher concentrations [4, 5]. One-half maximal inhibition of specific $[^3H]$(-)-alprenolol binding occurs at a concentration of (-)-alprenolol of approximately 10 nM. It should be pointed out that addition of (-)-alprenolol at concentrations above 10 μM does not lead to further inhibition of alprenolol binding (Table 1). This is because the nonspecifically bound $[^3H]$(-)-alprenolol is apparently bound to a set of nonsaturable sites which bind $[^3H]$(-)-alprenolol irreversibly. In the frog erythrocyte membrane system, ~80% of all the observed binding is displaceable or specific binding. If purified erythrocyte membranes are used, virtually no nonspecific binding is observed.

A wide variety of adrenergic agonist and antagonist compounds can be tested for their ability to compete with $[^3H]$(-)-alprenolol for occupancy of the receptor-binding sites. We have demonstrated that ability to compete for occupancy of these sites is directly related to the potency of the compounds tested as either beta-adrenergic agonists or antagonists. Compounds which are without beta-adrenergic activity have no ability to inhibit this binding [4-7]. Typical displacement curves for several antagonist and agonist drugs are shown in Figures 2 and 3.

V. POTENTIAL PITFALLS

Despite our successes in using $[^3H]$(-)-alprenolol to label apparent beta-adrenergic receptors in a number of tissues, it should not be automatically

TABLE 1. Inhibition of [^{3}H] (-)-Alprenolol Binding to Frog Erythrocyte Membranes by (-)-Alprenolol[a]

Addition	[^{3}H] (-)-alprenolol bound (cpm/pellet)		Specific binding
	Observed	Mean	
None (control)	1316 1327	1321	1062
(-)-Alprenolol 10^{-9} M	1112 1110	1111	852
10^{-8} M	744 736	740	481
10^{-7} M	440 380	410	151
10^{-6} M	297 291	294	35
10^{-5} M	247 270	259	0
10^{-4} M	265 255	260	0

[a]Specific binding is calculated by subtracting the cpm in the pellet in the presence of 10 μM (-)-alprenolol from the observed value.

assumed that in any system studied, the binding of [^{3}H] (-)-alprenolol will reflect, in its entirety, binding at the level of the adrenergic receptors. We have previously demonstrated that with other ^{3}H-labeled beta-adrenergic antagonists of somewhat lower specific radioactivity ([^{3}H]propranolol), binding to sites other than beta-adrenergic receptors occurs [37]. A number of other authors have also observed nonspecific binding of [^{3}H]propranolol to a variety of membrane sites [38, 39].

When we attempted to study the binding of [^{3}H] (-)-alprenolol to membranes derived from human leukocytes, we observed findings quite different from those reported above [40]. Thus, the dissociation constant for the alprenolol binding sites in these membranes was quite high, in the range of 10^{-4} to 10^{-5} M, and binding was nonstereospecific. Thus, the sites labeled with [^{3}H] (-)-alprenolol in these membranes were not the beta-adrenergic receptors. Accordingly, an important caution in performing studies of this type is to carefully delineate the properties of binding of

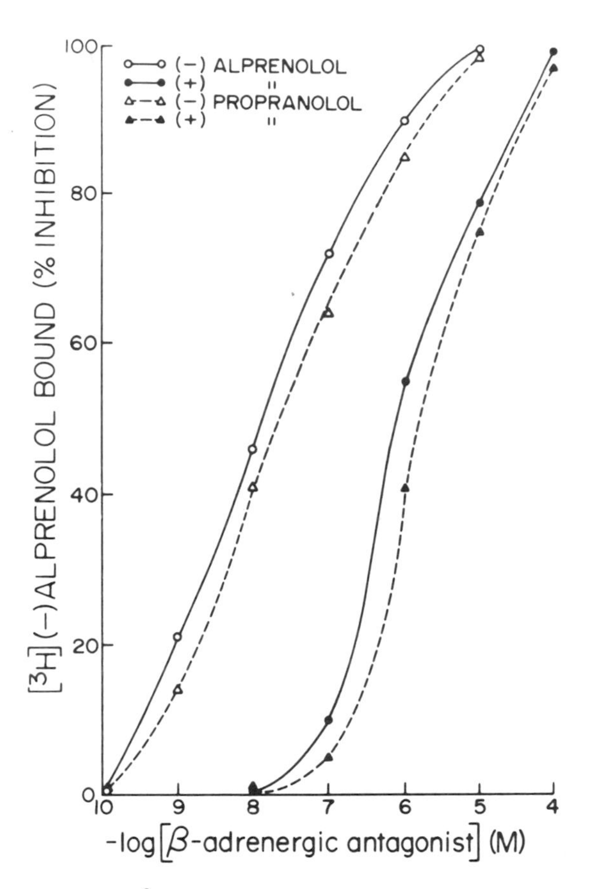

FIG. 2. Inhibition of [^{3}H] (-)-alprenolol binding to frog erythrocyte membranes by optical isomers of alprenolol and propranolol. From [5].

[^{3}H] (-)-alprenolol in each system in which it is used. The minimum criteria for equating the binding sites observed with the adenylate cyclase-coupled or physiological beta-adrenergic receptors would appear to be: (1) sufficiently rapid forward and reverse rates of binding, (2) stereospecificity of binding for both agonists and antagonists, and (3) appropriate binding specificity and affinity for a series of beta-adrenergic agonists and antagonists which directly parallels the biological potency of these compounds.

Another potential pitfall, of a more technical nature, relates to the extremely rapid rates of association and dissociation of [^{3}H] (-)-alprenolol binding to the beta-adrenergic receptors. The dissociation rates are very rapid (t 1/2 $<$ 15 sec in heart membranes). Accordingly, methods for

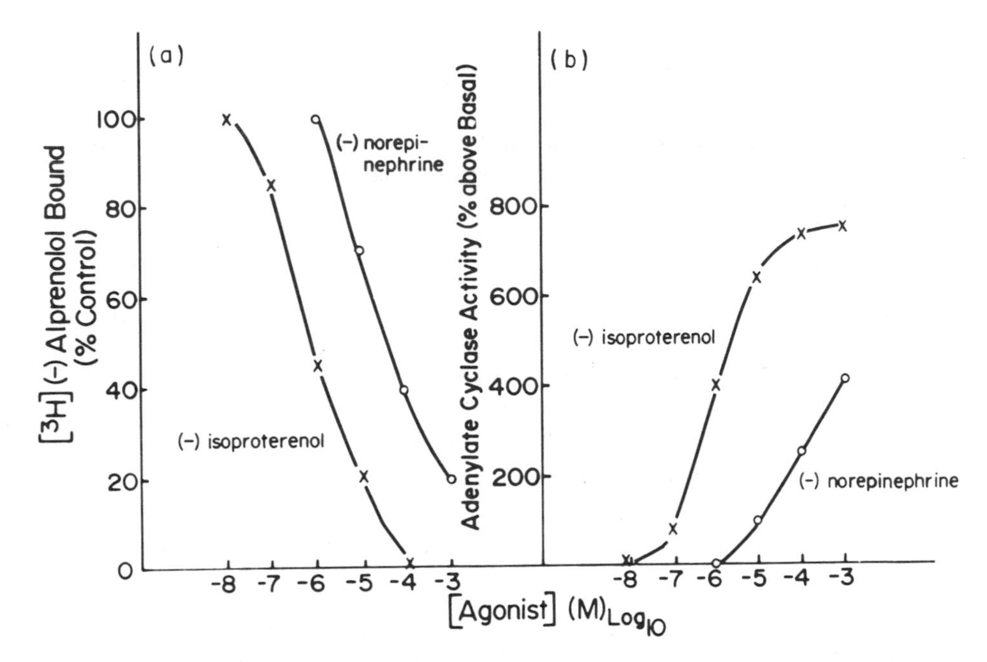

FIG. 3. (a) Inhibition of [^{3}H] (-)-alprenolol binding to frog erythrocyte membranes by (-)-isoproterenol and (-)-norepinephrine. (b) Activation of frog erythrocyte membrane adenylate cyclase by (-)-isoproterenol and (-)-norepinephrine. From [4].

separating membrane-bound [^{3}H] (-)-alprenolol from free [^{3}H(-)-alprenolol which are either slow (1-2 min), or which involve extensive washing of the membranes, are likely to lead to extensive dissociation of bound [^{3}H] (-)-alprenolol. The result will be gross underestimation of the amounts of labeled material bound to receptor sites. In our hands, as described, the microfuge has provided a very rapid and convenient way of effecting such separation. In some systems, however, the purified plasma membranes which can be used for such studies will represent very light fractions, which will not be sedimented by the relatively low g forces developed by these table-top micorfuge centrifuges. In such cases, equilibrium dialysis may be used effectively. However, this is more cumbersome and involves relatively prolonged periods of time for equilibration. However, the success with which equilibrium dialysis methods have been applied to studies of nicotinic and muscarinic cholinergic receptors suggests that they may well be applicable here as well.

VI. COMPARISON OF [^{3}H](-)-ALPRENOLOL BINDING WITH OTHER METHODS

As we have noted, attempts to identify beta-adrenergic receptors by using ^{3}H-labeled agonist compounds for direct binding studies met with a number of difficulties. The binding sites identified in these studies appeared to have time courses for interaction with the adrenergic compounds which were significantly slower, both in terms of rate of association and of dissociation, than might have been expected for the beta-adrenergic receptors. In addition, there were significant discrepancies between binding specificity and the specificity of the adenylate cyclase-coupled beta-adrenergic receptors present in the same membrane fractions. For these reasons, ^{3}H-labeled beta-adrenergic agonists appear to be less useful than compounds such as [^{3}H](-)-alprenolol for binding studies of this type.

A recent study by Levitzki et al. [41] has described the use of [^{3}H]-propranolol to label sites in turkey erythrocyte membranes, which seem to have properties very similar to those identified in frog erythrocyte membranes with [^{3}H](-)-alprenolol. Since the affinity of alprenolol and propranolol for beta-adrenergic receptors are very comparable, it should not be surprising that propranolol can be used for this type of study as well. However, there appear to be a number of problems associated with the use of [^{3}H]propranolol as a ligand for direct binding studies which are not associated with [^{3}H](-)-alprenolol. First, propranolol appears to have a much greater propensity to bind to nonspecific binding sites than does alprenolol. A number of laboratories have reported binding of [^{3}H]-propranolol to a variety of non-beta-receptor sites in membrane fractions from several tissues [37-39]. The nonspecific sites to which propranolol binds are apparently present in tissues in much higher concentrations than the beta-adrenergic receptors themselves [37, 38]. In Levitzki's studies, specific binding of propranolol to beta-adrenergic receptors appears to represent only a small fraction of the total binding observed, 20% or less in the experiments reported [41]. The large amount of nonspecific binding observed makes it very difficult to quantitate the much smaller amount of specific receptor binding which also occurs. Nonspecific binding with [^{3}H](-)-alprenolol appears to be less of a problem, although the extent to which nonspecific binding is observed has varied considerably from one tissue to another. The specific activity of ^{3}H-labeled propranolol is also somewhat less than than which can be obtained by catalytic reduction of (-)-alprenolol.

Recently, Aurbach and his co-workers have reported on a conceptually very similar approach to that described here, which utilizes a ^{125}I-labeled beta-adrenergic antagonist which possesses a high affinity for the beta-adrenergic receptors [42]. The studies reported were performed in turkey erythrocyte membranes. The beta-adrenergic antagonist used was specially synthesized for these studies and was hydroxybenzylpindolol. The compound bears a hydroxyphenyl moiety substituted on a branched-chain substituent on the amino nitrogen group. This hydroxyphenyl moiety can be iodinated to high specific activity. The specific activity of the compound used in the studies reported by Aurbach et al. was ~200 Ci/mmol, well above that of the [^{3}H] (-)-alprenolol described here. The data reported with this compound appear to be qualitatively very similar to those obtained with [^{3}H] (-)-alprenolol. Although the very high specific radioactivity of this compound would seemingly represent a potential advantage, such high specific activity does not appear to be necessary for studies of this type. In general, in performing binding studies, the labeled material is best used in a concentration range which is roughly equivalent to its K_D for the receptors under study. With iodinated compounds, incubations can be performed with the labeled material at concentrations as low as 10^{-11} M. However, if very low concentrations of radioactively labeled drug are used, only a very small fraction of the relevant receptors may be labeled, a situation which might tend to favor the development of experimental artifacts. Further, iodinated derivatives are likely to be more unstable than ^{3}H-labeled alprenolol and have much shorter half-lives.

The much greater efficacy of the beta-adrenergic antagonists, as opposed to the agonists, as ligands for studying direct binding to beta-adrenergic receptors is, in itself, a point of some interest. This is presumably a reflection of the much higher affinity of the potent antagonist compounds (K_D's in the range of 10^{-9} to 10^{-8} M) as opposed to the agonists (K_D's on the order of ~10^{-6} M for potent agonists such as isoproterenol in most adenylate cyclase systems).

It seems likely that the methods described here should be widely applicable for the study of beta-adrenergic receptor interactions in membranes from many tissues.

VII. APPROACHES TO BETA-ADRENERGIC RECEPTOR PURIFICATION

A. Solubilization

As yet, the techniques described here for binding of [^{3}H] (-)-alprenolol to beta-adrenergic receptors have not been applied to solubilized preparations. However, on the basis of previous studies performed with [^{3}H]-norepinephrine [25], there is sufficient precedent to describe the types of methods that are likely to be successful. Any complete understanding of adrenergic receptors and their relation to adenylate cyclase will require that both the receptors and the enzyme be freed from the plasma membrane, so that they may be examined in purified soluble form. Once the receptor-binding sites are solubilized, however, somewhat different methods must be developed for assay of binding activity. The usual methods which are applicable to particulate preparations, such as rapid centrifugation and Millipore filtration, cannot be used when the receptor-binding sites themselves are in solution.

Solubilization may be achieved by the use of one of a variety of ionic or nonionic detergents. Likely candidates are Triton X-100, Lubrol-PX, Lubrol-WX, and sodium deoxycholate. Membrane fractions or intact pieces of tissue are suspended in a buffered solution of one of the detergents, generally present at a final concentration of 0.1 to 1%. The suspension is homogenized, e.g., with a mechanically driven teflon-tipped pestle. The resultant homogenate is then centrifuged for 1 hr, at 105,000 g to remove particulate material. The supernatant can be assayed directly for receptor binding or enzyme activity. Procedures are best performed at 0 to 4°C.

As noted, somewhat different assay methods are required for the direct measurement of binding to soluble preparations. Methods such as ammonium sulfate precipitation of receptor-bound ligand or comparable methods are not likely to be successful because of the very rapid dissociation of reversible ligands from the beta-adrenergic receptor (documented earlier in this paper). One method, which may be applicable, utilizes rapid separation of protein-bound from free radioactively labeled adrenergic ligand on small, Sephadex G-25 fine columns [25].

Two-milliliter columns of preswelled Sephadex G-25 are prepared in the barrels of 3-ml, plastic disposable syringes. Four hundred-microliter aliquots of incubation mixtures which have come to equilibrium are placed onto the surface of these columns. After the level of solution has fallen below the gel surface, 500 μl of an appropriate buffer are applied. The columns are then placed into disposable glass tubes and centrifuged for 10 min at 1000 g. As controls, aliquots of radioactively labeled adrenergic ligand, e.g., [^{3}H](-)-alprenolol, can be incubated in buffer alone, or with soluble fractions to which a high concentration (i.e., 10 μM of propranolol has been added to block all specific binding sites. The column eluates, in the glass centrifuge tubes, will contain receptor-bound radioactive adrenergic ligand. Free ligand remains within the columns. Column eluates can then be added to scintillation fluid and counted. In many systems, this simple method will effect a rapid and very complete separation of protein-bound and free hormone or drug.

Alternatively, with soluble preparations, binding studies can be performed by conventional techniques, such as equilibrium dialysis or equilibrium Sephadex gel filtration [43].

B. Affinity Chromatography

In recent years, the techniques of affinity chromatography have been widely applied. Their use in the purification of hormone receptors has been somewhat limited thus far. Nonetheless, they have been used with some success to purify cholinergic receptor-binding sites [44], insulin binding sites [45], and [^{3}H] norepinephrine binding sites [25]. The presence of a primary or secondary amine function in most catecholamine agonists and antagonists is convenient for the preparation of such derivatives. Catecholamines such as norepinephrine can generally be covalently bound to agarose beads via a hydrocarbon side chain. Detailed methods for such reactions have been published elsewhere [46, 47]. Catecholamines can be coupled to such agarose derivatives via a number of different reactions which produce different bonds between the catecholamine and the agarose. Coupling can be effected through the amino nitrogen so as to form an amino, amido, or guanidino bond [47]. Catecholamines may also be coupled via their catechol rings, thus leaving the amino nitrogen free [47].

Methods for preparation of catecholamine-agarose affinity columns have been detailed elsewhere [25, 46]. Several cautions are worth reiterating: (1) Ionic strength of buffers used to equilibrate columns should generally be kept at 0.2 M or above to minimize ion-exchange effects and nonspecific adsorption of proteins. (2) The presence or absence of detergent may affect the adsorption of the binding proteins to the agarose-catecholamine derivatives and should be evaluated in each individual case. (3) The nature of the chemical linkage of the catecholamine to the agarose will undoubtedly be crucial for determining efficacy for any particular application. In particular, since the configuration about the amino nitrogen of catecholamines and beta-adrenergic antagonists appears to be crucial for biological activity, those derivatives which have the least distortion of this group are likely to be the most effective for purification of receptor proteins.

ACKNOWLEDGMENTS

This work was supported by grant HL-16037 from NIH and a grant-in-aid from the American Heart Association, with funds contributed in part by the North Carolina Heart Association. The author is an Established Investigator of the American Heart Association. The author wishes to thank the following associates and colleagues who were involved in the development of the binding methods described in this chapter: Drs. C. Mukherjee, M. Caron, L. Limbird, R. W. Alexander; and Mr. L. T. Williams and Mr. M. Coverstone.

Note Added in Proof: Since this article was originally written we have demonstrated that:

1. Rapid filtration using glass-fiber filters is the preferred method for separating receptor-bound from free radioligand.
2. Binding of $[^3H]$ (−)-alprenolol to beta-receptors in human lymphocytes can be easily studied.

REFERENCES

1. A. S. V. Burgen and L. L. Iverson, Brit. J. Pharmacol., 25:34 (1965).
2. U. S. VonEuler, Circ. Res., Suppl. 3, 20:5 (1967).
3. J. Axelrod and R. Tomchick, J. Biol. Chem., 233:702 (1958).
4. R. J. Lefkowitz, C. Mukherjee, M. Coverstone, and M. Caron, Biochem. Biophys. Res. Commun., 60:703 (1974).
5. C. Mukherjee, M. G. Caron, M. Coverstone, and R. J. Lefkowitz, J. Biol. Chem., 250:4869 (1975).
6. R. J. Lefkowitz, L. Limbird, M. Caron, and C. Mukherjee, Biomembrane Reviews, in press (1976).
7. R. W. Alexander, L. T. Williams, and R. J. Lefkowitz, Proc. Natl. Acad. Sci. U.S.A., 72:1564 (1975).
8. R. P. Ahlquist, Am. J. Physiol., 153:586 (1948).
9. C. E. Powell and R. H. Slater, J. Pharmacol. Exptl. Ther., 122:480 (1958).
10. A. M. Lands, A. Arnold, J. P. McAuliff, F. L. Luduena, and T. G. Brown, Nature, 214:597 (1967).
11. A. Arnold, J. P. McAuliff, D. F. Colella, W. V. O'Connor, and T. G. Brown, Arch. Int. Pharmacodyn., 170:451 (1968).
12. D. Dunlop and R. G. Shanks, Brit. J. Pharmacol., 32:201 (1968).
13. M. A. Wasserman and B. Levy, J. Pharmacol. Exptl. Ther., 182:256 (1972).
14. R. J. Lefkowitz, Circulation, 49:783 (1974).
15. J. F. Moran, M. May, H. Kimelberg, and D. J. Triggle, Mol. Pharmacol., 3:15 (1967).
16. M. S. Yong and M. Nickerson, J. Pharmacol. Exptl. Ther., 186:100 (1973).

17. G. A. Robison, R. W. Butcher, and E. W. Sutherland, Cyclic AMP, Academic Press, London, 1971, pp. 145-231.

18. H. P. Bar, Mol. Pharmacol., 10:597 (1974).

19. R. J. Lefkowitz, Biochem. Pharmacol., 24:1651 (1975).

20. J. P. Bilzekian and G. Aurbach, J. Biol. Chem., 248:5774 (1973).

21. J. P. Bilzekian and G. Aurbach, J. Biol. Chem., 248:5584 (1973).

22. V. Tomasi, S. Koretz, T. K. Ray, J. Dunnick, and G. V. Marinetti, Biochim. Biophys. Acta, 211:31 (1970).

23. M. Schramm, H. Feinstein, E. Naim, M. Long, and M. Lasser, Proc. Natl. Acad. Sci. U.S.A., 69:523 (1972).

24. R. J. Lefkowitz and E. Haber, Proc. Natl. Acad. Sci. U.S.A., 68: 1773 (1971).

25. R. J. Lefkowitz, D. O'Hara, and E. Haber, Proc. Natl. Acad. Sci. U.S.A., 69:2828 (1972).

26. R. J. Lefkowitz, G. Sharp, and E. Haber, J. Biol. Chem., 48:342 (1973).

27. R. J. Lefkowitz, D. O'Hara, and J. Warshaw, Biochim. Biophys. Acta, 332:317 (1974).

28. P. Cuatrecasas, G. P. E. Tell, V. Sica, I. Parick, and K. J. Chang, Nature, 247:92 (1974).

29. G. P. E. Tell and P. Cuatrecasas, Biochem. Biophys. Res. Commun., 57:793 (1974).

30. R. J. Lefkowitz, Biochem. Biophys. Res. Commun., 58:1110 (1974).

31. B. M. Wolfe, J. A. Zirrolli, and P. B. Molinoff, Mol. Pharmacol., 10:582 (1974).

32. M. E. Maguire, P. H. Goldmann, and A. G. Gilman, Mol. Pharmacol., 10:563 (1974).

33. E. Ablad, M. Bogard, and L. Ek, Acta Pharmacol. Toxicol., 25, Suppl. 2:9 (1967).

34. C. Grunfeld, A. P. Grollman, and O. M. Rosen, Mol. Pharmacol., 10:605 (1974).

35. M. Caron and R. J. Lefkowitz, unpublished observations.

36. C. Mukherjee, M. Caron, L. T. Williams, R. W. Alexander, and R. J. Lefkowitz, unpublished observations.

37. D. Vatner and R. J. Lefkowitz, Mol. Pharmacol., 10:450 (1974).

38. L. T. Potter, J. Pharmacol. Exptl. Ther., 155:91 (1967).

39. A. Huunan-Seppala, Acta Chem. Scand., 26:2713 (1972).

40. L. Limbird, J. Mickey, and R. J. Lefkowitz, unpublished observations.

41. A. Levitski, D. Atlas, and M. L. Steer, Proc. Natl. Acad. Sci. U.S.A., 71:2773 (1974).

42. G. D. Aurbach, S. A. Fedak, C. J. Woodard, J. S. Palmer, D. Hauser, and F. Troxler, Science, in press.

43. W. H. Perlman and O. Crepy, J. Biol. Chem., 242:182 (1967).

44. H. W. Chang, Proc. Natl. Acad. Sci. U.S.A., 71:2113 (1974).

45. P. Cuatrecasas, Proc. Natl. Acad, Sci. U.S.A., 69:1277 (1972).

46. D. O'Hara and R. J. Lefkowitz, in Methods in Enzymology (S. P. Colowick and N. O. Kaplan, eds.) Academic Press, New York, in press.

47. M. G. Caron and R. J. Lefkowitz, submitted for publication.

Chapter 4

CHOLERA TOXIN RECEPTORS

Vann Bennett* and Pedro Cuatrecasas†

Department of Pharmacology and Experimental Therapeutics
The Johns Hopkins University School of Medicine
Baltimore, Maryland

---

*Current affiliation: The Biological Laboratories, Harvard University, Cambridge, Massachusetts.
†Current affiliation: The Wellcome Research Laboratories, Burroughs Wellcome Company, Research Triangle Park, North Carolina.

I. INTRODUCTION

Cholera enterotoxin (choleragen), an oligomeric protein of ~100,000 mol wt secreted by Vibrio cholerae, stimulates ubiquitously adenylate cyclase activity in mammalian tissues [1, 2], as well as avian [3, 4] and amphibian [4-6] erythrocytes. The initial event in the action of choleragen involves a high affinity binding to cell-surface receptor sites [4, 7-13] which are believed to be G_{M1} monosialogangliosides (see Fig. 1) [4, 7-12, 14-20]. G_{M1} competes very effectively for the binding of ^{125}I-labeled cholera toxin to liver membranes [7, 8] and it blocks the biological effects of the toxin [8, 14-20].

Although high concentrations of certain complex sugars and glycoproteins will also block binding (Table 1), choleragen exhibits a remarkable specificity for glycosphingolipids containing sialic acid (gangliosides), particularly G_{M1} (Tables 2 and 3). No binding competition is observed with sulfatides, sphingomyelin, psychosine, or crude brain cerebrosides (Table 2). The simple glycosphingolipids (such as glucose-ceramide, galactose-ceramide, and galactose-galactose-ceramide) are also relatively inert, although more complex derivatives formed by sequential addition of terminal galactose and N-acetylgalactosamine residues result in an increased potency in blocking the binding of [^{125}I]choleragen (Table 2). The addition of sialic acid, in 3,2 linkage with the galactose of galactose-glucose-ceramide to form G_{M3} monosialoganglioside, results in at least a 20-fold increase in inhibitory potency (Table 3). Sequential additions of N-acetylgalactosamine to G_{M3} to form G_{M2}, and of galactose to G_{M2} to form G_{M1}, result in an increase of 100-fold and 50-fold, respectively, in the apparent affinity for choleragen (Table 3). Further addition of N-acetylneuraminic acid to G_{M1} reduces the inhibitory ability by 40- to 100-fold. The approximate concentrations of gangliosides required for half-maximal inhibition of [^{125}I]choleragen binding are: G_{M1} (0.02 μg/ml), G_{D1a} (0.7 μg/ml), G_{M2} (1 μg/ml), G_{T1} (1.8 μg/ml), and G_{M3} (50 μg/ml) [7].

Exogenous G_{M1} can incorporate spontaneously into cell membranes, resulting in an increased capacity for binding of ^{125}I-labeled choleragen (up to tenfold), as well as an enhanced sensitivity to the biological effects of the toxin [8, 21]. The fact that reconstituted G_{M1} can function as a biologically effective membrane receptor strongly supports the notion that this glycolipid is the natural cell-surface binding site for choleragen. Further evidence is provided by a study of transformed mouse cell lines from the same parent strain which exhibit loss of specific enzymes involved in the de novo biosynthesis of membrane gangliosides [10]. The binding of [^{125}I]choleragen, and the sensitivity to stimulation of adenylate cyclase activity and inhibition of DNA synthesis, varies in parallel with the ganglioside composition of these cells. The SV40 transformed cell

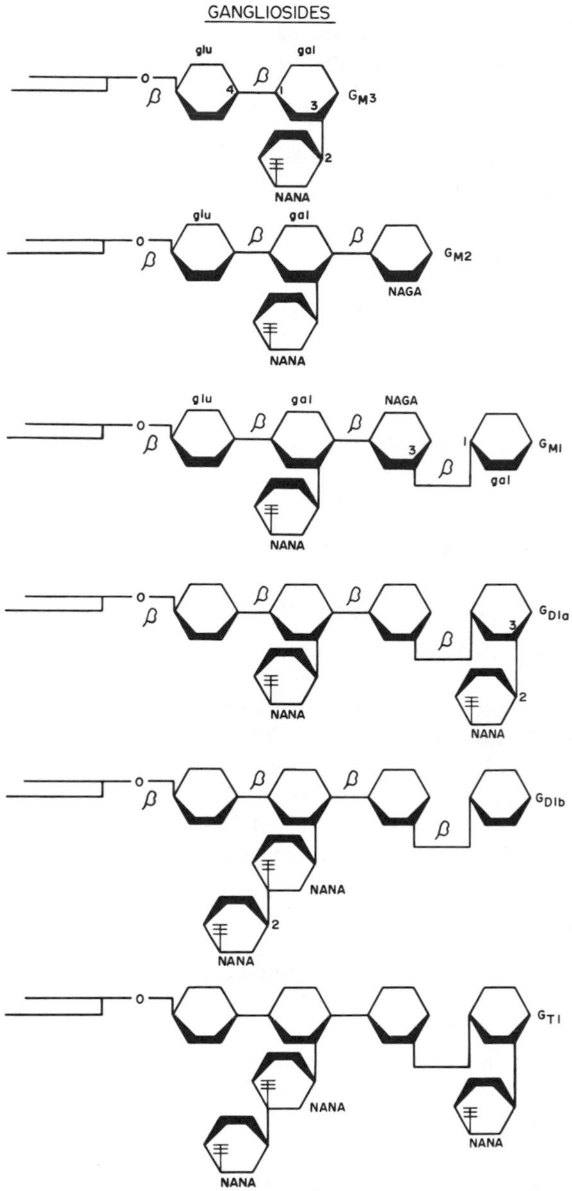

GANGLIOSIDES

FIG. 1. Structures of some commonly occuring gangliosides. The nomenclature is that of Svennerholm [56].

TABLE 1. Effect of Special Glycopeptides and Oliogosaccharides on the Specific Binding of Cholera Toxin to Liver Membranes[a,b]

Oligosaccharide	^{125}I-Labeled cholera toxin (% bound)
Fetuin glycopeptide I, 85 μmol	85
Fetuin glycopeptide II, 85 μmol	98
Thyroglobulin glycopeptide I, 85 μmol	86
Thyroglobulin glycopeptide II, 85 μmol	84
γG-Glycopeptide, 0.9 μmol	100
γM-Glycopeptide, 0.1 μmol	96
Lacto-N-tetraose (Gal $\xrightarrow{\beta 1,3}$ GlcNAc $\xrightarrow{\beta 1,3}$ Gal $\longrightarrow$ GlcNAc),[c] 1 mmol	80
Lacto-N-neotetraose (Gal $\xrightarrow{\beta 1,4}$ GlcNAc $\xrightarrow{\beta 1,4}$ Gal $\xrightarrow{\beta 1,4}$ GlcNAc),[c] 1 mmol	92

[a] ^{125}I-Labeled cholera toxin (50 ng/ml) was preincubated at 24°C for 60 min in Krebs-Ringer bicarbonate buffer containing 0.1% (wt/vol) albumin and the compound indicated in the table. Samples of these solutions were then incubated at 24°C for 15 min with Krebs-Ringer bicarbonate containing 0.1% (wt/vol) albumin and 40 μg/ml of liver membrane protein. Specific toxin binding was determined by filtration procedures as described in the text.

[b] From Cuatrecasas [7].

[c] Gal, galactose; GlcNAc, N-acetylglucosamine.

line which lacks chemically detectable G_{M2}, G_{M1}, and G_{D1a}, as well as UDP-N-acetylgalactosamine:G_{M3} N-acetylgalactosaminyltransferase activity, is the least sensitive to the biological effects and binds the least amount of choleragen (Fig. 2).

Agarose derivatives of gangliosides have been prepared which are capable of adsorbing choleragen nearly quantitatively at concentrations as low as 10^{-11} M [22]. Moreover, soluble ganglioside polymers prepared by coupling the terminal sialic acid residue to the amino groups of branched copolymers of lysine and alanine also prevent the binding of ^{125}I-labeled choleragen to liver membranes, and abolish the lipolytic

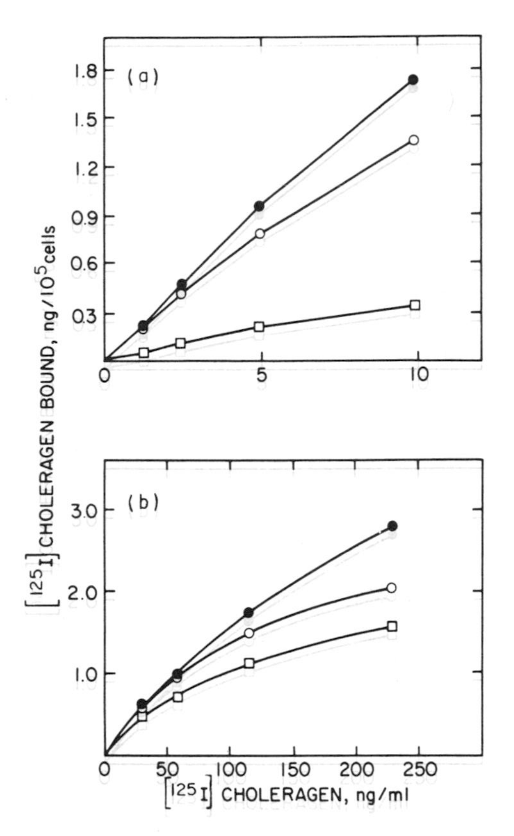

FIG. 2. Binding of cholera toxin to transformed Al/N cells. ○: TAL/N, P = 46; ●: TAL/N, P-269; □: SVS AL/N. (a) Binding to 8×10^4 cells in 0.4 ml of buffer. (b) Binding to 3.2×10^5 cells in 0.2 ml of buffer. From Hollenberg et al. [10].

activity of choleragen on fat cells [22]. These macromolecular derivatives of gangliosides apparently do not incorporate into membranes as do the free glycolipids, and thus may provide useful therapeutic tools for the management of clinical cholera [22].

Choleragen is composed of two major subunits, one of ~60,000 mol wt ("binding" subunit) which is responsible for the binding interaction with cell membranes, and another of ~36,000 mol wt ("active" subunit) which confers biological activity to the molecule [20, 22-25]. Choleragenoid, a 60,000-mol wt protein derived from choleragen [1], which contains only the binding subunit [20, 22-24], competitively inhibits the action of choleragen [4, 9, 18, 26] and has been shown to bind to the

TABLE 2. Effect of Various Glycosphingolipids and Other Glycolipids on the Specific Binding of ^{125}I-Labeled Cholera Toxin to Liver Membranes[a, b]

Compound[c]	Concentration in preincubation mixture (μg/ml)	^{125}I-Labeled cholera toxin (% bound)
None	-	100
Gal1 $\longrightarrow$ Cer	100	100
	5	100
Glc1 $\longrightarrow$ Cer	100	100
	5	100
Gal1 $\xrightarrow{\beta}$ 4Glc1 $\longrightarrow$ Cer	100	100
	5	100
Gal1 $\xrightarrow{\beta}$ 4Gal1 $\xrightarrow{\beta}$ 4Glc1 $\longrightarrow$ Cer	100	80
GalNac1 $\xrightarrow{\beta}$ 3Gal1 $\xrightarrow{\alpha}$ 4Gal1 $\xrightarrow{\beta}$ 4Glc1 $\xrightarrow{2}$ Cer	50	35
	20	44
	5	68

$HSO_3 \longrightarrow$ 3Gal1 $\longrightarrow$ Cer (sulfatide)	150	100
Gangliosides (crude, bovine brain)	50	0
	10	0
	1	2
	0.25	5
	0.50	73
Pychosine (Gal1 $\longrightarrow$ sphingosine)	100	100
Cerebrosides	500	100
Sphingomyelin	100	100

[a] ^{125}I-Labeled cholera toxin (0.1 μg/ml) was preincubated at 24°C for 60 min in 0.2 ml of Krebs-Ringer bicarbonate buffer containing 0.1% (wt/vol) albumin and the compound indicated in the table. Samples (50 μl) were then added to incubation mixtures consisting of 0.2 ml of Krebs-Ringer bicarbonate buffer, 0.1% (wt/vol) albumin, and 50 μg of liver membrane protein. After 20 min at 24°C, the specific binding of the toxin to the membranes was determined by filtration on cellulose-acetate filters. In the absence of liver membranes none of these compounds caused adsorption of the toxin to the filters. The final concentration of the various glycolipids in the membrane incubation mixture is five times lower than that indicated in the table.

[b] From Cuatrecasas [7].

[c] Cer, ceramide; Gal, galactose; Glc, glucose; GalNAc, N-acetylgalactosamine.

TABLE 3. Effect of Specific Gangliosides on the Binding of ^{125}I-Labeled Cholera Toxin to Liver Membranes[a,b]

Ganglioside[c]	Concentration in preincubation mixture (μg/ml)	^{125}I-Labeled cholera toxin (% bound)
None	–	100
G_{M3}(Cer $\longrightarrow$ 1Glc4 $\xrightarrow{\beta}$ 1Gal3 $\xrightarrow{\beta}$ 2NANA)	50	71
	25	95
	6	100
G_{M3}(Cer $\longrightarrow$ 1Glc4 $\xrightarrow{\beta}$ 1Gal3 $\xrightarrow{\beta}$ 2N-glycolyl-NA)	50	50
	20	60
	5	82
G_{M2}(Cer $\longrightarrow$ 1Glc4 $\xrightarrow{\beta}$ 1Gal4 $\xrightarrow{\beta}$ 1GalNAc); 2NANA $\xrightarrow{\alpha}$ 3 (on Gal)	5	16
	1.6	28
	0.5	65
	0.2	84
G_{M1}(Cer $\longrightarrow$ 1Glc4 $\xrightarrow{\beta}$ 1Gal4 $\xrightarrow{\beta}$ 1GalNAc3 $\xrightarrow{\beta}$ 1Gal)[d]; 2NANA $\xrightarrow{\alpha}$ 3 (on Gal4)	0.35	0
	70 ng/ml	18
	20 ng/ml	45

	6 ng/ml	74
	1 ng/ml	90
G_{Dla}(Cer → 1Glc4 $\xrightarrow{\beta}$ 1Gal4 $\xrightarrow{\beta}$ 1GalNAc3 $\xrightarrow{\beta}$ 1Gal3; Gal4 ← 3 α 2NANA; Gal3 ← α 2NANA	5	20
	1.6	34
	0.5	43
	0.2	70
G_{T1}(Cer → 1Glc4 $\xrightarrow{\beta}$ 1Gal4 $\xrightarrow{\beta}$ 1GalNAc3 $\xrightarrow{\beta}$ 1Gal3; Gal4 ← 3 α 2NANA8 $\xleftarrow{\alpha}$ 2NANA; Gal3 ← α 2NANA	10	43
	2.5	31
	0.8	85
	0.2	90

[a] These experiments were performed as described in Table 2.

[b] From Cuatrecasas [7].

[c] Nomenclature according to Svennerholm [56]. Cer, ceramide; Glc, glucose; Gal, galactose; GalNAc, N-acetylgalactosamine; NANA, N-acetylneuraminic acid.

[d] After digesting G_{M1} (0.1 mg/ml) with neuraminidase (20 μg/ml) for 3 hr at 37°C in 0.1 M sodium acetate (pH 6.2)–2 mM $CaCl_2$, the binding of toxin in the presence of 1 ng/ml of the ganglioside was 80%. Under similar conditions, digestion of G_{Dla} and G_{T1} resulted in a very large enhancement of inhibitory activity.

same membrane sites and with an affinity identical to choleragen [9]. No biological effects have yet been attributed unequivocally to pure choleragenoid.

Choleragen and choleragenoid are, thus, potentially useful as probes for detection and quantitation of a special class of cell-surface glycolipids which most likely are G_{M1} monosialogangliosides. These membrane molecules are of particular interest since the complement of cell-surface gangliosides can be altered dramatically following virus- or carcinogen-induced transformation [27-30] or extensive culturing of spontaneously transformed cell lines [31, 32]. Choleragen and choleragenoid can be radioiodinated to specific activities as high as 10^7 cpm/pmol, and therefore provide a sensitivity greatly exceeding (by 10^3 to 10^4) that of the chemical methods now available. The extremely high affinity of the binding of these proteins to their membrane sites ($K_D < 10^{-10}$ M) and the fact that they do not penetrate cells in detectable amounts [5] also permits their use as plasma membrane markers during membrane isolation procedures [33].

Choleragen is not toxic to cells. Despite the profound stimulation of adenylate cyclase activity, cultured cells exposed to choleragen remain viable [10, 34, 35], and the enzyme activity eventually returns to normal [5, 34]. Choleragenoid offers a potential advantage over choleragen as a membrane marker in that it is apparently metabolically inert.

Evidence has been reported recently suggesting that, following the initial binding interaction, choleragen, or some portion of the molecule, forms a detergent-stable complex directly with adenylate cyclase [36, 36a] Activation of adenylate cyclase by choleragen is fully retained upon conversion (with Lubrol-PX) to a nonsedimentable form (30 min at 250,000 g) [36, 36a, 37], and after gel filtration [36]. Furthermore, a peak of [^{125}I]-choleragen is associated with adenylate cyclase activity on gel filtration, provided the enzyme has been activated by labeled choleragen, and antisera directed against choleragen or its active subunit will specifically immunoprecipitate solubilized toxin-stimulated adenylate cyclase activity. These findings suggest that the ultimate "receptor" for choleragen is adenylate cyclase itself, and indicate that [^{125}I]choleragen may provide a means of labeling the enzyme.

II. RADIOIODINATION OF CHOLERAGEN AND CHOLERAGENOID

Although it is possible to label these molecules by other methods, such as reaction with [^{3}H]acetic anhydride [1], the specific activities achieved with iodination are superior by several orders of magnitude, and the ^{125}I-labeled proteins suffer no loss in biological activity [7]. Before use, cholera toxin and choleragenoid (5 mg in 0.2 ml) are chromatographed on a G-75 Sephadex column (1 cm × 20 cm) equilibrated at 24°C with 0.25 M sodium phosphate, pH 7.4. This step removes salts such as NaN_3 or Na-EDTA (which frequently are included and may themselves affect adenylate cyclase) and separates choleragen from any free active subunit which may have dissociated during storage. Iodination is performed essentially as described previously [7], using chloramine-T as an oxidant [38]. Twenty microliters of chloramine-T (0.5 mg/ml in 0.25 M sodium phosphate, pH 7.4) are added to a glass tube containing 50 μl of choleragen in 0.25 M sodium phosphate, pH 7.4, optical density at 280 nm of 1.5 to 2.2, and 10 μl of carrier-free $Na^{125}I$ (0.3-0.4 mCi/μl) freshly prepared by Union Carbide. After 30 to 45 sec at 24°C, the reaction is terminated by addition of 20 μl of sodium metabisulfite (1 mg/ml in 0.25 M sodium phosphate, pH 7.4). After 10 sec, 0.1 ml of 0.05 M sodium phosphate-0.1% (wt/vol) bovine serum albumin, pH 7.4, is added, and the sample is applied on a G-50 fine Sephadex column (0.6 × 30 cm) equilibrated at 24°C with 0.1 M sodium phosphate-0.1% (wt/vol) bovine serum albumin.

The radioactive material in the void volume of the column is pooled, diluted to 0.2 to 0.4 mCi/ml, and stored in aliquots at -20°C. The specific activity of the iodinated material is determined by diluting an aliquot of the reaction mixture with 0.1 M sodium phosphate, pH 7.4, containing 0.5% (wt/vol) bovine serum albumin, and measuring the percent of the total radioactivity which is precipitable in ice-cold 5% (wt/vol) trichloroacetic acid. This value gives the proportion of iodine incorporated into protein, and varies from 30 to 60%.

The specific activity of $[^{125}I]$choleragen prepared in this way varies from 1 to 2.5 μCi/pmol, assuming a molecular weight of ~90,000 [22, 23, 39] and $A_{1cm}^{1\%}$ (280 nmol of 11.42 [39]. This corresponds to 0.5 to 1 atoms of ^{125}I per molecule of protein; derivatives containing up to 5 atoms of ^{125}I per molecule (10 μCi/pmol) can be prepared by including more $Na^{125}I$ in the reaction mixture. The incorporation of ^{125}I varies with the particular lot of $Na^{125}I$ and is best with colorless preparations.

Ninety to ninety-six percent of the labeled protein is precipitated by ice-cold 5% (wt/vol) trichloroacetic acid in 0.1 M sodium phosphate buffer, pH 7.4, with 0.5% (wt/vol) albumin as a carrier protein. Seventy to eighty-five percent of the freshly prepared radioactive toxin will specifically bind to liver microsomal membranes [7], and is adsorbed by ganglioside-agarose columns [22]. Approximately 90% comigrates with choleragen or choleragenoid on sodium dodecyl sulfate disk gel electrophoresis [22]. The portion which does not adsorb to liver membranes will not bind to ganglioside columns and apparently represents a contaminating protein or damaged molecules. Nearly 80% of the iodine in ^{125}I-labeled choleragen is localized on the 36,000-mol wt active subunit

[22]. The percent of radioactivity which will bind to membranes and ganglioside columns decreases with the age of the iodinated material; progressive loss of activity is detectable after one week of storage, and after three weeks only 50% of the radioactivity will bind to liver membranes. The binding behavior of the radiolabeled material which will bind to membranes and ganglioside-agarose columns is not appreciably altered during storage, which suggests that only a certain proportion of the molecules is being inactivated. Thus, even aged iodinated choleragen and choleragenoid can be used for certain purposes.

The biological activity of choleragen is not affected by incorporation iodine. Freshly prepared [^{125}I]choleragen, with an average of approximately one atom of iodine per molecule (22 μCi/μg), activates adenylate cyclase of isolated rat fat cells in an identical fashion to native choleragen (Fig. 3). The lipolytic activity of the iodotoxin measured with isolated fat cells [40-42] also is indistinguishable from the native toxin [7].

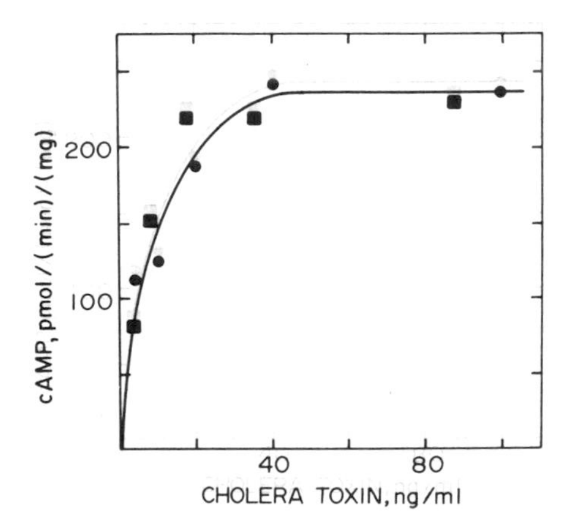

FIG. 3. Effect of preincubating rat fat cells with increasing concentrations of either native cholera toxin (●) or of ^{125}I-labeled cholera toxin (■) on the membrane adenylate cyclase activity. Isolated epididymal fat cells from 12 150-g rats were divided into 11 equal portions; and each of these was incubated at 37°C in 10 ml of Krebs-Ringer bicarbonate, 2% (wt/vol) BSA, pH 7.4, in the presence of various concentrations of either native cholera toxin or ^{125}I-cholera toxin (22 μCi/μg). After 3 hr, the cells were homogenized in tris-HCl (50 mmol, pH 7.7), and the 40,000-g membrane pellet assayed for adenylate cyclase activity (15 min at 30°C) as described [5].

III. QUANTITATIVE ASSAY OF THE INTERACTION OF ^{125}I-LABELED CHOLERAGEN AND CHOLERAGENOID WITH CELLS AND CELL MEMBRANES

A. Methods

The essential problem in measuring the binding of ^{125}I-labeled ligands to cells and cell membranes is to quantitatively separate bound and unbound ligand. Two methods are now available which permit the rapid assay of a large number of samples (up to 200 per day): filtration across Millipore filters [43] and centrifugation of membranes or cells through an oil phase in a Beckman microfuge [44]. Millipore filtration has been used in the past to measure the binding to fat cells and cell membranes of iodinated derivatives of insulin [43], glucagon [45], and plant lectins [46]. Cells or membrane fractions are incubated for 20 to 30 min at 24°C in 0.2 ml of Krebs-Ringer bicarbonate buffer (pH 7.4) containing 0.1% (wt/vol) bovine serum albumin and the iodinated choleragen or choleragenoid. Three milliliters of the same buffer (ice-cold) are added to the tube, and the contents are poured onto 25-mm cellulose acetate Millipore filters [EGWP (0.25-μ pore size) for membranes; EAWP (1-μ pore size) for cells], which are positioned under vacuum in a multiple-sample filtration manifold. The filters are washed under negative pressure with 10 ml of ice-cold Krebs-Ringer bicarbonate buffer containing 0.1% (wt/vol) albumin, and then removed and counted for ^{125}I in a well-type gammacounter. The steps of dilution, filtration, and washing can be completed in 10 to 15 sec.; the dissociation of bound choleragen is so slow, particularly at 0°C, that the loss of bound material during this time is negligible.

In the absence of cells or membranes, <0.2 to 1% of freshly prepared iodinated choleragen or choleragenoid is adsorbed to cellulose acetate Millipore filters. Mixed cellulose filters (HAWP), however, bind up to 30 to 40% of the applied ligands. Binding data should be expressed as "specific" binding, which refers to the amount of ^{125}I-labeled protein bound to cells or membranes which can be "displaced" by adding 10 to 20 μg/ml of native toxin 5 to 10 min before addition of the iodinated material. Usually, the contribution of nondisplaceable binding to the total quantity of radioactive uptake is very small ($<3\%$). It is also important to include control samples containing no cells or membranes, since under certain circumstances, [^{125}I]choleragen will bind or be adsorbed to proteins which themselves adsorb to the filters, and thus increase the background radioactivity [7].

Advantages of this filtration method include the rapidity and potential accuracy of the procedure, and the fact that large numbers of samples can be processed. Disadvantages arise from the initial cost of the filter manifold, and the expense of the Millipore filters. Furthermore, situations may arise when the nonspecific adsorption of iodinated material to the filter will present problems. The oil flotation (microfuge) method [44] provides an alternative procedure which circumvents some of these difficulties. Bound and free ligands are separated by centrifugation of the membranes or cells through an oil phase with a density less than that of the particulate material, but greater than that of the medium. The pelleted material is thus "washed" during its passage through the oil, and is well-separated from the buffer containing the free radioactive proteins. In the case of fat cells, which float in aqueous solutions, an oil phase of dinonyl phthalate with a density greater than the cells, but less than the medium, is layered over the cells, and centrifugation forces them upwards through the oil. The density of the oil phase is adjusted by mixing appropriate amounts of dinonyl phthalate and dibutyl phthalate. The correct proportion must be determined for each tissue: for rat liver microsomal membranes, 2:1 dibutyl phthalate-dinonyl phthalate is effective, while erthrocytes will sediment with undiluted dibutyl phthalate. The assay of [^{125}I]choleragen or choleragenoid binding with the microfuge technique requires incubation of cells or membranes for 20 to 30 min at 24°C in 0.1 to 0.2 ml of Krebs-Ringer bicarbonate buffer containing 0.1% (wt/vol) albumin and 10^{-12}-10^{-9} M [^{125}I]choleragen or choleragenoid. A 0.1 ml aliquot is transferred to a soft-plastic microfuge tube (0.4-ml capacity), containing 0.25 ml of the appropriate mixture of dibutyl phthalate and dionyl phthalate. Following centrifugation for 3 to 4 min in a microfuge, the tubes are removed and the tips cut off and assayed for ^{125}I. The instrument will accommodate 20 samples per run, and large numbers of tubes can be processed. The microfuge is relatively inexpensive, and the microfuge tubes can be obtained for less than $0.01 each. Difficulties with this technique arise when small amounts of membranes are assayed, since incomplete pelleting may occur; if the oil density is not adjusted properly, significant quantities of membranes may remain at the oil-water interface. If excessive quantities of cells or membranes are assayed, significant trapping of the medium may occur and raise the background. It is important here, as with the Millipore assay, to express binding as the amount of ^{125}I-labeled material bound to cells or membranes which is displaceable by previous addition of 10 to 20 μg/ml of native toxin or choleragenoid.

It may be useful to have on hand a supply of standardized membranes which can be used to test various preparations of iodoproteins. Liver microsomal membranes are especially practical in this regard [7]; these are easily prepared, and retain this binding property for at least two years when stored at -20°C. The procedure [43] involves homogenization of minced rat livers in 20 vol of ice-cold 0.25 M sucrose with a Brinkman polytron (90 sec at a setting of 2.5), followed by differential centrifugation for 10 min at 600 _g_ and 10 min at 10,000 _g_. The 10,000-_g_ supernatant is made 0.1 M in sodium chloride and 0.5 mmol in magnesium chloride, and centrifuged for 45 min at 35,000 _g_; the resulting membrane pellet is washed once, and finally resuspended at a concentration of ~5 mg/ml of membrane protein in 50 mmol tris-HCl, pH 7.6. These membranes are stored in aliquots at -20°C, and are thawed and suspended by brief sonication before use.

B. Interpretation of Binding Data

Two useful parameters can be derived from studies of the binding of choleragen and choleragenoid to cell membranes: the apparent affinity of the process, and the total number of binding sites per cell or per milligram of protein. A reasonable estimation of these values can be obtained from plots of ligand bound versus ligand concentration, provided that the assays are determined such that, at equilibrium, < 10-20% of the functional radiolabeled molecules are adsorbed to cell membranes. It is, therefore, necessary to perform preliminary assays such as those described in Figure 4, in order to determine the correct concentration of cells or membranes. This is particularly important when the displacement of labeled ligand, or loss of binding sites, is being examined, as would be the case if an unlabeled derivative of choleragen were being tested for the ability to compete for [^{125}I]choleragen binding. If membrane-binding sites are present in large excess ofer the number of [^{125}I]choleragen molecules, these additional sites would be filled by the derivative with no displacement of membrane-bound radioactivity. It would then be concluded incorrectly that the derivative could not compete for choleragen binding, or that it did so with very low affinity.

The true affinity of the choleragen-membrane interaction is difficult to measure, since, following the initial binding event, a portion of the membrane-bound choleragen becomes permanently adsorbed to the membrane [42]. The irreversible nature of this second process invalidates estimates of affinity based on the assumption of an equilibrium between bound and free ligand. This property of choleragen also requires that, in displacement or competition experiments, the unlabeled ligand be added prior to the [^{125}I]choleragen. The apparent K_m for choleragen binding is still a useful number, although it cannot be interpreted in a formal manner.

C. Application of [^{125}I]Choleragen and Choleragenoid Membrane-Binding Assays

The rapidity and accuracy of the binding assays with ^{125}I-labeled choleragen and choleragenoid make this procedure particularly attractive. Binding competition for [^{125}I]choleragen sites on liver membranes could serve as a sensitive assay for native choleragen [7] (Fig. 5), and has been used for detection of _E. coli_ enterotoxin [47], which binds to similar

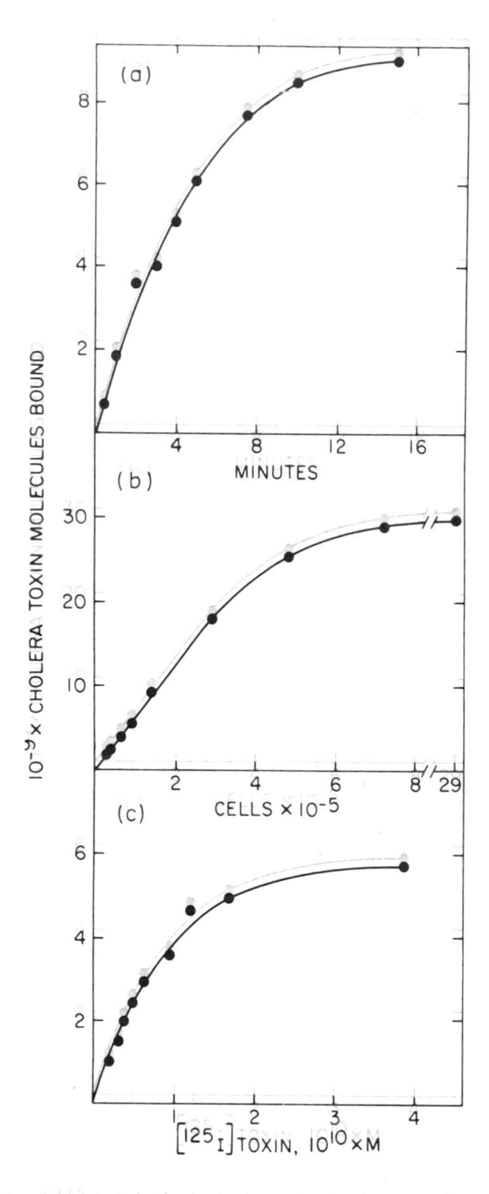

FIG. 4. Binding of ^{125}I-labeled cholera toxin to intact toad erythrocytes. Binding was performed at 0°C with thoroughly washed erythrocytes suspended in amphibian-Ringer, 0.1% (wt/vol) BSA, pH 7.5. The time course of binding of ^{125}I-toxin (2.6×10^{-10} M, 1.3 Ci/μmol) was measured in the presence of 1.4×10^5 cells in a volume of 0.24 ml (a). The effect of cell number on the binding of ^{125}I-toxin (2.6×10^{-10} M) was determined after incubating for 45 min (0°C) in a volume of 0.24 ml (b). The effect of increasing the concentration of ^{125}I-toxin was measured (40 min, 0°C) using 7.2×10^3 cells in a volume of 0.5 ml (c). From Bennett and Cuatrecasas [5].

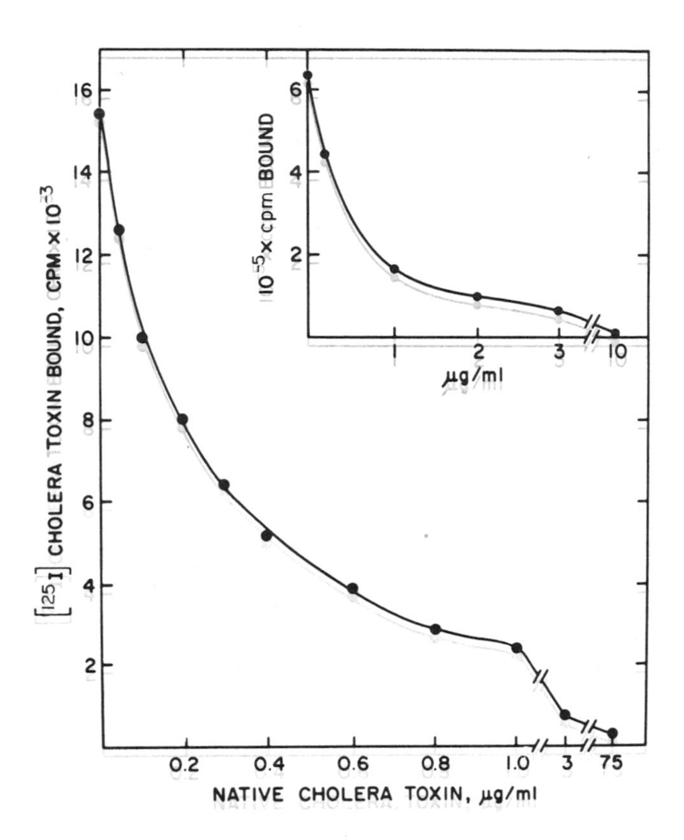

FIG. 5. Effect of native cholera toxin on the binding of ^{125}I-labeled cholera toxin to liver membranes. The indicated concentration of native toxin was added to 0.2 ml of Krebs-Ringer bicarbonate buffer containing 0.1% (wt/vol) albumin and 50 μg of membrane protein. After incubating for 5 min at 24°C, ^{125}I-labeled cholera toxin (15 ng/ml, 9 μCi/μg) was added and the samples were incubated for 15 min at 24°C. Binding to the membranes was determined by filtration over cellulose-acetate Millipore filters as described in the text. The inset describes a similar experiment in which the concentration of ^{125}I-labeled cholera toxin in the incubation medium was increased by more than 20-fold (0.4 μg/ml); the concentration of protein in this experiment was 0.75 mg/ml. From Cuatrecasas [7].

membrane sites as choleragen [26]. The binding assay also provides a means of quantitatively following the titer of choleragen antisera in immunized animals [47], since [^{125}I]toxin complexed with antibody will not bind to membranes. The possibility of using [^{125}I]choleragen binding to monitor the ganglioside composition of cell lines during transformation has already been mentioned [10].

IV. USE OF ^{125}I-LABELED CHOLERAGEN AND CHOLERAGENOID AS PLASMA MEMBRANE MARKERS

The unusual rapidity of the binding of choleragen and choleragenoid to cell membranes, the stability of the resulting membrane-ligand complex, and the fact that binding sites are ubiquitous and relatively abundant in most cell surfaces permit the use of these molecules as general plasma membrane markers [33].

The basic approach of this procedure is to label surface gangliosides (G_{M1}) with low concentrations of radioiodinated protein (10-50 ng/ml, 10,000-20,000 cpm/ng) before cell homogenization by a 5- to 10-min incubation with intact cells at 4°C, or by in situ perfusion of an organ such as liver. The unbound, ^{125}I-labeled choleragen or choleragenoid is removed by washing at 4°C, and the distribution and the recovery of plasma membrane-bound label are followed quantitatively during subsequent homogenization and fractionation procedures. The spontaneous dissociation of membrane-bound iodotoxin is negligible at 4°C, and prolonged manipulations are possible without significant loss of label. Other radiolabeled ligands such as insulin and the plant lectin, wheat germ agglutinin, also interact with high affinity to cell-surface receptors, and are useful as plasma membrane markers [33].

The use of [^{125}I]choleragen as a labeling agent is restricted to intact cells or whole tissues, since the nuclear fraction of cell homogenates binds a significant amount of radioactivity (Fig. 6). The distribution of membrane-bound [^{125}I]choleragen from labeled fat cells is nearly identical on linear sucrose gradients to that of [^{125}I]insulin, which has been shown to bind to receptors localized exclusively at the cell surface [48, 49]. Moreover, no intracellular label can be detected even after prolonged incubation with intact cells [5]. Choleragen is not known to perturb membranes in any manner other than activation of adenylate cyclase, and this process does not occur significantly at temperatures < 10°C or incubation periods less than 30 min [5, 42]. Choleragenoid binds to cells in an identical fashion as choleragen, and at this time has no known biological effects.

The application of these ligands as plasma membrane markers presents significant advantages over the conventional use of enzymatic

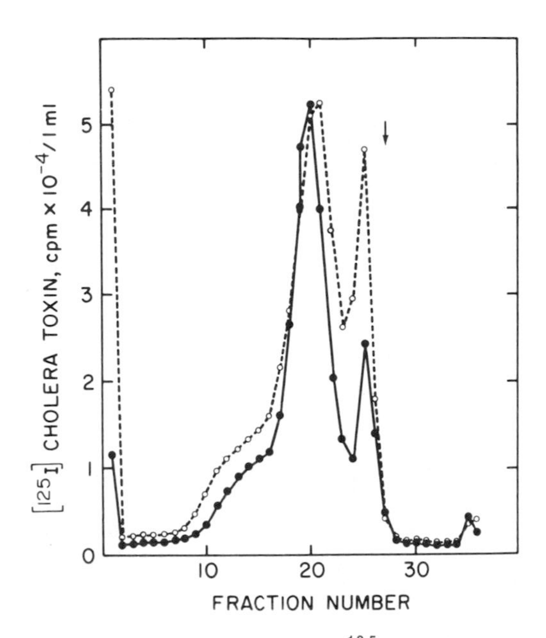

FIG. 6. Comparison of the distribution of [^{125}I]cholera toxin on sucrose density centrifugation when membrane labeling is performed in intact fat cells and in whole particulate preparations from homogenates. Isolated fat cells from eight rats (150 to 175 g) were divided into two equal parts. One part (A) was incubated with [^{125}I]cholera toxin (5.1×10^6 cpm) for 5 min at 24°C in 5 ml of KRB-1% albumin. The other (B) was treated the same way, but without [^{125}I]cholera toxin. Both sets were washed and homogenized. A total particulate pellet (40,000 g, 30 min) was resuspended in 5 ml of KRB-1% albumin with a Polytron (set at 2.6, 10 sec). [^{125}I]Cholera toxin was added to (B), incubated for 5 min at 24°C and washed twice by centrifugation. The distribution of [^{125}I]cholera toxin from the samples labeled in the intact cells (●) and in the particulate fractions (○) are determined on 27-ml linear sucrose gradients (54.1% to 27.6% wt/vol) following centrifugation at 25,000 rpm for 60 min at 0°C in a Beckman SW27 rotor. The same amount of protein material was applied to each gradient. Arrow indicates the interphase of linear gradient and isotonic sucrose-tris buffer. From Chang et al. [33].

activities. Seldom have these enzymes been demonstrated to be localized exclusively in the plasma membrane [50, 51]; different patterns of localization may occur in different tissues or species [50, 52], and changes may occur during the cell cycle [53]. Another difficulty with marker enzymes is that their activity cannot be accurately measured in the intact cell before homogenization. Furthermore, this procedure may damage or alter the catalytic properties of these enzymes due to the action of proteases, removal of activators or inhibitors, or by exposure to unphysiological conditions. The measurements of enzymatic activity may be difficult, especially in crude homogenates, where competing reactions may occur. For example, Na^+- K^+-activated ATPase is difficult to measure accurately in many tissues, and in rat fat cell membranes, this enzyme represents no more than 10% of the total ATPase activity [54].

In a typical experiment involving labeling of adipocyte plasma membranes, isolated epididymal fat cells from eight rats ($\sim 10^7$ cells) are suspended in 10 ml of Krebs-Ringer bicarbonate, pH 7.4, containing 0.1% (wt/vol) bovine serum albumin and 50 ng/ml of [^{125}I]choleragen (10,000 to 20,000 cpm/ng). The cells are incubated for 5 min at room temperature, and subsequently washed three times with 10 vol each of cold Krebs-Ringer bicarbonate buffer (pH 7.4). Treatment of adipocytes under these conditions results in incorporation into the plasma membranes of $\sim$200,000 cpm, or 10^3 molecules of [^{125}I]choleragen per cell. The recovery of plasma membranes during homogenization, centrifugation, and other procedures can be easily quantitated in terms of the complement of membranes initially present in the intact cells.

V. INTERACTION OF [^{125}I]CHOLERAGEN WITH MEMBRANE COMPONENTS FOLLOWING SOLUBILIZATION WITH NONIONIC DETERGENTS

The choleragen-induced activation of adenylate cyclase persists following dissolution of the membrane with the nonionic detergent, Lubrol-PX [36, 37], and membrane-bound [^{125}I]choleragen remains complexed in a macromolecular form in the solubilized state [36]. The chromatographic patterns of adenylate cyclase activity and of membrane-bound ^{125}I-labeled choleragen from solubilized membranes on Sepharose 6B columns equilibrated with detergent are shown in Figure 7. Fat cells were exposed to 5×10^{-10} M [^{125}I]choleragen at 37°C for either 10 min—which is sufficient to form the initial, inactive choleragen-membrane complex—or for 3 hr, at which time nearly maximal activation of adenylate cyclase has occured [7, 42]. The cells were then homogenized in 50 mM tris-HCl, 0.5 mM NaEDTA (pH 7.7), with a Brinkman polytron (30 sec at a setting of 3.0) and the whole particulate fractions collected by centrifugation for 30 min at 35,000 g. The membranes were suspended at a concentration of 20 mg of protein per ml in about 1 ml of 50 mM tris-HCl, 2% Lubrol-PX, sonicated three times for 1 sec each, and 50 μl of 0.1 M $MgCl_2$, 3% (wt/vol) bovine serum albumin, 2 mM GTP were added. These suspensions were centrifuged for 30 min at 250,000 g, and the supernatants were applied to Sepharose 6B columns (1 × 57 cm) equilibrated at 4°C with

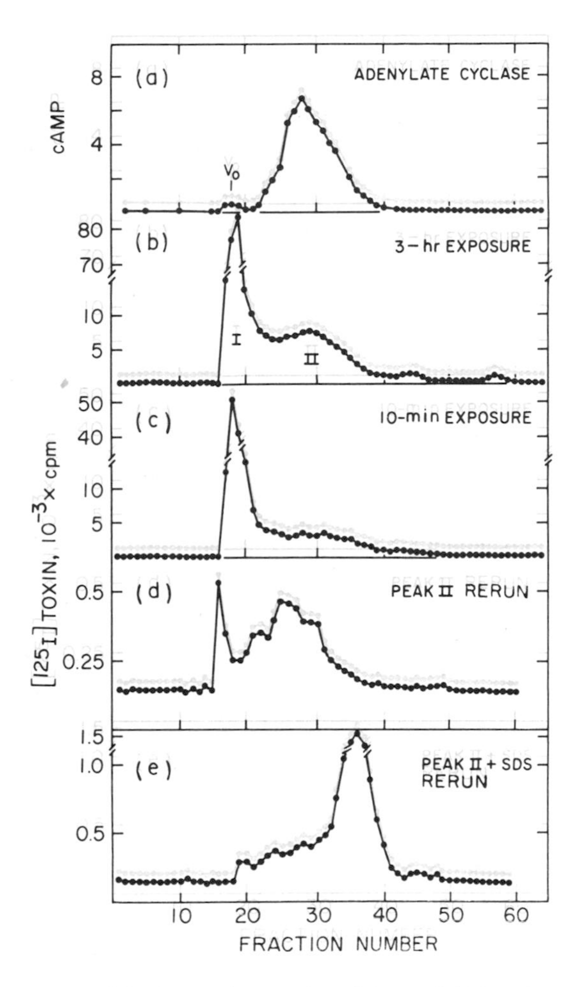

FIG. 7. Sepharose 6B chromatography of Lubrol PX-solubilized adenylate cyclase from fat cells incubated with ^{125}I-labeled cholera toxin. Cells from 25 rats (200 g) were incubated at 37°C for 3 hr. ^{125}I-labeled cholera toxin (2.4 μg, 9.4 μCi/μg) was added either at the beginning of the incubation or for the final 10 min. Samples (0.75 ml) of the 250,000 g membrane-detergent extracts were applied to Sepharose 6B columns (1 × 57 cm; flow rate 10 ml/hr; 0.6-ml fractions), and cyclase activity (a) was determined. Panels (a) and (b) refer to cells exposed to toxin for 3 hr while panel (c) is for the 10-min exposure to ^{125}I-labeled toxin. The peak of cyclase in panel (a) (fractions 28–30) was rechromatographed with [panel (e)] or without [panel (d)] heating the sample in 1% sodium dodecyl sulfate (60 min at 40°C). From Bennett, O'Keefe, and Cuatrecasas [36].

$MgCl_2$ (4 mmol), bovine serum albumin (0.15%, wt/vol), Lubrol-PX (0.2%, vol/vol), and tris-HCl (50 mmol, pH 8).

A major portion of the membrane-bound ^{125}I-labeled choleragen appears in the void volume, and this is relatively independent of the time or temperature of preincubation with choleragen. A small radioactive peak is found associated with the adenylate cyclase activity, provided the cells have been incubated at 37°C for 3 hr with the ^{125}I-labeled toxin. This small peak, which corresponds to ~1,000 molecules of toxin per cell, appears in the same position after rechromatography, although heating in 1% sodium dodecyl sulfate shifts the peak from 1.5 V_0 to ~2 V_0. Analysis of this material on sodium dodecyl sulfate disk gel electrophoresis (not shown) indicates that the radioactivity primarily represents the active 36,000-mol wt subunit of choleragen, although the presence of the 60,000-mol wt subunit cannot be excluded.

It is of interest that if [^{125}I]choleragen is added directly to solubilized fat cell membranes, some radioactivity still appears in the excluded volume, although no peak is observed in the position of adenylate cyclase. If native toxin is added to the membranes or solubilized proteins before [^{125}I]toxin, both radioactive peaks disappear. [^{125}I]Choleragen alone elutes in a complicated pattern in Lubrol-PX, but the largest-molecular-weight species appears at about 2 V_0 on Sepharose 6B. The high molecular weight of the major peak of membrane-bound [^{125}I]choleragen and that of [^{125}I]choleragen added to the solubilized membranes suggest that cell-surface gangliosides may be associated with membrane proteins or other macromolecules in a relatively stable complex.

The direct association of at least a portion of the small radioactive peak with adenylate cyclase is supported by the finding that sera directed against choleragen will specifically immunoprecipitate solubilized choleragen-stimulated adenylate cyclase activity [36, 36a]. The fact that ~90% of the membrane-bound choleragen is not associated with adenylate cyclase on gel filtration (Fig. 7) is not surprising; the apparent K_m for stimulation of adenylate cyclase in erythrocytes and melanocytes [5, 34] occurs with 1,500 to 3,000 molecules of toxin bound per cell, whereas specific binding is half-maximal at ~40,000 or more molecules per cell. Studies using choleragenoid to "decrease" the number of toxin-binding sites suggest that the excess binding sites represent equivalent "spare" receptors [36].

[^{125}I]Choleragen may, thus, be a useful marker for adenylate cyclase in detergent extracts of cell membranes, provided this fraction can be easily separated from the major portion of membrane-bound radioactivity. In any event, this iodoprotein may provide a sensitive tool for elucidating on a molecular level the disposition of cell-surface gangliosides. It should, in principle, be possible, using [^{125}I]choleragen as a label, to isolate and characterize the membrane macromolecule(s) which complex these glycolipids in detergent extracts, and presumably in the intact membrane.

VI. SUMMARY

Choleragen and choleragenoid are proteins secreted by Vibrio cholerae, which bind with an unuaually high affinity to a special class of membrane glycosphingolipids which most likely is G_{M1} monosialogangliosides. These ligands can be radiolabeled with carrier-free ^{125}I to specific activities as high as 10^7 cpm/pmol, and are the most sensitive reagents currently available for detection of cell-surface gangliosides. These molecules may thus provide important tools in the elucidation of the role of glycolipids in membrane structure and function. Little is known of the mechanism at this time, although gangliosides may be involved in the processes of transformation [27-30] and contact inhibition of cell growth [32].

The cell-surface receptors for choleragen are ubiquitous in eukaryotic cells, the membrane complex is quite stable at 4°C, and no intracellular label can be detected even after prolonged incubation with intact cells. These properties permit the labeling of the plasma membranes of intact cells with [^{125}I]choleragen or [^{125}I]choleragenoid with the formation of a complex which persists during cell homogenization and membrane isolation procedures [33]. The ease and accuracy of determination of membrane-bound [^{125}I]ligand provides a method for quantitation of recovery of labeled plasma membranes which has significant advantages over the usual enzymatic markers. Direct evidence has been presented which shows that incubation of fat cells with choleragen results in the formation of a detergent-stable complex between adenylate cyclase and some portion of the toxin molecule [36]. Treatment of cells with ^{125}I-labeled choleragen prior to solubilization of the membrane may, thus, be a method for labeling soluble adenylate cyclase.

An area of future interest is the use of choleragen or choleragenoid as morphological markers for localization of gangliosides on cell surfaces. Rhodamine and fluoroscein derivatives of choleragen have been prepared recently which can be visualized by fluorescence light microscopy [55]. Such fluorescent ligands can be detected on cells when these are labeled at average densities of 50 to 100 molecules per square micron and have been useful in examining the distribution of membrane-bound choleragen at a resolution of 0.2 to 0.5 μ [55]. It should also be possible to prepare bulky, or electron-dense, derivatives of choleragen which would be visible by transmission and scanning electron microscopy. These molecules would permit localization of cell-surface gangliosides with resolutions as high as 50 Å. Such morphological tools would add another dimension to the present methodology, which provides only quantitative information about populations of molecules.

ACKNOWLEDGMENTS

This research was supported by grants from National Institutes of Health (AM 14956), The American Cancer Society, and The Kroc Foundation. Vann Bennett is supported by the Home Life Insurance Company. Pedro Cuatrecasas is recipient of a USPHS RCDA (AM 31464).

REFERENCES

1. R. A. Finkelstein, CRC Crit. Rev. Microbiol., 2(4):553 (1973).
2. G. W. G. Sharp, Ann. Rev. Med., 24:19 (1973).
3. M. Field, Proc. Natl. Acad. Sci. U.S.A., 71(8):3299 (1974).
4. V. Bennett and P. Cuatrecasas, J. Memb. Biol., 22:1 (1975).
5. V. Bennett and P. Cuatrecasas, J. Memb. Biol., 22:29 (1975).
6. V. Bennett and P. Cuatrecases, Federation Proc., 33(5):1357, Abstract (1974).
7. P. Cuatrecasas, Biochemistry, 12:3547 (1973).
8. P. Cuatrecasas, Biochemistry, 12:3558 (1973).
9. P. Cuatrecases, Biochemistry, 12:3577 (1973).
10. M. D. Hollenberg, P. H. Fishman, V. Bennett, and P. Cuatrecasas, Proc. Natl. Acad. Sci. U.S.A., 71:4224 (1974).
11. J. Holmgren, L. Lindholm, and I. Lonnroth, J. Exptl. Med., 139: 801 (1974).
12. J. M. Boyle and J. D. Gardner, J. Clin. Invest., 53:1149 (1974).
13. W. A. Walker, M. Field, and K. J. Isselbacher, Proc. Natl. Acad. Sci. U.S.A., 71:320 (1974).
14. W. E. van Heyningen, C. C. J. Carpenter, N. F. Pierce, and W. B. Greenough, J. Infect. Dis., 124:415 (1971).
15. C. A. King. and W. E. van Heyningen, J. Infect. Dis., 127:639 (1973).
16. J. Holmgren, I. Lonnroth, and L. Svennerholm, Infect. Immunity, 8:208 (1973).

17. J. Holmgren, I. Lonnroth, and L. Svennerholm, Scand. J. Infect. Dis., 5:77 (1973).

18. N. F. Pierce, J. Exptl. Med., 137:1009 (1973).

19. W. E. van Heyningen, Nature, 249:415 (1974).

20. S. van Heyningen, Science, 183:656 (1974).

21. B. Beckman, E. O'Keefe, and P. Cuatrecasas, manuscript in preparation.

22. P. Cuatrecasas, I. Parikh, and M. D. Hollenberg, Biochemistry, 12:4253 (1973).

23. I. Lonnroth and J. Holmgren, J. Gen. Microbiol., 76:417 (1973).

24. R. A. Finkelstein, M. Boesman, S. H. Neoh, M. K. La Rue, and R. Delaney, J. Immunol., 113:145 (1974).

25. S. van Heyningen and C. A. King, Biochem. J., 146:269 (1975).

26. J. Holmgren, Infect. Immunity, 8:51 (1973).

27. F. A. Cumar, R. O. Brady, E. H. Kolodny, V. W. McFarland, and P. T. Mora, Proc. Natl. Acad. Sci. U.S.A., 67:757 (1970).

28. H. Den, B.-A. Sela, S. Roseman, and L. Sachs, J. Biol. Chem., 249:659 (1974).

29. P. H. Fishman, R. O. Brady, R. M. Bradley, S. A. Aaronson, and G. J. Todaro, Proc. Natl. Acad. Sci. U.S.A., 71:298 (1974).

30. R. O. Brady and P. T. Mora, Biochim. Biophys. Acta, 218:308 (1970).

31. P. H. Fishman, R. O. Brady, and P. T. Mora, in Tumor Lipids: Biochemistry and Metabolism. (R. Wood, ed.), American Oil Chemists Society, Champaign, Ill., 1973, p. 250.

32. S. Hakomori, Proc. Natl. Acad. Sci. U.S.A., 67:1741 (1970).

33. K.-J. Chang, V. Bennett, and P. Cuatrecasas, J. Biol. Chem., 250:488 (1975).

34. E. O'Keefe and P. Cuatrecasas, Proc. Natl. Acad. Sci. U.S.A., 71:2500 (1974).

35. M. D. Hollenberg and P. Cuatrecasas, Proc. Natl. Acad. Sci. U.S.A., 70:2964 (1973).

36. V. Bennett, E. O'Keefe, and P. Cuatrecasas, Proc. Natl. Acad. Sci. U.S.A., 72:33 (1975).

36a. N. Sahyoun and P. Cuatrecasas, Proc. Natl. Acad. Sci. U.S.A., 72:3438 (1975).

37. B. Beckman, J. Flores, P. Witkum, and G. W. G. Sharp, J. Clin. Invest., 53:1202 (1974).

38. W. M. Hunter and F. C. Greenwood, Nature, 194:495 (1962).

39. J. J. LoSpalluto and R. A. Finkelstein, Biochim. Biophys. Acta, 257:158 (1972).

40. M. Vaughan, N. F. Pierce, and W. B. Greenough, Nature, 226:658 (1970).

41. W. B. Greenough, N. F. Pierce, and M. Vaughan, J. Infect. Dis., 121:5111 (1970).

42. P. Cuatrecasas, Biochemistry, 12, 3567 (1973).

43. P. Cuatrecasas, Proc. Natl. Acad. Sci. U.S.A., 68:1264 (1971).

44. J. Gliemann, K. Osterlind, J. Vinten, and S. Gammeltoft, Biochim. Biophys. Acta, 286:1 (1972).

45. B. Desbuquois, F. Krug, and P. Cuatrecasas, Biochim. Biophys. Acta, 343:102 (1974).

46. P. Cuatrecasas, Biochemistry, 12:1312 (1973).

47. P. Agre and P. Cuatrecasas, manuscript in preparation.

48. P. Cuatrecasas, J. Biol. Chem., 247:1980 (1972).

49. P. Cuatrecasas, Federation Proc., 32:1838 (1973).

50. C. J. Lauter, A. Solyom, and E. G. Trams, Biochim. Biophys. Acta, 266:511 (1972).

51. C. C. Widnell, J. Cell. Biol., 52:542 (1972).

52. A. R. Oseroff, P. W. Robbins, and M. M. Burger, Ann Rev. Biochem., 42:647 (1973).

53. H. B. Bosmann, Biochim. Biophys. Acta, 203:256 (1970).

54. J. B. Modolell and R. O. Moore, Biochim. Biophys. Acta, 135:319 (1967).

55. S. Craig and P. Cuatrecasas, Proc. Natl. Acad. Sci. U.S.A., 72:3844 (1975).

56. L. Svennerholm, J. Neurochem., 10:613 (1963).

Chapter 5

FOLLICLE-STIMULATING HORMONE: MEASUREMENT BY A RAT TESTES TUBULE TISSUE RECEPTOR ASSAY

Leo E. Reichert, Jr.

Department of Biochemistry
Division of Basic Health Sciences
Emory University
Atlanta, Georgia

I. INTRODUCTION

Tissue receptor assays have become important tools for measurement of biologic activity of proteins and nonprotein hormones, and for assessing the consequences of structural modifications on the biologic activity of hormone derivatives. We have been concerned for some time with development and application of tissue receptor assays for pituitary luteinizing hormone (LH) [1] and follicle-stimulating hormone (FSH) [2]. Our application of such assays has been primarily in monitoring effectiveness of various purification sequences, in determining the characteristics of gonadotropin subunit association or dissociation phenomena [3, 4], in structure-function studies, and in studies on the comparative properties of gonadotropins from a variety of species [5, 6]. The receptor source utilized in our studies has been the mature rat testes. As the radioligand, we have chosen human LH or human FSH. There are, of course, other analytical procedures available for measurement of gonadotropic activity. Of these, the most commonly used and, generally speaking, the most sensitive, has been radioimmunoassay. Perhaps the most significant application of radioimmunoassay has been for measurement of hormone concentrations in human and animal plasma or serum. Whole animal bioassays, the traditional means for hormone measurement, are also available for FSH and LH, but their lack of sensitivity and relatively poor precision render then unsuitable for many types of investigations. While radioimmunoassays have given results similar to those obtained with bioassays in most instances, the caveat always remains that immunologic, rather than biologic, activity is being measured.

We have exhaustively evaluated the results of tissue receptor assay measurements for LH and FSH biologic activity, and compared these with those obtained through use of accepted and specific whole animal bioassays. In general, good agreement in relative potency estimates was observed [2, 7]. However, it should be noted that the tissue receptor assay for FSH which we describe here is a membrane-binding assay, and that membrane binding may not necessarily reflect biologic activity in all situations for at least two reasons: First, it is possible that hormone derivative or metabolite such as desialylated FSH may bind avidly to its receptor, but yet have low biologic activity in the whole animal due to factors related to plasma survival time. Second, it is also possible that a hormone derivative or metabolite may bind to a specific membrane receptor, but fail to elicit the hormone-responsive sequela (such as adenyl cyclase activation) necessary for expression of biologic activity, perhaps as a consequence of a subtle defect in the hormone-receptor interaction. As with all assay systems, it is important to bear in mind shortcomings, as well as positive attributes, if the technique is to be applied in a meaningful way to various types of research problems. The FSH tissue receptor assay described in this report has been extensively validated in terms of a generally accepted specific in vivo bioassay for FSH [2], and has been shown to be a useful tool in a variety of structure-function studies on FSH [4, 23].

II. PREPARATION OF THE RADIOLIGAND

A. Chloramine-T Iodination Procedure

Human FSH is utilized as the radioligand in our tissue receptor assay. It cannot be assumed that FSH from other species will behave in a manner similar to human FSH when iodinated by the chloramine-T procedure described. Highly purified hFSH suitable for iodination can be obtained upon application to the Hormone Distribution Officer, NIAMDD, NIH, Bethesda, Maryland. The method of preparation of hFSH made available through this program has been described in detail elsewhere [8, 9].

1. Solutions Required for Iodination

Chloramine-T solution: 1 mg chloramine-T in 1.5 ml of 0.05 M phosphate buffer, pH 7.5. Sodium metabisulfite: 3.25 mg in 1 ml of 0.05 M sodium phosphate buffer, pH 7.5. Note: Each of these solutions must be prepared fresh, immediately prior to use. Transfer solution: 100 mg of KI and 1 mg of bromphenol blue in 10 ml of 16% sucrose. Rinse solution: 100 mg KI and 1 mg of bromphenol blue in 10 ml of 8% sucrose. Note: The latter two solutions may be stored in the freezer and need not be prepared fresh for each iodination.

2. Procedure for Iodination

For the following iodination procedure, it is convenient to utilize disposable micropipets. The sizes employed are 5 μl, 10 μl, 20 μl, 50 μl, and 100 μl. Note that all pipets, columns and other equipment utilized in subsequent steps must be saved for calculation of percent iodination, described in Sec. II. A. 5. For convenience, we radioiodinate 50 μg of hFSH each time, but lesser amounts may be used, provided proportions of all reagents are scaled down accordingly.

The iodination reaction is performed in the cold (2-3°C). Fifty micrograms of hFSH are dissolved in 50 μl of glass-distilled water in a 6 × 50-mm disposable culture tube. To this is added 20 μl of 0.5 M sodium phosphate buffer, pH 7.5. The 0.5 M phosphate buffer is needed to bring the pH of the mixture to the range required for iodination to take place.

Approximately 1 mCi of $Na^{125}I$ (IMS-300; Amersham-Searle) is added to the hormone for iodination. The $Na^{125}I$ is usually received from the manufacturer at a concentration of 10 mCi in about 15 μl. This amount is too small for accurate pipetting, so that it is diluted to ~50 μl with 1×10^{-4} M NaOH to give a final solution with a pH of ~10.0. Five microliters, containing 1 mCi, is then added to the hFSH solution.

The iodination reaction is initiated by addition of 10 μl of chloramine-T solution, and allowed to continue for 30 sec with gentle shaking. The reaction is then quenched by addition of 50 μl of sodium metabisulfite solution. If the iodination reaction is allowed to proceed for longer than 30 sec, there will be a greater incorporation of label into hFSH, but with a concomitant loss of biologic activity. One hundred microliters of transfer solution are added to the reaction tube, mixed thoroughly, and the reaction tube contents transferred to a column of Sephadex G-100 or Bio-Gel P-60. The reaction tube is then washed with 70 μl of rinse solution, and the contents again transferred to the column for separation of the bound from free iodine, as described in Sec. II. A. 3.

3. Separation of Bound from Free Iodine.

Sephadex G-100 or Bio-Gel P-60 are usually utilized for this separation. The Bio-Gel P-60, a polyacrylamide gel, can be purchased prepacked in disposable 0.7 × 20-cm columns (Bio-Rad, Richmond, Calif.). It is not difficult, however, to prepare one's own columns for this purpose. Glass tubing of ~1.2 × 25-cm dimension is a convenient size. Draw a one-eighth in. tip on one end of the column and attach rubber tubing with a clamp. Insert a small piece of glass wool into the column for gel support. The column is then packed by pouring, using a slurry of Sephadex G-100 or Bio-Gel P-60 pre-equilibrated with 0.05 M phosphate buffer, pH 7.5. One milliliter of a 2% solution of egg albumin is passed through the column to saturate nonspecific binding sites capable of adsorbing the iodinated hFSH.

When using columns of the dimensions specified, it is convenient to collect about 20 individual 1 ml fractions in 12 × 75-mm disposable tubes. Prior to collection of the column fraction, each tube is marked at the 2-ml level, and 1 ml of 0.05 M phosphate buffer, pH 7.5, containing 0.1% crystallized egg albumin, is added to each tube. After the contents of the initial reaction tubes (including transfer and rinse solutions, total about 305 μl), have been placed on the column, the column is developed with phosphate buffer until an appropriate number of tubes have been collected; that is, until peaks of radioactivity corresponding to bound and free ^{125}I have been passed through the column, usually about 20 tubes. Note that the original reaction tube for iodination, as well as the gel column, must be saved for counting. In the case of the gel column, usually ~10 ml Bio-Gel P-60 or 12 to 15 ml Sephadex G-100 in total volume, the exact total gel volume is measured using a graduated cylinder, and then a 1-ml aliquot is pipetted into a culture tube for counting.

4. Counting of Radioactivity

Tubes are counted in an appropriate spectrometer. In addition to the tubes collected from the column, it is necessary to count the pipets used for transfer of the $Na^{125}I$, the pipets used for chloramine-T, sodium metabisulfite, transfer and the rinse solutions, the tube in which the iodination was performed, and an aliquot of the gel used for separation of $[^{125}I]$hFSH from free iodine (Table 1).

5. Calculation of Specific Radioactivity

The results of the gel filtration of the reaction mixture of iodinated hormone and free iodine are visualized by plotting cpm/tube vs tube number. Two clearly separate peaks should be observed. When using Bio-Gel P-60 (0.7 $\times$ 20 cm), the labeled hormone emerges about tube 3, and the free iodine about tube 9. When Sephadex G-100 (1.2 $\times$ 25 cm) is used, as illustrated in Table 1, the bound hormone will emerge about tube 7 or 8, and the free hormone about tube 14 to 15. It is assumed that the total counts, which includes the radioactivity contained in all 20 tubes, as well as in the pipets, gels, etc., represents the counts present in the 1 mCi (or whatever amount is used) of ^{125}I initially added to the FSH. The percent iodination is calculated by dividing the total number of counts in the $[^{125}I]$hFSH (Table 1-D) by the total counts in 1 mCi of added $^{125}I \times 100$. The specific activity of the iodinated hormone is calculated by multiplying 1,000 μCi (the amount of ^{125}I added initially) by the percent iodination, and then divided by the μg of hFSH used for the iodination. For use in the tissue receptor assay, we routinely select only those tubes comprising the leading edge of the $[^{125}I]$hFSH peak. This is to further minimize any likelihood of contamination of the labeled hormone with free ^{125}I. To determine the μg of protein per ml of column effluent, simply divide the μg of $[^{125}I]$hFSH in the pooled tubes (Table 1-H) by its volume (Table 1-I).

The iodinated hormone is appropriately diluted with 0.01 M phosphate buffer, pH 7.5, containing 0.1 M sucrose, 5 mM $MgCl_2$, and 0.1% egg albumin (2X crystallized), to a concentration of 5 ng/50 μl, and stored frozen until needed, usually in aliquots of 5 ml, which is normally adequate for an experiment consisting of 100 assay tubes.

6. Stability of the Radioligand

Some properties of iodinated hFSH are described in Table 2. Under the conditions of the tissue receptor assay as described in Sec. IV, specific binding of $[^{125}I]$hFSH ($\sim$10 μCi/μg) will be approximately 80% of counts bound to the receptor, while nonspecific uptake will be about 20%. Specific uptake will be considered later, but is defined as the uptake of radioligand which can be prevented by preincubation of the receptor with large amounts of unlabeled FSH. Nonspecific uptake is defined as that uptake of label which cannot be prevented by preincubation of the receptor with large amounts of unlabeled FSH. Note that, due to the inherent imprecision of whole animal bioassays, it is extremely difficult to quantitate loss of FSH biologic activity due to the iodination procedure when this is less than ~50% of the initial activity present. Because of this, specific uptake is generally considered to reflect retention of biologic activity by the iodinated hormone.

TABLE 1. Results of Iodination and Calculation of Specific Activity

Sephadex G-100 Eluates (2 ml/tube)	
Tube No.	cpm/tube[a-d]
1	-
2	-
3	-
4	-
5	75,000
6	78,000
7	1,383,000
8	1,524,000
9	820,000
10	602,000
11	311,000
12	599,000
13	945,000
14	1,474,000
15	1,302,000
16	475,000
17	51,000
18	31,000
19	17,000
20	14,000

TABLE 1 (continued)

	Procedure	
A.	Mass of protein iodinated	50 μg
B.	μ Ci of $Na^{125}I$ added	1,000
C.	Total counts [Tubes 1 through 20, (a)-(d)]	10,742,000
D.	Total counts in 50 μg of hormone (A). Includes counts in tubes 5 through 11 plus (b), (c), and (d)[e]	5,230,000
E.	Percent iodination	D/C × 100 = 48.69%
F.	Specific activity	$\frac{(B)(D/C)}{(A)}$ = 9.74 μCi/μg
G.	Counts in tubes 5 through 8[f]	3,060,000
H.	μg of $[^{125}I]$hFSH in pooled tubes 5 through 8	G/D × A = 29.25 μg
I.	Volume tubes 5 through 8	2 ml × 4 = 8 ml
J.	μg $[^{125}I]$hFSH/ml	H/I = 3.65 μg/ml
K.	Dilute the solutions to 5 ng/50 μl and store in appropriate aliquots	

[a] ^{125}I pipet = 604,000 cpm.

[b] Other pipets = 165,000 cpm.

[c] Reaction vessel = 184,000 cpm.

[d] 1-ml Sephadex aliquot × 14 = 88,000 cpm.

[e] Calculations based on the method of Greenwood et al. [26] which assumes that the counts present in (a), (b), and (c) are due to nonspecific adsorption of $[^{125}I]$hFSH. The counts in (D), therefore, are considered to represent those present in 50 μg of protein, the mass of hormone iodinated as given in (A).

[f] Tubes 5 through 8 represent leading edge of elution profile of $[^{125}I]$hFSH.

TABLE 2. Conditions for Iodination of hFSH

Description		Comment
μmol of hFSH	1.5×10^{-3}	Mol wt assumed to be 33,000
Molar ratio (hFSH:chloramine-T)	1:15	-
Temperature	2–4°	-
Reaction time	30 sec	Before quenching with sodium metabisulfite
mCi $Na^{125}I$ added	1	-
Specific activity of labeled hFSH (μCi/μg)	9.5 ± 2	-
Percent specific binding	80	$(Bt-Bn)/Bt \times 100$
Percent nonspecific binding	20	100-specific binding
Percent biological activity retained after iodination	100	As determined by bioassay [22]
Percent of added [^{125}I]hFSH bound to receptor under conditions of the assay	3–6	-

When stored in the frozen state, there is a gradual increase in nonspecific binding of [^{125}I]hFSH from 20% to ~30% over a period of one month.

[^{125}I]hFSH which exhibits nonspecific binding in excess of 30% should be used with some caution, although such higher levels of nonspecific binding do not necessarily preclude use of the radioligand, assuming proper controls are run and adequate statistics are applied to the resulting data (Sec. II. A. 5).

7. Comments on Measurement of Specific Radioactivity

It should be noted that the specific activity determined earlier (Sec. II. A. 5) is a reasonable approximation, but is not quantitatively accurate in an absolute sense. This is because of the various assumptions inherent in the calculations. For use as the radioligand in tissue receptor assays, this estimate of specific radioactivity represents an acceptable degree of accuracy. For other applications, however, a more precise estimate of specific activity and mass of labeled hormone may be necessary. Finally, at this low level of iodination (~10 μCi/μg), not every molecule of hormone can be assumed to be labeled. This may limit use of such lightly iodinated hormones in applications related to binding kinetics or Scatchard analysis.

B. Lactoperoxidase Iodination Procedure

The chloramine-T iodination procedure described in Sec. II. A has been successfully applied to iodination of highly purified rat FSH, prepared as described elsewhere [10], in a series of preliminary experiments. However, we do not yet have sufficient experience with rat FSH to allow unequivocal recommendation of the chloramine-T approach as the preferred method for iodination of that or other nonhuman species of FSH. Good success, however, has been had in iodination of rat FSH through a variation of the lactoperoxidase method of Miyachi et al. [11, 13]. In this method, 5 to 10 μg of previously lyophilized hormone was dissolved in 30 μl of 0.1 M sodium phosphate buffer (pH 6.9) in a 1-ml reaction vial capped with a serum stopper. The contents of the vial were adjusted to 50 μl with 0.1 M sodium phosphate buffer (pH 6.9) after adding 0.5 mCi of [^{125}I]sodium iodide (carrier free) and 5 μl of lactoperoxidase (160 ng/μl; 412/280 nm ratio >0.6). Four-microliter aliquots of hydrogen peroxide (30 ng/μl) were added three times at 5-min intervals and the contents of the vial were mixed continuously at room temperature.

[^{125}I]rFSH was separated from free ^{125}I by transferring the contents of the reaction vial to a Whatman CF-11 column, followed by elution with bovine serum albumin in phosphate buffered saline [12]. It was found useful to repurify the rat FSH after iodination by electrophoresis on polyacrylamide gel, essentially as described elsewhere [24]. The iodinated hormone is eluted from the appropriate gel segments and then diluted as required for tissue receptor assay. The specific activity of the [^{125}I]-rFSH, usually about 3 μCi/μg, can be determined by use of the tissue receptor assay with [^{131}I]rFSH as the radioligand [13]. This is a somewhat more precise approach than that described in Table 1, and may be more desirable for applications other than FSH quantitation by TRA.

III. PREPARATION OF RAT TESTES TUBULE HOMOGENATE

We use mature rats (250-300 g) of the Sprague-Dawley strain. The rats are sacrificed by dry-ice asphyxiation immediately prior to the assay. Testes are taken and weighed to the nearest milligram. The tunica albuginea is removed and the remaining tissue chilled and homogenized in the cold (3-5°C) using a hand homogenizer with loose-fitting pestle. The ratio of buffer per gram of tissue, wet weight, is 2 ml/g. The buffer utilized in preparation of the testes receptor is 0.01 M phosphate, pH 7.5, containing 5 mM $MgCl_2$, 0.1 M sucrose, and 0.1% egg albumin (2X crystallized). Homogenization is carried out with 15 to 20 strokes of the pestle, and the resulting homogenate is filtered through a single layer of fine-mesh cheesecloth. The extent of homogenization must be determined by experience; but it should be noted that if this is overly severe, excessive fragmentation of the tubule tissue will occur, so that they will not be retained by the cheesecloth. The tubules retained by the cheesecloth are washed with 1 vol of buffer (2 ml/g of starting tissue) and then transferred to a motor driven, teflon-pestle tissue grinder. To this is added buffer (1 ml/g of starting tissue) and the tubules are gently homogenized by a single upward and downward stroke of the pestle. For this purpose a teflon pestle with radial serrations at the tip has been found most suitable. The homogenate is then centrifuged at 1,500 g for 10 min (4°C) in a tared centrifuge tube, and the supernatant discarded. The resulting pellet is recentrifuged at 12,000 g for 10 min (4°C) to remove additional buffer, and therby allow accurate weighing of the receptor pellet. The weighed pellet is dispersed by vortexing in the sample buffer to make a final concentration of 1 g of pellet per 10 ml, and this is refiltered through a single layer of fine-mesh cheesecloth to remove any large-sized fragments that might remain. This precaution is necessary in order to avoid difficulties in accurate aliquoting of the receptor preparation for the tissue receptor assay. The wet weight of these fragments was found to be negligible in comparison to the wet weight of resuspended tissue ($< 0.1\%$). This final filtrate, designated RTTH, is then used in performance of the tissue receptor assay (TRA). Note that, after the initial homogenization, the tubule receptors are retained by the cheesecloth. However, after the second homogenization using the teflon pestle, the tubule receptors, now more finally dispersed, will pass through the cheesecloth.

IV. PROCEDURE FOR THE TISSUE RECEPTOR ASSAY (TRA)

The conditions recommended for use in the TRA were selected on the basis of extensive preliminary studies of the [^{125}I]hFSH-RTTH interaction [14]. The following additions are carried out in the cold (2°C) using 12 × 75-mm Kimble-type, flint glass, disposable centrifugation tubes.

To each tube is added an aliquot of 0.01 M phosphate buffer, pH 7.5, containing 5 mM $MgCl_2$, 0.1 M sucrose, and 0.1% egg albumin (2X crystallized), followed by aliquots of the assay standard and unknown, to make a final volume in each tube of 400 μl. To this is added chilled RTTH (450 μl), and then 5 ng of [^{125}I]hFSH in 50 μl (usually ~70,000 cpm) to make a final incubation mixture of 0.9 ml.

An alternate procedure is to add the [^{125}I]hFSH to the RTTH, and then add the radioligand-receptor mixture to the standards and unknowns. In this modification, sufficient [^{125}I]hFSH is added to the homogenate preparation to give a final concentration of 5 ng of labeled hormone in 500 μl of RTTH. This admixing is done immediately prior to performance of the bioassay. It has the advantage of eliminating one pipetting step and, in our hands, tends to increase the precision of the assay, although at a slight loss in sensitivity. It is absolutely essential in this application that all reagents and containers be kept cold, especially when mixing the labeled hormone and the receptor. If this is not done, nonspecific uptake will appear markedly elevated, since considerable specific uptake of iodinated hFSH may occur prior to addition of excess cold hormone to the nonspecific binding tubes (Table 3).

The tubes are incubated in a reciprocal water-bath shaker, 100 oscillations/min, at 37°C for 3 hr. After incubation, the tubes are centrifuged at 1,500 g for 10 min, and the supernatant discarded. The tissue pellet is resuspended in the cold in 1 ml of chilled phosphate buffer (vide supra) using a vortex mixer, and then centrifuged for an additional 10 min. One such wash is sufficient to remove interfering amounts of unbound radioactivity. The supernatant from the second centrifugation is decanted, the tips of the tubes wiped off with Kimwipes (mundane but essential), and the tubes then counted in an autogamma spectrometer.

V. CALCULATION OF ASSAY RESULTS

A detailed consideration of procedures for statistical analysis of TRA data is beyond the scope of this report. Approximate potency estimates may be

TABLE 3. Tube Arrangements and Contents for FSH-TRA[a]

		Total counts	Control	Nonspecific binding	Std. LER-1575C (150 ng/ml)					Unknown No. (2,500 ng/ml)				
Tube number		1	2	3	4	5	6	7	8	9	10	11	12	13
Buffer	(μl)	400	400	200	375	350	300	200	-	375	350	300	200	-
Cold hormone	(ng)	-	-	10,000	3.75	7.5	15	30	60	62.5	125	250	500	1000
	(μl)	-	-	200	25	50	100	200	400	25	50	100	200	400

[a]0.5 ml-RTH homogenate containing 5 ng LER-1575C ^{125}I in all tubes (see text Sec. IV).

obtained by simple graphical analysis of the results, but this gives no indication of the precision of the data, can be seriously misleading, and is not recommended as the basis for decision making. Computer programs for analysis of data from competitive protein-binding assays have become generally available and can be found elsewhere [15, 25]. In addition, most commercially available, programmable desk-top electronic calculator systems offer ready-made programs for quite sophisticated analysis of data, using accepted and reliable statistical procedures [16-18]. The following sections are for scientists with limited experience in this area and should allow for handling of data from the FSH-TRA while greater expertise in data processing and analysis is being pursued.

A. Definition of Specific and Nonspecific Binding

The percent specific binding is defined as $(Bt-Bn)/Bt \times 100$, where Bt is total counts bound to the receptor and Bn is total counts bound to the receptor in tubes designated for estimation of nonspecific binding. Non-Specific binding is determined by addition to the receptor preparation of a 1,000-fold excess of unlabeled hormone prior to addition of the radioligand. As indicated in Table 2, for hFSH having a specific activity of about 10 μCi/μg, approximately 3 to 6% of the total counts added to the system will be taken up by the receptor, and of this, about 80% will be specific uptake.

B. Design of the Assay and Processing of Data

Each category of tubes is run at least in duplicate with the average count used in the calculations listed below. A typical arrangement of tubes for an FSH-TRA is given in Table 3 with the total volume in each tube equal to 900 μl.

1. Total Count Tubes

Each tube contains 500 μl of RTTH-[^{125}I]hFSH (see Sec. IV) and 400 μl of 0.01 M phosphate buffer, pH 7.5. Their purpose is to allow an estimate of the total amount of radioactivity added to each reaction tube. These tubes are not centrifuged, but counted directly.

2. Control Tubes

These tubes contain 500 μl of RTTH-[^{125}I]hFSH plus 400 μl of buffer. The tubes are processed as described in Sec. IV. Their purpose is to allow measurement of the total amount of radioactivity bound to the receptor pellet in the absence of unlabeled hormone.

3. Nonspecific Binding Tubes

Each contains 500 μl of RTTH-[^{125}I]hFSH plus a 1,000-fold excess of cold hormone in 200 μl of buffer, plus buffer (200 μl) to make a volume of 900 μl. The value for nonspecific binding is subtracted from all reaction tubes.

4. Reaction Tubes

These are the tubes in which the competitive hormone binding occurs. Each reaction tube will contain 500 μl of RTTH-[^{125}I]hFSH mixture plus assay standards or unknowns in a volume not exceeding 400 μl. A typical assay design is given in Table 3.

5. Total Percent of Counts Bound

This value is calculated as control tube/total count tube × 100. This will usually range from 3 to 6% of the total counts added to the system.

6. Nonspecific Binding

The uptake of approximately 20% of the counts bound will not be prevented by the 1,000-fold excess of unlabeled hormone and will represent nonspecific binding. This figure will vary somewhat among assays since it is affected by the procedure utilized to prepare the rat testis tubule receptor preparation. An increase in nonspecific binding is a reflection of hormone damage, receptor damage, or both.

7. Calculation of Results

As indicated earlier, a simple, graphical calculation will allow a rapid assessment of assay results, although this should be followed by a more sophisticated statistical analysis of the data. If the correct dosage of standards and unknowns were chosen, the radioligand uptake-inhibition curve should be linear between 80% counts bound (20% uptake inhibition) and 20% counts bound (80% uptake inhibition). The range of linearity may be extended somewhat by converting the percent bound values to their corresponding logit value by the equation

$$\text{logit \% bound} = \log_e (\text{\% bound}/100\text{-\% bound}).$$

Figure 1 gives the results of a typical FSH-TRA plotted in this manner. The data may then be processed by a variety of standard statistical procedures as already described.

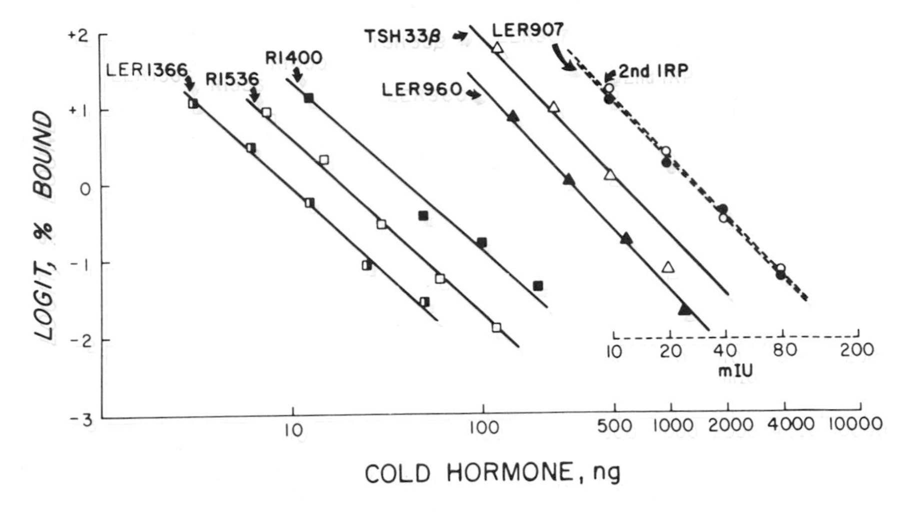

FIG. 1. Results of a tissue receptor assay for FSH. Each point is the mean of duplicate determinations. LER-960: highly purified hLH; TSH-33B: highly purified hTSH; LER-1366: highly purified hFSH. 2nd IRP: 2nd International Reference Preparation for Gonadotropins. Taken from [2].

VI. CHARACTERISTICS OF THE ASSAY: CHOICE OF A REFERENCE PREPARATION

The result of a TRA assay of a series of human pituitary hormone preparations is given in Figure 1. We have previously demonstrated that FSH from ovine, bovine, porcine, rabbit, rat, and equine pituitary glands will compete with [^{125}I]hFSH for receptors in rat testes [6]. Apparently FSH from all the above sources have structural features which can be recognized by rat testes in a manner similar to that seen with rat ovaries [19, 20]. Because of this, pituitary FSH activity from each of these sources can be quantitated using the heterologous hFSH-rat testes system, assuming a proper reference preparation is chosen. A wide range of such reference materials is available upon request from the Hormone Distribution Officer, NIAMDD. A list of some of these, together with the dose ranges found suitable in the FSH-TRA, is given in Table 4. Characteristics of the [^{125}I]hFSH uptake-inhibition curves obtained when heterologous unlabeled FSHs are used are summarized in Table 5.

TABLE 4. Useful Ranges of Various FSH Reference Preparations in the Receptor Assay

Preparation	Species	Comments	Dose for [^{125}I]hFSH uptake inhibition (ng)	
			20%	80%
LER-907	Human	Crude pituitary reference preparation, 20 IU/mg	500 (10 mIU)	8,000 (160 mIU)
LER-1575C	Human	Typical preparation of pure hFSH, 3,608 IU/mg[a]	3.75	60
NIH-FSH-S10	Ovine	Equivalent in potency to NIH-FSH-S1 and other members of this series[a]	30	480
NIH-FSH-B1	Bovine	Equivalent to 0.49 NIH-FSH-S1 μ/mg by whole animal bioassay	100	1,600
NIH-FSH-P1	Porcine	Equivalent to 0.76 NIH-FSH-S1 units/mg by whole animal bioassay	50	800
NIAMD-RP-1	Rat	Crude rat pituitary reference preparation	15	240

[a] 1 mg NIH-FSH-S1 = 25 IU human pituitary FSH [21].

TABLE 5. Characteristics of the [^{125}I]hFSH Uptake-Inhibition Curve for FSH from Several Sources

Species	Slope[a]	Standard deviation	95% confidence limit		Lambda[b]
			Lower limit	Upper limit	
Ovine	-2.182	0.148	-2.533	-1.831	0.086
Bovine	-2.121	0.166	-2.542	-1.700	0.078
Porcine	-1.985	0.090	-2.193	-1.777	0.061
PMSG	-2.260	0.174	-2.919	-1.601	0.037
Rabbit	-2.095	0.132	-2.411	-1.779	0.074
Rat	-2.277	0.113	-2.548	-2.006	0.065
Horse	-4.106	0.236	-4.763	-3.450	0.035
Human	-2.062	0.150	-2.449	-1.674	0.072

[a] Slopes are from plots of Logit percent bound vs log of hormone concentration.

[b] Lambda = mean index of precision (taken from Ref. 6).

VII. APPLICATIONS OF THE TISSUE RECEPTOR ASSAY

We have previously shown that FSH from numerous sources will compete with iodinated hFSH for receptors in homogenates of rat tubules or whole testes [6]. Therefore, the TRA has an obvious use as a rapid method for assessing the effectiveness of procedures designed to purify FSH from pituitary glands of different species. Usually, the great dilutions required for preparation of samples of TRA eliminate the possible interference of buffer or elution salts on the assay, although the likelihood of this occurring should not be overlooked. In addition, the cautions mentioned in Sec. I should always be kept in mind. It is at least possible for a fraction having a high in vitro activity to be purified, but found subsequently to have low in vivo activity when tested in accepted and specific whole animal bioassays. For this reason, final testing in such an in vivo system [20] is recommended if at all possible. The greater sensitivity of the TRA compared to whole animal bioassays should make it particularly suitable in studies on purification of FSH from scarce species of pituitary glands.

An important potential use of the FSH-TRA is in measurement of plasma levels of the hormone. This is a feasible application, although as this is written, problems related to interference by presumably nonhormonal factors in plasma have only been potentially solved [27, 28].

The TRA is well-suited to studies on problems related to structure-function properties of FSH. We have utilized this approach to study the kinetics of combination of the beta-subunit of hFSH with the slpha-subunit of a number of other glycoprotein hormones [14]. The subunits themselves have no receptor-binding activity in this system, so that appearance of $[^{125}I]$hFSH uptake inhibition is considered a reflection of subunit annealing and dimer formation. Conversely, the technique may be used to study dissociation of hFSH [23]. A decrease in radioligand-uptake inhibition upon incubation with urea, for example, can be considered a reflection of hFSH dissociation into its constituent alpha- and beta-subunits. The application of TRA in determining the consequences of group-specific modification on association of FSH subunits, or FSH-receptor binding activity are obvious.

ACKNOWLEDGMENTS

This study was supported by grants AM-03598 (NIAMDD) and HD-08228 (NICHD). This is Publication 1230 from the Division of Basic Health Sciences, Emory University. It is a pleasure to acknowledge the able technical assistance of Mrs. Rosemary Ramsey, Ms. Eloise Carter, and Mrs. Nancy Shih in various phases of this work.

REFERENCES

1. L. E. Reichert, Jr., F. Leidenberger, and C. G. Trowbridge, Recent Progr. Hormone Res., 29:497 (1973).

2. L. E. Reichert, Jr. and V. K. Bhalla, Endocrinology, 94:483 (1974).

3. L. E. Reichert, Jr., G. F. Lawson, F. L. Leidenberger, and C. G. Trowbridge, Endocrinology, 93:938 (1973).

4. L. E. Reichert, Jr., C. G. Trowbridge, V. K. Bhalla, and G. F. Lawson, J. Biol. Chem., 249:6472 (1974).

5. F. Leidenberger and L. E. Reichert, Jr., Endocrinology, 92:646 (1973).

6. L. E. Reichert, Jr. and V. K. Bhalla, Gen. Comp. Endocrinol., 23:111 (1974).

7. F. Leidenberger and L. E. Reichert, Jr., Endocrinology, 91:901 (1972).

8. L. E. Reichert, Jr., in Gonadotropins (B. B. Saxena, C. G. Beling, and H. M. Gandy, eds.), Wiley, New York, 1972, p. 107.

9. L. E. Reichert, Jr., in Methods in Investigative and Diagnostic Endocrinology, Vol. 2A: Peptide Hormones (S. A. Berson and R. S. Yalow, eds.), American Elsevier, New York, 1970, p. 504.

10. L. E. Reichert, Jr., in Methods in Enzymology: Hormones and Cyclic Nucleotides (J. C. Hardman and B. W. O'Malley, eds.), Academic Press, New York, 1975, Vol. 37, Part B, p. 360.

11. Y. Miyachi, J. L. Vaitukaitis, E. Nieschlag, and M. B. Lipsett, J. Clin. Endocrinol. Metabol., 34:23 (1972).

12. A. R. Midgley, A. J. Zeleznik, H. J. Rajaniemi, J. S. Richards, and L. E. Reichert, Jr., in Gonadotropins and Gonadal Function (N. R. Moudgal, ed.), Academic Press, New York, 1974, p. 416.

13. C. Desjardin, A. J. Zeleznik, A. Rees Midgley, Jr., and L. E. Reichert, Jr., Endocrinology, in press, 1976.

14. V. K. Bhalla and L. E. Reichert, Jr., J. Biol. Chem., 249:43 (1974).

15. W. D. Odell and W. H. Daughaday (eds.), Principles of Competitive Protein Binding Assays, Lippincott, Philadelphia, 1971.

16. J. W. McArthur and T. Colton (eds.), Statistics in Endocrinology, MIT Press, Cambridge, Mass., 1970.

17. D. J. Finney (ed.), Statistical Methods in Bioassay, Charles Griffin, London, 1964.

18. C. I. Bliss (ed.), Statistics of Bioassay, Academic Press, New York, 1952.

19. A. F. Parlow and L. E. Reichert, Jr., Endocrinology, 73:740 (1963).

20. L. E. Reichert, Jr., Endocrinology, 80:1180 (1967).

21. A. Albert, E. Rosemberg, G. T. Ross, C. A. Paulsen, and R. J. Ryan, J. Clin. Endocrinol. Metabol., 28:1214 (1968).

22. S. Steelman and F. M. Pohley, Endocrinology, 53:604 (1953).

23. L. E. Reichert, Jr. and R. B. Ramsey, J. Biol. Chem., 250:3034 (1975).

24. A. R. Midgley, Jr., G. D. Niswender, V. L. Gay, and L. E. Reichert, Jr., Recent Progr. Hormone Res., 27:235 (1971).

25. D. Rodbard, Clin. Chem., 20:1255 (1974).

26. F. C. Greenwood, W. M. Hunter, and J. S. Glover, Biochem. J., 89:114 (1963).

27. L. E. Reichert, Jr., R. B. Ramsey and E. B. Carter, Endocrinology, 41:634 (1975).

28. K. W. Cheng, Biochem. J., 149:123 (1975).

Chapter 6

ISOLATION OF GLUCAGON RECEPTOR PROTEINS FROM RAT LIVER PLASMA MEMBRANES

Steven Goldstein and Melvin Blecher

Department of Biochemistry
Georgetown University Medical Center
Washington, D. C.

I. INTRODUCTION

At the time this chapter is being written, it would appear to be unnecessary to discuss in detail the development of the concept that interaction of hormones with specific receptors represents the primary step in their mechanism of action. Equally well-established, and well-reviewed [1-3], is the concept that many polypeptide hormones, such as glucagon, interact with specific cell-surface (plasma membrane) receptors; and that a result of this interaction is the activation of adenylate cyclase, also located in the plasma membrane, but situated, perhaps, on the inner aspect of this membrane.

A corollary concept is that glucagon-plasma membrane-receptor interactions should be amenable to assessment either by direct binding assays or, indirectly, by quantifying the degree of activation of adenylate cyclase. The direct binding assay can be used not only to assess the quality of the labeled ligand (e.g., [^{125}I]glucagon) and the integrity of the receptor, but also is amenable to being made the basis of a quantitative radioreceptor assay for glucagon in biological fluids and elsewhere, as has been accomplished for ACTH [4], hGH [5], hFSH [6], HCG [7], and prolactin [8]. The indirect method can serve also as a bioassay for the amount and quality of the glucagon (radiolabeled or native), and is useful for establishing the integrity of the entire receptor-adenylate cyclase complex.

A long series of elegant, direct and indirect studies on the interaction of glucagon with isolated liver plasma membranes has been carried out over the past three years by Rodbell's laboratory [9]; the Marinetti and Barnebei groups have also contributed significant studies using a similar system [10]. Such studies have produced valuable information on the properties and characteristics of this interaction: kinetics; specificity; saturability; reversibility; binding sites and binding constants; nucleotide effects; and the relationships between binding, activation of adenylate cyclase, and the glucagonase activity. The preparation of [^{125}I]glucagon and of Neville rat liver plasma membranes, and methods for the direct and indirect estimation of glucagon-particulate receptor interactions, are described in this chapter, as well as in Dr. Pohl's contribution, Chapter 8.

Although invaluable, the data obtained with particulate membranes is essentially descriptive in nature, and reveals relatively little about the actual molecular mechanisms of glucagon interaction with its receptor and activation of the cyclase. A direct approach to this problem is to solubilize and purify the membrane system, and then to examine it by conventional means of studying ligand-protein interactions.

In order to undertake solubilization experiments, one must stipulate what is to be solubilized and what is to be assayed. Ideally, one would like to solubilize the entire complex, and then to dissect out and isolate separately the components of this complex. This problem has been approached using a variety of ingenious techniques, but, unfortunately, with little success. Levey [11] has devised techniques to "solubilize" (the preparation used was a 12,000-g supernatant fluid derived from a Lubrol-tissue homogenate) adenylate cyclase(s) from cat myocardium, which are sensitive to glucagon and epinephrine following addition of specific phospholipids to the preparation; unfortunately, these reports have not yet been confirmed, nor have the methods (which are described by Levey, Fletcher, and Ramachandran in Chap. 7) been applied successfully to other tissues. More recently Ryan and Storm [12] described the preparation from rat liver plasma membranes of a soluble (100,000-g supernatant fluid derived from membranes extracted with Triton X-305) adenylate cyclase which is very sensitive to fluoride, glucagon, and epinephrine, without further treatment or additions; no attempt was made to separate and isolate binding and enzyme activities, nor has this work yet been confirmed.

An alternative to solubilizing the entire receptor-adenylate cyclase system from plasma membranes is separately to solubilize and isolate each macromolecule. This approach has the advantage, initially at least, in not having to be concerned with the integrity of both components during the isolation procedures for either one. This approach has met with more success. Using Lubrol-PX, Pohl et al. [13] achieved a 15% solubilization of adenylate cyclase from Neville rat liver plasma membranes; the enzyme did not respond to glucagon even when phospholipids were added to the extract. Qualitatively similar results were described by Swislocki et al. [14], who used 0.1 M Lubrol-PX to achieve a 50% solubilization (165,000-g supernatant fluid) of the enzyme; and by Blecher and Johnson [15], who found fluoride-sensitive, glucagon-insensitive adenylate cyclase in the 0.5% Lubrol-PX extracts (105,400-g supernatant fluid) of Neville rat liver plasma membranes from which they isolated the glucagon-receptor protein (vide infra). Tomasi et al. [10] reported preliminary experiments in which they detected macromolecules which bound glucagon and epinephrine in deoxycholate extracts of rat liver plasma membranes which were preincubated with the labeled hormone. There is, however, some doubt as to whether the binding proteins were soluble or particulate, since no attempt was made to clarify the extract; furthermore, this preliminary report has neither been confirmed by others nor extended by the authors beyond the original observation.

Among the methods to be described here are our procedures for the solubilization, purification, and assay of a glucagon-binding macromolecule which has many of the characteristics of the glucagon receptor present in the starting Neville rat liver plasma membranes. Since the nonionic detergent used (Lubrol-PX) has such profound effects on the glucagon receptor and on the chromatographic behavior of radiolabeled glucagon, methods are given for the determination of the critical micellar concentration of this detergent. Since studies on the specificity of both the particulate and soluble receptors require the examination of other polypeptide hormones, e.g., ACTH, insulin, secretin, and "gut glucagon" (vasoactive intestinal polypeptide), which may interact with the same binding sites, we present methods for their iodination.

II. ISOLATION OF RAT LIVER PLASMA MEMBRANES

Rat liver plasma membranes are isolated by the method of Neville [16] up to step 11 of the procedure. The entire isolation procedure of Neville is not followed since Pohl et al. [13] have determined that the response of adenylate cyclase to both glucagon and fluoride is identical in either these partially purified membranes or in those isolated after a second sucrose gradient step. Additionally, the yield of glucagon-sensitive adenylate cyclase activity is 14 times greater in the partially purified membranes.

Male Sprague-Dawley derived rats (160-200 g; Flow Laboratories, Dublin, Virginia) were equilibrated to our stock diet for at least 24 hr before being used. Typically, 38 animals are obtained and are used in two batches of 19 on successive days.

Following decapitation of the rats, livers are rapidly excised and stored in a chilled container until all are collected. Mincates of the liver are then prepared as follows. The pooled livers, carefully trimmed of fat, are placed on a prechilled glass plate and chopped finely with a stainless-steel blade (the type used in a Stadie-Riggs microtome); this type of preparation is more easily homogenized in large batches than is tissue minced with a scissors. All of the following manipulations are carried out at ice-bath temperatures. Ten-gram portions of the mincate, suspended in 35 ml of 1 mM $NaHCO_3$ (medium), are homogenized with eight excursions in a loose-fitting Dounce homogenizer (Type A, Kontes or Blaessig Glass Co.). At this point, an aliquot of the homogenate may be removed and saved for determination of marker enzyme activities (adenylate cyclase and 5'-nucleotidase) at a later time. Two homogenates are combined, brought to 500 ml with medium, stirred magnetically for 5 min, and finally filtered once through two layers of cheesecloth. The filtrates are distributed among 250-ml flat-bottom polycarbonate centrifuge bottles, and centrifuged at 1,500 g for 10 min (2,600 rpm in International PR-2 Centrifuge, #287 rotor). Supernatant fluids are aspirated to waste using a serum pipet attached to a water pump. In this way, virtually all of the supernatant fluid can be removed without disturbing the loosely pelleted material. Homogenizations and centrifugations are continued until all of the tissue has been processed. The pooled pellets (~70 ml), suspended in a Dounce homogenizer using three excursions of the pestle, are transferred to a 250-ml plastic beaker. The suspension is brought to 44.0 ± 0.1% (wt/wt) sucrose (Abbé Refractometer) by the addition of a 70% (wt/wt) sucrose solution. Generally, 80 ml of the 70% sucrose solution are added directly to the suspension, and the final adjustment of the sucrose concentration is made with water.

Sucrose gradients are prepared in the 1 × 3.5-in. tubes of a Beckman SW.27 rotor. These tubes have a 20% greater volume than those usable in a SW.25 rotor, and six gradients can be run at a time instead of three. Twenty-six milliliters of the tissue suspension are pipetted into each tube, followed by an overlay of 13 ml of 42.3 ± 0.1% (wt/wt) sucrose (Abbé Refractometer). Centrifugation is carried out in a Beckman L2-65B ultracentrifuge at 27,000 rpm (95,100 g_{av}) for 2 hr. The float containing the plasma membranes can be removed easily by first pinching the tube slightly below the float, then lifting off the float with a spoon-shaped teflon-coated spatula. These floats are transferred to a preweighed plastic centrifuge tube (50 ml), chilled medium is added to bring the volume to ~45 ml, and the well-mixed (plastic stirring rod) suspension centrifuged at 40,000 g for 30 min in a Sorval SS-34 rotor. Following aspiration of the supernatant fluid to waste, the tubes are reweighed in order to estimate the yield of plasma membranes. After the addition of an equal volume of medium, the pellet is aspirated repeatedly through a Luer-lok 20-gauge needle fitted to a plastic syringe. The plasma membrane suspension is distributed (0.2- or 1.0-ml aliquots) in Vanguard screwcap plastic vials for storage in liquid nitrogen. The plasma membranes are characterized by electron microscopy [17] and by marker enzymes: glucagon and fluoride-stimulated adenylate cyclase activities [18] and 5'-nucleotidase activity. Highly stimulable adenylate cyclase activity should be present if the membranes are to be useful for further studies. Typically, from a basal activity of about 0.4 nmol cAMP formed per mg protein per 10 min, glucagon (10^{-7} M) stimulated this activity 10-fold and 10 mM NaF, 5-fold. Although other liver membrane preparations [19] contain epinephrine-stimulable adenylate cyclase, concentrations of up to 55 μmol epinephrine fail to activate the present preparation. Plasma membrane 5'-nucleotidase activity was enriched 12-fold relative to the activity in the whole homogenate (717 nmol P_i produced per mg protein per 10 min.

III. PREPARATION AND CRITERIA OF PURITY OF RADIOACTIVE LIGANDS

A. Introduction

For most of our studies, the chloramine-T method of Hunter and Greenwood [20] has been used to iodinate polypeptide hormones. Glucagon, secretin, and vasoactive intestinal peptide (VIP) have been iodinated according to one modification of the method, while ACTH and insulin have been iodinated by a second modification. Additionally, we have recently iodinated secretin with a lactoperioxidase-hydrogen peroxide system in order to produce a labeled species with a specific activity higher than could be achieved with the chloramine-T method; this procedure is described in Sec. III. C. In all of the following procedures the iodination is carried out in a well-ventilated fume hood.

B. Chloramine-T Method for Iodination

1. Iodination of Glucagon, Secretin, and VIP

a. Reagents

0.6 M potassium phosphate, pH 7.0 (B-1 buffer)

Chloramine-T, 0.84 mg/ml in B-1 buffer

Sodium metabisulfite, 0.57 mg/ml in B-1 buffer

Hormone solution, approximately 1 μmol/ml in 1 mM Na_2CO_3

$Na^{125}I$, carrier free, high specific activity (>300 mCi/ml) (ICN Corp., Irvine, Calif.)

Potassium phosphate, pH 7.4, 0.1 M

Potassium phosphate, pH 7.4, 10 mM containing 1% BSA (B-2 buffer)

50% aqueous ethanol

b. Procedure

To decrease the amount of damage to the polypeptide chain(s) during, and subsequent to, iodination, the hormones are only lightly iodinated. For example, an average of about 0.6 atoms of ^{125}I are incorporated per mole of glucagon by using one mole each of glucagon and $Na^{125}I$, and 20 nmol of chloramine-T, under the iodination conditions described.

A 1-ml, polystyrene, concial centrifuge tube (Fisher Scientific Cat. No. 4-978-145) has been found to be a convenient vessel in which to carry out the iodination. The components of the assay can be pipetted onto different areas of the tube and easily combined by vortexing. B-1 buffer (10 μl) is placed at the bottom of the tube. The hormone to be iodinated (10 μl) is pipetted onto the wall of the tube near its bottom, as is the appropriate volume of $Na^{125}I$ necessary to deliver one natom of the isotope. This latter volume is calculated by first multiplying the specific activity of the stock $Na^{125}I$, in mCi/μl, by a factor of 0.472 natom/mCi, and then taking the inverse of the product. Chloramine-T solution (10 μl) is added last, also to the side of the tube. The reaction is begun by tapping the tube gently on a flat surface to mix the components. The tube is vortexed for 15 sec at room temperature, and the reaction terminated by the addition of 20 μl of sodium metabisulfite and 200 μl of ice-cold B-2 buffer. All of the following steps are done at 4°C. To a graduated 15-ml, conical, glass centrifuge tube is added 3 ml of B-2 buffer. The contents of the reaction tube are transferred to the glass tube with a Pasteur pipet, care being taken to express the mixture directly into the buffer. The reaction tube is rinsed three times with the same buffer, the rinses are added to the contents of the glass centrifuge tube, and buffer is added to bring the total volume to 8 ml. After mixing, an aliquot of the mixture is removed to determine the total number of ^{125}I cpm present.

Isolation of the iodinated hormone, free of damaged peptide and unincorporated ^{125}I, is achieved by a talc-adsorption method. A 50-mg pellet of talc (Gold Leaf Pharmacal Co., Englewood, N. J.) is crushed between two sheets of glassine paper, and the powder added to the glass tube. The contents are mixed by vortexing for 1 min, and the talc is pelleted by centrifugation at 2,000 rpm for 10 min in an International PR-2 centrifuge (#269 rotor). The supernatant fluid is carefully removed with a Pasteur pipet, and an aliquot is taken to determine the amount of unadsorbed radioactivity. The pellet is washed four times by suspending it each time in 6 ml of 0.1 M potassium phosphate, pH 7.4, vortexing the suspension for 30 sec, and centrifuging as before; the washings were sampled for determination of radioactivity in polypeptide fragments. The radioiodinated hormone is eluted from this talc by vigorous suspension of the pellet in 6 ml of 50% aqueous ethanol. Following removal of the solvent by evaporation under reduced pressure (Buchi-Rotovap) the iodohormone is extracted from the residue with 2 ml of B-2 buffer. The solution is clarified by filtering it through a Millipore AAWP membrane (which has been soaked with B-2 buffer), held in a plastic Swinney adapter attached to a 3-ml plastic syringe. Because of the high contamination of the syringe and Swinney adapter with radioactivity, no attempt is made to wash them; both are discarded after a single use. Iodinated hormones in

200-μl aliquot are stored at -20°C in plastic tubes, the same as those used for the iodination reaction. Since these hormones are subject to radiation damage even during short period of storage, no single preparation was used for more than one week. Although it is possible to salvage older preparations by reprocessing them through the talc procedure, we prefer to iodinate a new batch of glucagon when needed, the time required for the iodination reaction being so short.

2. Iodination of ACTH and Insulin

a. Reagents

Potassium phosphate, pH 7.4, 0.25 M (B-3 buffer)

Chloramine-T, 3.5 mg/ml in B-3 buffer

Sodium metabisulfite, 4.5 mg/ml in B-3 buffer

$Na^{125}I$, carrier free, > 300 mCi/ml

Potassium phosphate, pH 7.4, 0.1 M, 2% BSA (B-4 buffer)

HCl, 0.37 N containing 6% BSA

Hormones: (a) ACTH, 1 mg/ml in water; (b) insulin, 1 mg/ml in 0.01 N HCl

b. Procedure

ACTH and insulin are iodinated essentially as described in Sec. II. B. 1. Either 2.9 nmol of ACTH or 1.7 nmol of insulin (10 μl of each, 1 mg/ml of solution, respectively) are added to a plastic centrifuge tube which contains 100 μl B-3 buffer and 1 mCi of $Na^{125}I$. To start the reaction, 15 μl of chloramine-T solution (186 nmol) are pipetted onto the side of the reaction tube, and the mixture mixed for 15 sec at room temperature. To stop the reaction, 4.73 nmol sodium metabisulfite (20 μl of 4.5 mg/ml solution) are added, followed by 500 μl of B-4 buffer. The contents of the reaction tube are transferred to a 10 × 75-mm test tube which was previously siliconized with Dri-Film SC-87 (Pierce Chemical Co.). An additional 500-μl aliquot of B-4 buffer is used to rinse the reaction tube.

The purification of these radioiodinated hormones is by the talc procedure. A 50-mg talc tablet (crushed) is added to the tube, and thoroughly mixed with its contents; about 1 min of vortexing is sufficient to break down the large pieces of the talc and to adsorb completely all of the hormone. The iodo-hormone-talc complex is isolated by centrifugation in an International Model CL table-top centrifuge (# 809 rotor, rheostat position 7). After the supernatant is decanted, the talc pellet is washed twice with 2 ml of either water (for ACTH) or B-4 buffer (for insulin. [^{125}I]hormones are eluted from the talc with 2 ml of 0.37 N HCl containing 6% BSA. Storage of 200-μl aliquot of the iodinated material is at -20°C in 1-ml plastic centrifuge tubes.

C. Iodination of Secretin by the Lactoperoxidase Method

Because secretin could be iodinated in our laboratory to only a small extent by the chloramine-T method (typically, 22,000 cpm/pmol versus 450,000 cpm/pmol for glucagon), an alternative method of iodination was sought. Holohan et al. [21] have described a procedure for the enzymatic iodination of secretin, by the use of which we were able to increase the specific activity of the iodinated hormone at least threefold, compared with the chloramine-T method. [^{125}I]Secretin with a specific activity of 64,000 cpm/pmol has been prepared by the following method.

a. Reagents

0.2 M sodium phosphate, pH 6.0 (B-5 buffer)

Potassium phosphate, pH 7.4, 0.05 M containing 0.25% BSA (B-6 buffer)

Lactoperoxidase (Calbiochem, B Grade, 122 IU/mg), 50 μg/ml in B-5 buffer

Hydrogen peroxide, 0.86 mM

Sodium metabisulfite, 0.5 mg/ml in B-5 buffer

Potassium iodide, 31.1 μg/ml in B-5 buffer

$Na^{125}I$, carrier free

Secretin, 100 μg/ml in B-5 buffer

Ethanol, 0.7 N HCl (3:1 vol/vol)

b. Procedure

The iodination is carried out in a 1-ml, conical, plastic centrifuge tube (Fisher Scientific) at room temperature. To the tube ad 10 μl lactoperoxidase solution, 50 μl secretin solution, and $Na^{125}I$ (1 mCi). The reaction is initiated by the addition of 10 μl of 0.86 mM hydrogen peroxide, and vortexing for 30 sec. Following 15 min of incubation at room temperature, with occasional mixing, the reaction is ended by addition of 100 μl of sodium metabisulfite solution and 200 μl of potassium iodide solution. Transfer the contents of the reaction tube to a graduated, 15-ml, conical glass centrifuge tube. Rinse the reaction tube with a total of

3 ml B-6 buffer, adding the rinses to the contents of the glass tube; bring the total volume to 5 ml. Add 10 mg of microfine silica (Quso-32, Philadelphia Quartz Co., Chester, Pa.) and mix. Centrifuge for 10 min at 2,000 rpm in International PR-2 centrifuge (#269 rotor, 4°C). After decanting the supernatant, the silica pellet is washed three times by suspension in 2 ml of water, followed by centrifugation as before. The [^{125}I]secretin is eluted from the silica by 2 ml of ethanol: 0.7 N HCl (3:1, vol/vol). The iodinated secretin is obtained free of any silica particles by increasing the duration of the centrifugation to 15 min. The product is stored, as before, in 200-μl aliquots, in capped, 1-ml, plastic centrifuge tubes at -20°C.

D. Estimation of Recovery and Specific Radioactivity of ^{125}I-Labeled Hormones

Because the iodo-hormones must be stored in the presence of higher concentrations of BSA, it is obviously impossible to determine directly the mass of peptide recovered after iodination and purification. In order to estimate the mass recovered, bioassays must be performed where possible. In the case of radioiodinated glucagon, the estimation is based on the stimulation of adenylate cyclase activity in rat liver plasma membranes as prepared in Sec. II. The recovery of other hormones can be estimated in a similar manner if one has an enzyme preparation stimulable by that hormone.

An assay of adenylate cyclase activity is carried out in the presence of a range of concentrations of native glucagon. These concentrations are chosen to cover the entire dose-response curve for the hormone, from no stimulation of activity to maximal stimulation (5×10^{-11} g to 1×10^{-7} g). It is advisable to use at least two concentrations of the hormone within each order of magnitude of concentration. At lease three concentrations of the radioiodinated glucagon preparations are assayed; these should range from the undiluted hormone solution to approximately a 100-fold dilution of the preparation.

The results of the assay of glucagon standards are graphed on semi-log paper, with the concentrations of the unlabeled hormone plotted on the log axis and enzyme activity of the linear axis. The apparent concentrations of radioiodinated glucagon can be read directly from the graph and corrected for dilution. Not all of the enzyme activities observed in the presence of [^{125}I]glucagon will necessarily fall on the linear portion of the dose-response curve, because of either too much or too little dilution.

Such a method for the determination of recovery cannot detect and quantify large, biologically inactive, radioactive glucagon fragments which may follow [^{125}I]glucagon through the entire purification procedure, and which will contribute to the total radioactivity recovered. Until and unless it is rigorously determined by analytical means that the iodinated hormone is the only peptide species present, the values for the mass and specific radioactivity of [^{125}I]glucagon can be only a good approximation. In general, in freshly iodinated and purified preparations, the yield of biologically active [^{125}I]glucagon ranged between 37 and 71%, averaging 51%.

E. Assessment of Purity of Iodinated Hormones

Cellulose thin layer chromatography has been used to determine the purity of radioiodinated hormones.

1. Materials

Precoated, cellulose, thin layer plates (E. Merck, Darmstadt, Germany, obtained through Brinkmann Instruments, Inc., N.Y.): 20 × 20-cm glass support, 0.10-mm layer thickness

Desaga glass chromatography tank lined with Whatman No. 2 filter paper

Solvent system: n-butanol:pyridine:water:glacial acetic acid (15:10:12:13)

Hormone solutions: (a) native hormone, ~ 4 mg/ml; (b) iodinated hormone, prepared as described in Sec. III. A, B, or C

Na^{125}I, dilute carrier-free stock 1:1000 with water prior to use

2. Procedure

Prior to use of the chromatography tank, saturate the filter paper lining with the solvent system. Then add between 100 and 150 ml of the solvent system to the tank, cover tightly, and allow to stand while the plate is prepared. Score one thin layer plate 1 cm from a side; this marks the top of the plate. Along this line, divide the plate into ten 2-cm-wide lanes, and score lanes into the coating perpendicular to the first line. Make small marks next to the lines dividing the lanes 2 cm from the bottom of the plate; the samples will be spotted between these marks. Thus, the total distance each sample will travel during development of the chromatogram will be 17 cm. Typically, the following samples should be spotted in order to assess the purity of the iodinated hormone: (1) Na^{125}I, 400,000 cpm; (2) native hormone, 8 to 10 μg; and (3) iodinated hormone, ~200,000 cpm. After the samples are spotted and solvents allowed to evaporate under a stream of air, the plate is placed in the chromatography tank and the chromatogram developed. About 7 hr are required for the solvent front to reach the score mark across the top of the plate.

After the plate is removed from the tank, a stream of warm air (hair dryer) is used to help speed the evaporation of the solvents. Each lane is scanned in a thin layer radioactivity scanner (e.g., Brinkmann-Berthold Radioscanner), care being taken to note the origin and the solvent front on the recorder trace. Then the entire plate is sprayed with ninhydrin reagent and dried for 1 hr at 110°C to visualize peptides.

Such chromatography is carried out for all radioiodinated hormones: glucagon, VIP, secretin, ACTH, and insulin. In each case, the native hormone chromatographed as a single species in this solvent system, as demonstrated by the presence of a single ninhydrin positive spot. The R_f of each iodinated hormone was identical with that of the corresponding native hormones. In no case did we find more than one radioactive peak beyond the origin, although some radioactivity ($<$ 1-2%) did remain at the origin. As the samples age, a radioactive spot appears between the iodohormone and the solvent front. This material may represent iodide released during storage due to the damage that occurs even at -20°C. Such a peak also appears at the V_t when aged samples are chromatographed on small-pore Sephadex (G-10, G-15, etc.) columns.

IV. SOLUBILIZATION OF GLUCAGON RECEPTORS

A. Purification of [^{14}C] Lubrol-PX

The concentration of detergent in a solubilized preparation should be followed throughout the purification of the material. Artifacts can arise during the chromatography of proteins with detergent, or the detergent may inhibit hormone-binding assays if the detergent is present in appropriately high concentrations. The nonionic detergents may be removed from receptor solutions by dialysis or ultrafiltration, but reduction may not be complete due to a high affinity of the amphipathic detergents for certain membrane proteins. It is, therefore, possible that, as the solubilized proteins are concentrated by methods such as ultrafiltration, the detergent solution should be tagged with tracer amounts of radioactive detergents. Radioactive [^{14}C] Lubrol-PX generously supplied by Dr. Gerald Levey, University of Miami School of Medicine, was used for this purpose. More recently, New England Nuclear Corp. has prepared [^{3}H] Lubrol-PX for us by the Wilzbach technique [23].

Several solvents systems were tested for the purification of [^{14}C]-Lubrol on thin layer silica gel. Among these were: (1) $CHCl_3$-toluene-formic acid (80:17:3); (2) $CHCl_3$; (3) hexane:acetone (4:1); (4) ethyl ether: ethanol (99:1) saturated with water; (5) ethanol:water (4:1); and (6) n-butanol saturated with 1.5 N ammonium hydroxide. Of these, only the last solvent system was successful. The chromatogram developed with this system showed two peaks of radioactivity: One peak remained close to the origin while the other migrated with an R_f of 0.69, the same as exhibited by authentic Lubrol-PX (ICI America, Inc., Wilmington, Del.). In practive, the chromatography is carried out in a 2 × 13-cm strip of Bakerflex Silica Gel 1B (thin layer on a Mylar support). The strip is marked with a pencil 1 cm from each end; the total length of the developed chromatogram is, therefore, 11 cm. At the origin, spot 2 μl of [^{14}C]-Lubrol using a Drummond micropipet; rinse the pipet three times with 2 μl of water, and transfer the rinses to the [^{14}C] Lubrol spot. A hair dryer is necessary to speed evaporation of the water.

A 250-ml Erlenmeyer flask is used as the chromatograph tank. The shape of the vessel is important; in a rectangular tank (i.e., Desaga) or in a sandwich tank (Eastman), the authentic Lubrol [as visualized by spraying with methanol:H_2SO_4 (4:1) and heating at 120°C], as well as the [^{14}C] Lubrol, is smeared across the chromatogram; the reasons for this have not been explained. Twenty milliliters of the solvent system (n-butanol saturated with 1.5 N ammonium hydroxide) is poured into the flask; the thin layer strip is inserted, and the top covered with Parafilm. The chromatogram is developed until the solvent front reaches the upper limit line.

After the strip is dried, it is scanned with a thin layer radioactivity scanner. The [^{14}C] Lubrol zone is scraped free of the Mylar support and is vacuumed into a Pasteur pipet containing a glass wool plug in the constriction. Invert the pipet and elute [^{14}C] Lubrol with 0.4-ml aliquots of 0.1% Lubrol-PX in water. An aliquot of each fraction is counted in a liquid scintillation spectrometer to determine recovery in each fraction. Generally, 80% of the recovered detergent is present in the first three fractions. Pool the most active and store at -20°C until used.

B. Determination of Critical Micellar Concentration of Lubrol-PX

For the reasons noted at the beginning of Sec. IV. A., it is important to determine the critical micellar concentration (CMC) of detergent used to solubilize biological membrane. Benzonana [24] has described three techniques for the determination of the CMC of deoxycholate: spectrophotometry, interfacial tension, and light scattering. We have used the spectrophotometric technique to determine the CMC of Lubrol-PX.

The reagents used are:

Buffer: 20 mM tris-HCl, pH 7.6, containing 0.1 M NaCl

1.6 mM methyl orange in buffer

10 mM and 100 mM Lubrol-PX (mol wt 500) in buffer.

As noted by Benzonana [24], methyl orange is an azo dye which exhibits a spectral shift when in contact with deoxycholate micesses, the difference spectrum showing maximum variation at 4840 Å. The first step in determining the CMC of Lubrol-PX (or of other nonionic detergents) is, therefore, to determine whether methyl orange also shows a spectral shift when in contact with Lubrol micelles, and the wavelength at which the shift is maximal. Spectral shifts are measured in a Beckman Acta V recording spectrophotometer in cells of 10-mm optical length. All solutions are in a total volume of 4 ml. Methyl orange solution (50 μl) is diluted to 20 μmol with buffer (3.95 ml) and the spectrum of the dye is recorded between 550 and 370 nm against a blank consisting of buffer only. A second sample is prepared in buffer containing 20 μmol methyl orange and 25 mmol Lubrol, a concentration of the detergent expected to be greater than the CMC. The spectrum of this solution is also recorded against buffer. As can be seen in Figure 1, Curves A and B, the spectrum of methyl orange shifts toward the blue wavelengths in the presence of Lubrol micelles. If Lubrol is present at very low concentrations, <0.15 mmol, this shift is not seen. When the spectrum of the methyl orange solution is determined versus a blank solution containing Lubrol and methyl orange, a difference spectrum is obtained (Fig. 1, Curve C); the wavelength of greatest difference is 490 nm.

To determine the CMC, a series of samples is prepared in buffer containing 20 μM methyl orange and increasing amounts of detergent. As a first approximation, the detergent concentration can range from 0.02 to 0.50 mM. The absorbancy of these samples is obtained at 490 nm versus a blank solution containing methyl orange only. At concentrations of detergents equal to or greater than the CMC, a decrease in absorbance will occur. In order to determine the CMC more precisely, the range of detergent concentrations at which the absorbance is first affected is expanded. By this method, the CMC of Lubrol-PX was determined to be between 0.16 and 0.18 mM (Fig. 2). Neer [25] has also determined the CMC by the same method and has reported a value of 0.005% (0.10 mM, assuming a mol wt of 500).

C. Solubilization and Partial Purification of Glucagon-Binding Protein from Rat Liver Plasma Membranes

Solubilization of rat liver plasma membranes is performed in a 15-ml, Corex, round-bottom centrifuge tube. Five milliliters of frozen plasma membrane suspension are thawed in cold water and transferred to the

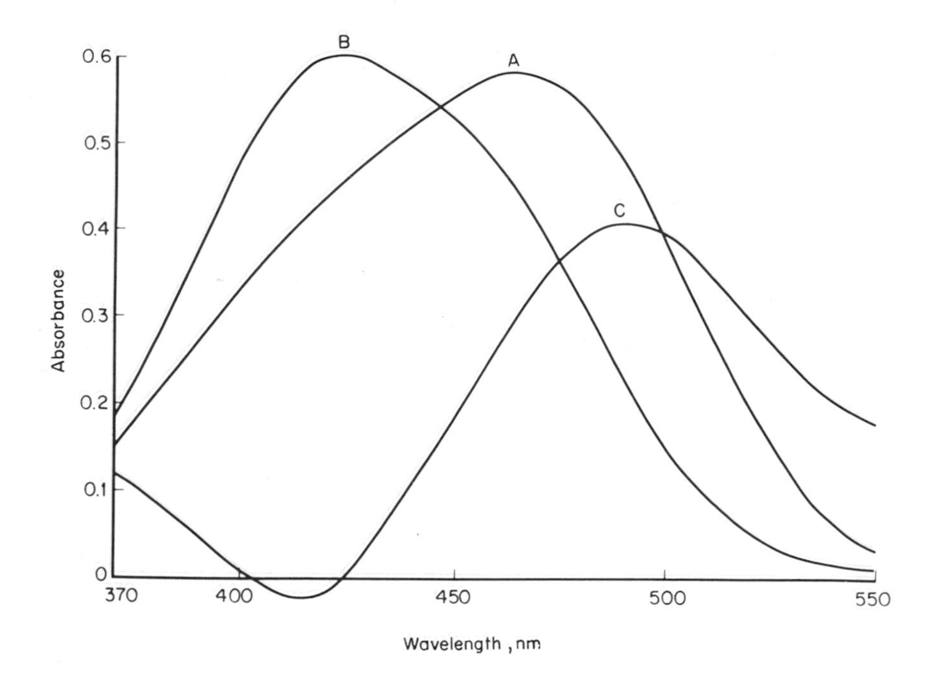

FIG. 1. Spectra of methyl orange in the presence and absence of Lubrol-PX. The blank in each case is 20 mM tris-HCl, pH 7.6, containing 0.1 M NaCl (buffer). Curve A, 20 μM methyl orange in buffer; Curve B, 25 mM Lubrol-PX and 20 μM methyl orange in buffer; and Curve C, the difference spectrum of methyl orange, i.e., 20 μM methyl orange in buffer read against 20 μM methyl orange and 25 μM Lubrol-PX in buffer.

centrifuge tube, which is placed in an ice bath. Five milliliters of a solution containing 20 mM HCl, pH 7.6; 1% Lubrol-PX; and 0.4 mM dithiothreitol (DTT) are added to the tube. The final concentration of Lubrol for the solubilization step is, therefore, 0.5%. [^{14}C] Lubrol (~700,000 cpm) in 0.1% unlabeled Lubrol-PX may be added if one is also following the detergent concentration; in this case, make an appropriate reduction in the amount of unlabeled Lubrol-PX added.

A small, magnetic stirring bar is placed in the centrifuge tube, and the mixture stirred for 1 hr. To increase the solubilization of membrane protein, the mixture is sonicated for 30-sec periods with a Bronson Sonifier microprobe (40 W). By keeping the tube in a salt water-ice bath, the temperature of the solubilization mixture will not rise above 4°C during each sonication period. Nevertheless, 1-min intervals for cooling are allowed between each sonication period.

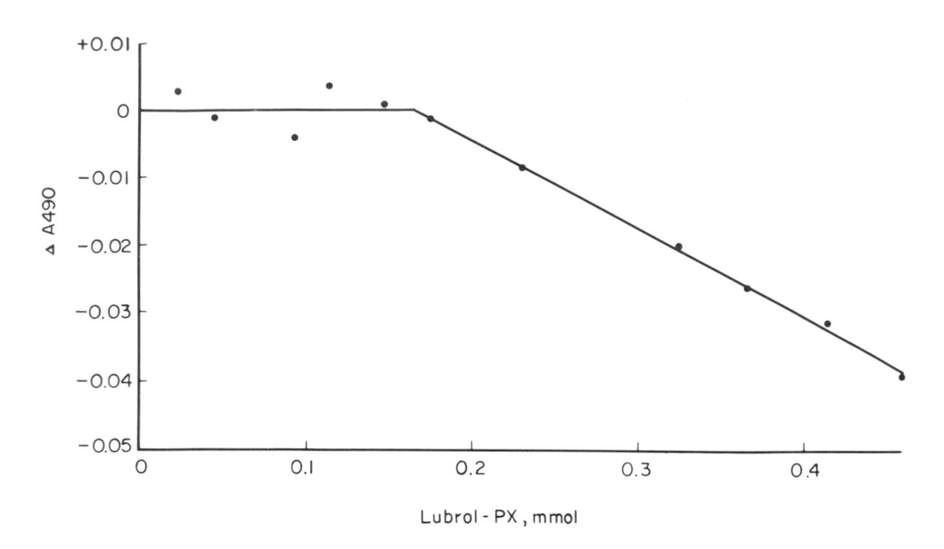

FIG. 2. Determination of the CMC of Lubrol-PX by the methyl orange method. To 5.0 ml (final volume) of 40 μM methyl orange in 10 mM tris buffer (pH 7.6)-0.1 M NaCl was added Lubrol-PX in the final concentrations shown in the figure. Absorbancies at 490 nm were determined against a blank containing all components except Lubrol-PX. Changes in absorbance are plotted as a function of the concentration of Lubrol-PX. The point at which the absorbance begins to fall represents the CMC of the detergent. Due to the electronic instability of the spectrophotometer and the small absorbance change at the CMC, the experiment should be repeated at least in triplicate and the CMC's averaged. We might note parenthetically that glucagon, at a concentration of 6×10^{-9} M, did not influence the CMC of Lubrol-PX.

The mixture is transferred to 5/8 × 3-in. screwcap centrifuge tubes (Oak Ridge type) and centrifuged in a Beckman type 40 rotor at 105,400 g_{av} for 2 hr (4°C). The supernatant is decanted and saved. The brownish pellet is suspended in 5 ml of ice-cold 1 mM sodium bicarbonate with the aid of a teflon stirring rod. This suspension is placed in a fresh 15-ml Corex centrifuge tube, 5 ml of solubilization medium are added, and the colubilization process repeated as described.

The two supernatant fluids (pooled volumes are ~20 ml) are dialyzed and ultrafiltered in order to reduce the concentrations of detergent. The pooled supernatants were dialyzed in a Visking bag against two changes of detergent-free buffer (2 liter each of 0.2 mM DTT in 10 mM Tris-HCl, pH 7.6). The dialysate is distributed among Amicon cones (Amicon CF-50

Centriflo Cones; 50,000 mol wt cutoff), and the volume of retentate is reduced by centrifugation (2,000 rpm in International PR-2 centrifuge, model 269 rotor) to ~1.5 ml in each cone.

Retentates are washed twice by centrifugation, following addition of 5 ml of buffer and vortexing of the cone to loosen adhering material. Finally, retentates are carefully removed with a fire-polished Pasteur pipet, the cones are rinsed with buffer, and the pooled retentates centrifuged at 105,400 g_{av} for 30 min in order to remove insolubilized proteins, etc. The supernatant fluid is called the low-Lubrol fraction.

An aliquot of the low-Lubrol membrane extract is chromatographed on a column (0.9 × 10 cm) of hydroxylapatite (Bio-Rad) equilibrated at 4°C with 0.2 mM DTT in 10 mM tris-HCl, pH 7.6. Apply the aliquot to the column and elute with the equilibration buffer. Protein elution is followed by determination of absorbancies at either 230 or 280 nm. An absorbance monitor (e.g., ISCO or LKB) connected to a microflow cell is useful for this purpose. The eluate is collected into preweighed tubes for determination of cumulative eluate volumes. This is necessary because the presence of Lubrol in the fraction causes a decrease in surface tension of water, thus reducing the size of the drops; therefore, determination of fraction size by drop counting is not accurate. Under these conditions, glucagon-binding proteins do not adsorb to the hydroxylapatite, while over 92% of the other membrane proteins do adsorb. The choice of eluting buffer is critical. At slightly higher ionic strengths, such as obtained by substituting 10 mM phosphate buffer for 10 mM tris in the eluting buffer, more extraneous proteins are eluted with the binding protein, yielding a lower specific binding activity.

The fractions containing the bulk of the nonadsorbed protein are pooled. Amicon Centriflo cones are again used to concentrate the binding proteins. Determination of Lubrol concentration in the hydroxylapatite eluate by tracer analysis showed [26] that the detergent is far below its critical micellar concentration when the aliquot os this partially purified glucagon-binding protein solution is diluted for use in the hormone-binding assay (see Sec. V. B. 4). Therefore, the remaining Lubrol should not interfere with quantification of hormone-binding activity.

An effect of Lubrol, which is apparent especially during the latter steps of purification of the glucagon-binding protein, is its interaction with the Folin-Ciocalteau reagent used in the protein analysis method of Lowry et al. [27]. This interaction produces a sediment which, after centrifugation of the sample for 5 min in a table-top centrifuge, does not interfere with the quantification. The sediment is most noticeable in protein analyses performed on solutions of purified receptor proteins, e.g., Centriflo concentrate, since a relatively larger aliquot ot this solution must be used in the Lowry assay.

V. HORMONE-BINDING ASSAYS

A. Binding to Plasma Membranes

The binding of iodinated hormones to particular rat liver plasma membranes is assessed by modifications of the microcentrifuge method outlined by Rodbell et al. [28]. We have used this method to quantify the binding of secretin, VIP, and insulin, as well as of glucagon.

A separate buffer system is prepared for each variation in the assay. All buffers contain, in 5 ml, 28 mM tris-HCl, pH 7.6; 1.4 mM EDTA; 0.14% BSA; and ^{125}I-labeled hormone ($\sim 4.2 \times 10^{-10}$ M). In earlier experiments, a nonspecific protease inhibitor, Trasylol (700 Kallikrein Inactivator Units per ml) was included in buffers as a means of preventing the possible destruction of hormone; however, no effect on glucagon binding was observed and so it was omitted from further experiments. Nonspecific binding of the ^{125}I-labeled hormone can be assessed by including a large excess of the unlabeled hormone (4.2×10^{-7} M) in the buffer. Similarly, competition experiments can be performed by including a large excess of other native hormones in the buffer.

The assay is carried out in 10 × 75-mm glass test tubes. Each binding-assay mixture contains 100 μl of the appropriate buffer, along with 40 μl of rat liver plasma membrane suspension (diluted in 1 mmol sodium bicarbonate to a concentration of 3 mg/ml). Therefore, the final concentrations of the components are: 20 mM tris-HCl, pH 7.6; 1 mM EDTA; 0.1% BSA; 3×10^{-10} M iodo-hormone; and 3×10^{-7} native hormone, when present. Blank tubes are also prepared in which 40 μl of sodium bicarbonate, in lieu of the membrane suspension, are added to the buffer. These blanks are used to correct for nonspecific binding of hormone to the centrifugation tubes.

The assay is started by the addition of the plasma membranes. After incubation for 30 min at 30°C, duplicate 50-μl aliquots of each incubation mixture are carefully layered over 300 μl of 2.5% BSA in 20 mmol tris-HCl, pH 7.6, contained in plastic 400-μl microcentrifuge tubes (Arthur H. Thomas Co., No. 2591-D15). The tubes are centrifuged for 5 min in a cold room in either a Beckman model 152 or a Coleman model 6-811 microcentrifuge. The supernatants are aspirated to waste with the aid of a Pasteur pipet connected to a water pump. It is convenient to use a Pasteur pipet in which the stem is drawn to a fine capillary end; the narrow tip provides very good control of the aspiration.

The pellet is washed with 300 μl of 10% sucrose solution, added without disturbing the pellet. The tubes are recentrifuged, and the supernatant aspirated to waste. The amount of radioactivity bound to the plasma membranes is determined in a gammacounter; the entire centrifuge tube is placed in the counting tubes. Aliquots of each binding-assay buffer are also counted to ensure that the concentration of radioactive ligand is the same in each assay condition. After subtraction of the blank counts from these counts obtained with the tubes containing plasma membranes, corrections are made for the aliquot taken for counting. Division of the net cpm bound by the specific radioactivity of the labeled hormone yields molar binding. Division of this volume by the milligrams of plasma membrane protein provides the specific-binding activity. At the concentrations of membranes and iodinated hormones used, specific-binding activities are in the order of 10^{-15} moles (femtomoles) bound per mg protein. Because of such low levels of binding activity, binding-assay incubations must be done in at least duplicate, and duplicate aliquots of each incubation mixture must be taken for membrane isolation.

B. Binding to Solubilized Hormone Receptors

1. Introduction

Glucagon, VIP, and secretin are polypeptide hormones containing a high percentage of hydrophobic amino acid residues. For example, of the 29 amino acids in glucagon, nine of these are hydrophobic in nature. This unique property creates special problems when using binding assays in which free hormone is separated from soluble receptor-bound hormone by filtration through cellulose or cellulose ester membranes (Millipore Corp., or Amersham-Searle's Oxoid filters). Pretreatment of such membranes with nonspecific proteins such as BSA, or with large amounts of the unlabeled hormone being tested, fail significantly to reduce the high blank values due to binding of free labeled hormone by the filters [15]. Furthermore, methods which use polyethylene glycol [29] or dextran-coated charcoal [30] to precipitate receptor-bound hormone were not useful, since a significant fraction of the free glucagon was precipitated as well. The ligand saturation-gel filtration technique of Hummel and Dreyer [31] was found to be amenable to the assay of soluble glucagon receptors, as well as to those for other polypeptide hormones.

2. Preparation and Calibration of Gel Filtration Columns

Disposable plastic chromatography columns having an inner diameter of 0.9 cm were purchased from Kontes Glass Co. (Chromaflex, Catalogue #K-420160). Circles of nylon screen (Nitex 380 mesh; purchased from D. C. Screen Process Supply Co., 501 13th St. N.W., Washington, D.C. 20004) are punched out using No. 2 and No. 3 cork borers. The smaller circles are placed in the tips of the columns to serve as bed supports. The columns are suspended from a Plexiglas rack, and 6.5 ml of a 50%

slurry of Sephadex G-50 (coarse) in water are pipetted into each column. After allowing the gel to settle, the bed is washed with 10 ml of 0.1 M sodium chloride (in order to reduce adsorption of radioiodinated polypeptides to the gel beads) and bed lengths of the columns are then equalized by the addition or subtraction of some of the Sephadex. The larger circles of nylon screen are then gently pushed onto the surface of the beds with a glass rod. Columns are again washed with 0.1 M sodium chloride.

Void volumes of the columns are then determined with Blue Dextran 2000. In this step, columns having void volumes far different from the average can be found and discarded; also, the recovery of proteins in the void volume can be estimated. A 0.2-ml aliquot of Blue Dextran 2000 (0.2%, wt/vol, in 0.1 M sodium chloride) is applied to several of the columns. (A short length of plastic tubing attached to the disposable tip of an Eppendorf pipet or similar device is useful in applying solutions just over the nylon screen covering the gel.) Successive 0.1-ml aliquots of 0.1 M sodium chloride are added to each column until the Blue Dextran begins to emerge from the top of the columns; at this point, elution is continued with an additional 0.5 ml of 0.1 M sodium chloride. If some Blue Dextran still remains within the column at this point, the void volume will be somewhat larger than the total of the sodium chloride solutions added to the midpeak point. In general, the void volumes of these Sephadex columns range between 0.67 and 0.75 ml. Once the void volume is determined, test all of the columns for <u>recovery</u> of Blue Dextran. Apply 0.2 ml of the Blue Dextran solution to each column. Then place the column over receiving tubes (13 × 100-mm test tubes), add 5 ml of 0.1 M sodium chloride, and collect the eluates. Dilute the eluates to 1 ml with 0.1 sodium chloride, and prepare a Blue Dextran standard by diluting 0.2 ml of the Blue Dextran with 0.8 ml of 0.1 M sodium chloride. The absorbancy of each tube is read at 625 nm against a blank of 0.1 M sodium chloride, and the percent recovery of the Blue Dextran from each column is calculated. A value between 85 and 90% recovery is common under these conditions. Columns with recovery values outside this range are discarded.

3. Hummel-Dreyer Binding Assay

As with the particulate-binding assay, individual buffers are prepared for each variation in the assay. The composition of the basic column buffer (prepare 30 ml of each) is: 1.0 mM EDTA-0.1% BSA in 20 mM tris-HCl, pH 7.6, containing 1.0 ng/ml of ^{125}I-labeled hormone. Variations include the basic column buffer plus 1.0 mg/ml of the same native hormone (to determine specific binding) or of another hormone (to test for competition). After prewashing the columns with 0.1 M sodium chloride, the gel is equilibrated at 4°C with 6 ml of the appropriate column buffer. A solution of hormone-binding protein (2-50 μl containing 0.25 to 1.5 μl of protein) is incubated with 0.5 ml of column buffer for 30 min at 30°C in a shaking water bath. This period of time is sufficient for saturation to be achieved.

The incubation mixture is rapidly cooled to 4°C in an ice-water bath, and 0.2-ml aliquots of the mixture are applied to duplicate gel columns. After the solution enters the gel, a volume of column buffer equal to the void volume is added; the column eluate is discarded. Protein-bound hormone is eluted with the next 0.5 ml of column buffer. This fraction is collected in 1-dram glass vials (Kimble No. 60975-L) which serve as counting vials. For quantification of radioactivity by liquid scintillation spectrometry, 3.3 ml of a scintillation system composed of 0.4% 2,5-diphenyloxazole (PPO)-0.01% 1,4-bis-2-(5-phenyloxazolyl) benzene (POPOP) in toluene-Triton X-100 (2:1 vol/vol) are added. The vials are capped, and a single-phase system results after shaking for 10 sec; no phase separation occurs even after the samples are cooled in the counter. These vials are placed in standard, glass, 20-ml counting vials for use in the scintillation counter.

In this method, ^{125}I-labeled hormone which is complexed to the binding protein appears as an increase in radioactivity relative to the basal radioactivity of the column buffer. Therefore, the binding capacity is calculated from the specific radioactivity of the iodinated hormone, and the amount of radioactivity in 0.5 ml of the columns eluate containing the binding protein, less the radioactivity in an equal volume of column eluate not containing binding protein.

4. Critique of Binding Assay

A well-known property of the nonionic detergents used in solubilizing hormone-binding proteins is the formation of micellar aggregates when concentrations are sufficiently high, i.e., above the critical micellar concentration. Studies on the solubilization of the glucagon-binding protein demonstrated that glucagon, perhaps because of its amphipathic structure, could form high-molecular-weight complexes with these micelles. Such complexes will appear in the void volume of the Sephadex G-50 columns, and will yield erroneously high values for hormone binding. It is, therefore, imperative that the critical micellar concentration of the detergent be reduced below that value before using this binding assay for hormones which associate with detergent micelles.

[^{125}I]Glucagon was found to adsorb to Sephadex beads. However, if the gel filtration medium is equilibrated with 0.1% BSA, only between 5 and 7% of the hormone is adsorbed to the gel. Since protein-bound radioactivity ranges between two to four times that of the column buffer, this amount of nonspecific adsorption is insignificant.

This method provides an accurate means of measuring hormone-binding activity of solubilized binding proteins, and is particularly useful for ligand-receptor complexes which are readily dissociable. It is not, however, very economical, since large amounts of polypeptide hormones (often difficult to obtain, and expensive) are required, because of the necessity of including the hormone in the column buffer. In addition, the time taken for the complete assay (incubation and column steps) precludes the obtainment of the early time points required for the calculation of binding-rate constants. Furthermore, because of the manipulations required and the fact that two columns are required for each experimental variable, it is difficult for a single operator to do as many binding assays as can be performed by other techniques.

ACKNOWLEDGMENTS

The authors gratefully acknowledge the contributions of Drs. Nicholas A. Giorgio and Carl B. Johnson to the development of methods presented in this review. Our contributions were supported by grant AM-05475 and GRS funds from the National Institutes of Health.

REFERENCES

1. G. A. Robison, R. W. Butcher, and E. W. Sutherland, Cyclic AMP, Academic Press, New York, 1971.
2. J. Roth, Metabolism, 22:1059 (1973).
3. P. Cuatrecasas, Ann. Rev. Biochem., 43:169 (1974).
4. R. J. Lefkowitz, J. Roth, and I. Pastan, Science, 170:633 (1970).
5. (a) T. Tsushima and H. G. Friesen, J. Clin. Endocrinol. Metab., 37:334 (1973); (b) P. Gordon, M. A. Lesniak, C. M. Hendricks, and J. Roth, Science, 182:829 (1973).
6. (a) L. E. Reichert, Jr., and V. K. Bhalla, Endocrinology, 94:483 (1974); (b) L. E. Reichert, Jr., this work, Chap. 5.
7. B. B. Saxena, S. H. Hasan, F. Haour, and M. Schmidt-Gollwitzer, Science, 184:793 (1973).
8. (a) R. P. C. Shiu, P. A. Kelly, and H. G. Friesen, Science, 180: 968 (1973); (b) this work, Chap. 0.

9. (a) M. Rodbell, Federation Proc., 32:1854 (1973); (b) S. L. Pohl, this work, Chap. 8.

10. V. Tomasi, S. Kortez, T. K. Ray, J. Dunnick, and G. V. Marinetti, Biochim. Biophys. Acta, 211:31 (1970); A. Rethy, V. Tomasi, A. Trevisani, and O. Barnebei, Biochim. Biophys. Acta, 290:58 (1972).

11. (a) G. S. Levey, Recent Progr. Hormone Res., 29:361 (1973); (b) G. S. Levey, M. A. Fletcher, I. Klein, E. Ruiz, and A. Schenk, J. Biol. Chem., 249:2665 (1974); (c) G. S. Levey, M. A. Fletcher, and S. Ramachandran, this work, Chap. 7.

12. J. Ryan and D. R. Storm, Biochem. Biophys. Res. Commun., 60: 304 (1974).

13. S. L. Pohl, L. Birnbaumer, and M. Rodbell, J. Biol. Chem., 246: 1849 (1971).

14. N. I. Swislocki, J. Tierney, and M. Sonenberg, Biochem. Biophys. Res. Commun., 53:1109 (1973).

15. M. Blecher and C. B. Johnson, unpublished.

16. C. M. Neville, Jr., Biochim. Biophys. Acta, 154:540 (1968).

17. N. A. Giorgio, C. G. Johnson, and M. Blecher, J. Biol. Chem., 249:428 (1974).

18. C. B. Johnson, M. Blecher, and N. A. Giorgio, Biochem. Biophys. Res. Commun., 46:1035 (1972).

19. C. B. Johnson, Ph.D. Dissertation, Georgetown University, Washington, D.C., 1974.

20. W. M. Hunter and F. C. Greenwood, Nature, 194:495 (1962).

21. K. N. Holohan, R. F. Murphy, R. W. J. Flanagan, K. D. Buchanan, and D. T. Elmore, Clin. Chim. Acta, 45:153 (1973); K. N. Holohan, R. F. Murphy, R. W. J. Flanagan, K. D. Buchanan, and D. T. Elmore, Biochim. Biophys. Acta, 322:178 (1973).

22. M. Brenner and A. Niederwieser, in Methods in Enzymology (C. H. W. Hirs, ed.), Vol. 11, Academic Press, New York, 1967, p. 39.

23. K. E. Wilzbach, J. Am. Chem. Soc., 79:1013 (1957).

24. G. Benzonana, Biochim. Biophys. Acta, 176:836 (1969).

25. E. J. Neer, J. Biol. Chem., 249:6527 (1974).

26. M. Blecher, N. A. Giorgio, and C. B. Johnson, in Advances in Enzyme Regulation (G. Weber, ed.), Vol. 4, Pergamon Press, New York, 1974, p. 289.

27. O. H. Lowry, N. J. Rosebrough, A. L. Farr, and R. J. Randall, J. Biol. Chem., 193:265 (1951).

28. M. Rodbell, H. M. J. Krans, S. L. Pohl, and L. Birnbaumer, J. Biol. Chem., 246:1861 (1971).

29. P. Cuatrecasas, Proc. Natl. Acad. Sci. U.S.A., 69:318 (1972).

30. S. G. Korneman, Endocrinology, 87:1119 (1972).

31. J. P. Hummel and W. J. Dreyer, Biochim. Biophys. Acta, 63:530 (1962).

Chapter 7

METHODS TO CHARACTERIZE THE CARDIAC GLUCAGON RECEPTOR

Gerald S. Levey
Mary A. Fletcher

Divisions of Endocrinology, Metabolism and Immunology
Department of Medicine
University of Miami School of Medicine
Miami, Florida

S. Ramachandran

Applied Science Laboratories, Inc.
Department of Production, Research and New Products
State College, Pennsylvania

I. INTRODUCTION

Many hormones increase the concentration of adenosine 3',5'-monophosphate (cAMP) in their target tissues [1]. Understanding the nature of the processes whereby hormones bind to specific cell-surface receptors and how they subsequently induce activation of the membrane-bound enzyme adenylate cyclase, which catalyzes the production of cAMP, is fundamental to the elucidation of the mechanism of action of these hormones. Many studies have delineated the critical role served by membrane lipids in the activation process [2-6]. Over the past several years, we have utilized solubilized preparations of cat myocardial adenylate cyclase in an attempt to understand the interactions of the various components of the hormone-responsive, membrane-bound adenylate cyclase [6-10]. Using highly purified, acidic phospholipids, we have found that monophosphatidylinositol is critical for the catecholamine activation of myocardial adenylate cyclase, and phosphatidylserine for the glucagon and histamine activation. Furthermore, these lipids did not appear to be necessary for catalytic activity of the enzyme under the conditions tested, or for the binding of [^{125}I]glucagon or [^{3}H]norepinephrine. The specific methods for the solubilization of myocardial adenylate cyclase with Lubrol-PX, the removal of the detergent by DEAE-cellulose chromatography, and the use of the purified acidic phospholipids in the adenylate cyclase assay have been described in detail elsewhere [11].

Recently we have reported the binding of [^{125}I]glucagon to solubilized preparations of myocardial adenylate cyclase and the evidence for a dissociable glucagon-binding site which may be important in the activation-inactivation of the glucagon-responsive enzyme [12, 13]. The present methodological report will serve to present, (1) the specific means by which the acidic phospholipids (used for the activation studies) are prepared from bovine brain; (2) the methods for iodination of glucagon and assessment of binding of the [^{125}I] glucagon; (3) the methods used to dissociate the glucagon receptor and determine the molecular weight of the dissociated receptor.

II. ISOLATION OF PHOSPHATIDYLSERINE AND PHOSPHATIDYLINOSITOL FROM BOVINE BRAIN

Phosphatidylserine and phosphatidylinositol are the two most difficult phospholipids to isolate in the pure state from bovine brain. These particular phospholipids which have been used in our studies of the hormone-responsive adenylate cyclase were from freshly slaughtered bovine brains obtained in less than 12 hr from the commercial slaughterhouses. The brains were freed from blood by rinsing with ice water. Solvents such as chloroform, methanol, and acetic acid (glacial) were all distilled in an all-glass distillation system and stored.

Whole beef brain (2 kg) is cut into 1-in. chunks and homogenized in a Waring Blendor with 4 liters of acetone for 10 min. The homogenate is filtered and the cake reextracted with another 4 liters of acetone and filtered again. Acetone filtrates contain large amounts of water and neutral lipids. Small amounts of phosphoglycerides are lost during acetone extraction. The residue from acetone extraction is suspended in 4 liters of absolute ethanol and homogenized for 10 min. The homogenate is filtered, and the filtrate contains mainly phosphatidylcholine and small amounts of other phosphoglycerides. Next, the ethanol-extracted residue is further treated with 4 liters of petroleum ether, homogenized for 5 min, and filtered. The petroleum ether filtrate is saved and the residue extracted again with another 4 liters of petroleum ether. Petroleum ether extracts contain mainly phosphatidylethanolamine, phosphatidylserine, phosphoinositides, cerebrosides, and sphingomyelin. This fraction is usually referred to as "cephalin" fraction.

A. Phosphatidylserine

The petroleum ether extract is concentrated to dryness under a Rotovap and the residue is resuspended with 25 ml of chloroform.

A column containing 120 g of DEAE cellulose (acetate form) prepared according to the method of Rouser et al. [14] is packed using chloroform. Crude "cephalin" from petroleum ether concentrate is dissolved in chloroform as described, and is charged to the column. The column is eluted with chloroform until no more neutral lipids are eluted from the column as monitored by thin layer chromatography (TLC). Further successive elutions are performed with the following solvents: chloroform:

methanol (9:1) to remove ceramides, phosphatidylcholine, lysophosphatidylcholine, and sphingomyelin; chloroform:methanol (7:3) to remove phosphatidylethanolamine and free fatty acids; methanol to remove salts; chloroform:acetic acid (3:1) to remove pigments; and finally with glacial acetic acid to remove phosphatidylserine and small amounts of protein. The acetic acid eluate contained most of the phosphatidylserine and is concentrated to dryness on a Rotovap at 37° C using a water vacuum. Final traces of acetic acid are removed using a high-vacuum pump and the residue dissolved in 100 ml chloroform. The chloroform layer is washed with distilled water three times and concentrated to dryness. This residue is dissolved in 25 ml $CHCl_3$ and charged to a HiFlosil (silicic acid) column packed in chloroform and eluted with chloroform; chloroform:methanol (90:10), chloroform:methanol (80:20), and chloroform:methanol (70:30). Sixteen-ounce fractions are obtained from the column and each fraction monitored by TLC. Pure phosphatidylserine is eluted from the column when a cloroform:methanol (90:10) and chloroform:methanol (80:20) solvent mixture is used. Fractions containing pure phosphatidylserine are combined and evaporated to dryness in a Rotovap. Samples are then weighed and dissolved in distilled benzene and stored in sealed glass ampules under nitrogen.

B. Phosphatidylinositol

Isolation of phosphatidylinositol from bovine brain is very difficult. The crude "cephalin fraction" obtained from petroleum ether extract as described in Sec. II. A is further purified by the following precipitation technique. For every 1 g of cephalin, 8 ml of chloroform is added to solubilize the crude cephalin fraction. To this chloroform solution, 11.8 ml ethanol is added, and a white precipitate formed which is stirred and centrifuged. The supernatant contains mostly phosphatidylserine and phosphatidylethanolamine. The white precipitate is dissolved in 22 ml chloroform, and 44 ml of methanol is added. Phosphoinositides are precipitated and centrifuged. Solubilization in chloroform and precipitation with twice the volume of methanol is repeated at least three more times. The final white precipitate obtained by this method is enriched with "crude phosphoinositides."

One hundred twenty grams of freshly washed DEAE (acetate) is slurried in chloroform and packed in a glass column. Crude phosphoinositides are dissolved in 20 ml of chloroform and charged to the DEAE column. The column is eluted with the following solvent systems: chloroform:methanol (9:1); chloroform:methanol (7:3); methanol, acetic acid, methanol; and finally with chloroform:methanol:ammonium acetate (0.6 M) (20:9:1). A final solvent system containing chloroform:methanol; ammonium acetate (0.6 M) (20:9:1) eluted the phosphoinositides from the

DEAE column. Fractions containing the phosphatidylinositol are combined and evaporated to dryness, and the residue dissolved in 100 ml of chloroform and washed with distilled water to remove salts. The chloroform layer is separated and concentrated to yield phosphatidylinositol. If found impure, it is subjected to a silicic acid chromatography column and eluted in chloroform/methanol mixtures, 70:30 to 60:40. Crude phosphoinositides can also be purified by silic acid (HiFlosil) chromatography and eluted in chloroform/methanol mixtures 70:30 and 60:40. Fractions containing the pure phosphatidylinositol are combined and evaporated to dryness. Samples are dissolved in chloroform and stored in sealed glass ampules under nitrogen before use.

Phosphatidylinositol obtained from DEAE and silicic acid chromatography columns were found to be extremely labile and unstable. During column elution they appeared to be homogeneous. However, when concentrated, they either decomposed or gave multiple spots on TLC. Phosphoinositides can be associated with divalent ions such as Ca^{2+} or Mg^{2+}. Insufficient extraction from tissue, nonhomogeneity during precipitation techniques, and the various fatty acid substituents at the acyl moiety could account for the labile nature of phosphoinositides isolated from beef brain. This instability is of great importance when utilizing them in experiments dealing with hormone responsiveness of solubilized adenylate cyclases.

III. IODINATION OF GLUCAGON

Glucagon is iodinated by a modification of the procedure of Hunter and Greenwood [27]. The following solutions are added successively into a 10 × 75-mm flint glass tube: 20 μl $NaPO_4$, 0.6 M, pH 7.4; 10 μl (35 μg) crystalline glucagon; 3.5 mg/ml, 0.01 N HCl or 0.01 M tris-HCl, pH 8.7; ^{125}I, 2.5 to 3.0 mCi; 10 μl chloramine-T, 3.5 mg/ml; 0.05 M $NaPO_4$, pH 7.0; and 25 μl Na metabisulfite, 2.4 mg/ml; 0.05 M $NaPO_4$, pH 7.0. Five microliters of the iodination mixture is then added to a solution of 100 μl of Veronal 0.1 M, pH 8.6, and 15 μl plasma containing bromphenol blue. This mixture is applied to Whatman 3M paper, and electrophoresed for 90 min at 25°C, dried and strip scanned on a Nuclear Chicago Actigraph to determine the relative efficiency of glucagon iodination. These iodinations are usually 90 to 95% complete, the remainder being free iodine. $[^{125}I]$Glucagon is purified on a column of cellulose powder as described by Rodbell et al. [15], by applying the $[^{125}I]$glucagon to a 2.5-cm cellulose column prepared in a Pasteur pipet of 0.6-cm diameter and prewashed with 1% albumin in 10 mmol sodium phosphate, pH 7.5. The column is then washed with 3.0 ml of a solution of 1% albumin in 10 mmol sodium phosphate adjusted to pH 7.5, and the $[^{125}I]$glucagon eluted with 0.6 ml of the same solution adjusted to pH 10.0 with concentrated ammonium hydroxide. The $[^{125}I]$glucagon eluted in this manner is biologically active, as determined by its ability to activate both the particulate myocardial adenylate cyclase and the solubilized, detergent-free adenylate cyclase in the presence of phosphatidylserine [13].

IV. [^{125}I]GLUCAGON-BINDING ASSAY

Solubilized enzyme is incubated at 37°C in a final volume of 100 μl containing 1.0% albumin in 10 mM tris-HCl, pH 7.7, and [^{125}I]glucagon (0.370 μCi/pmol). The binding reaction is linear for ~40 min. After the appropriate time interval, generally 60 min, the incubation mixture is added to dry, 2.5-cm cellulose columns in disposable Pasteur pipet with an inside diameter of 0.6 cm and washed with 1.4 ml of 1% albumin in 10 mM tris-HCl, pH 7.7. Bound [^{125}I]glucagon is not adsorbed to the column, in contrast to free (unbound) [^{125}I]glucagon, which binds to the cellulose. The eluate is then counted in a Nuclear Chicago Autogamma. The method removes more than 90% of the free (unbound) [^{125}I]glucagon, as determined by the number of counts found in the control samples of identical composition, incubated simultaneously using 10 mmol tris-HCl, pH 7.7, in place of the enzyme fraction. Similar control values are obtained when Lubrol-PX is included in the incubation mixture. All experimental values are corrected by subtracting the blank values obtained from incubations in the absence of the solubilized binding fraction. Four X-crystalline bovine serum albumin must be used at all steps in the iodination and binding experiments, since we have observed that blank values are lower and stability with storage vastly superior to bovine serum albumin (fraction V).

V. DISSOCIABLE GLUCAGON-RECEPTOR SITE

Two hundred microliters of the 12,000-g supernatant derived from the original Lubrol-PX homogenate of heart muscle is applied to a 6 × 96-mm Sephadex G-100 column, poured as a slurry, and equilibrated with 10 mM tris-HCl, pH 7.7, at room temperature. The column fractions are eluted with 10 mM tris-HCl, pH 7.7; 0.25-ml fractions are collected dropwise under visual monitoring; and a 25-μl aliquot of each fraction assayed for [^{125}I]glucagon-binding activity as described in Sec. IV. The maximum-binding fractions (0.9-1.4 ml) are pooled, a 200-μl aliquot rechromatographed on the Sephadex G-100 [Fig. 1(a)], and 0.25-ml fractions collected dropwise again under visual monitoring. Twenty-five-microliter aliquots of each of these fractions are assayed for [^{125}I]glucagon binding and fluoride-stimulated adenylate cyclase activity. Bound and free [^{125}I]-glucagon are next separated by cellulose chromatography as described, and the total cellulose effluent (1.0 ml), which contains bound [^{125}I]glucagon, counted in a autogamma. In a separate set of experiments [Fig. 1(b)],

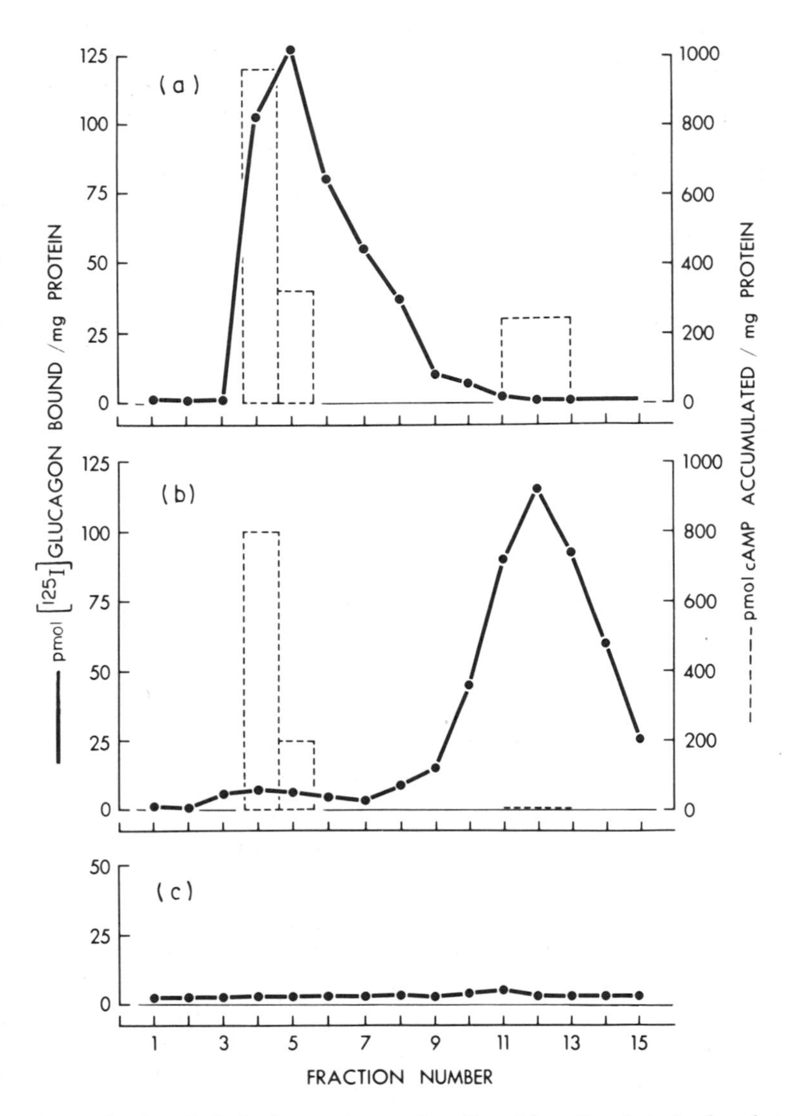

FIG. 1. Sephadex G-100 chromatography; fluoride-stimulated adenylate cyclase activity and [^{125}I]glucagon binding. See discussion in the text. (Reprinted with permission of the American Society of Biological Chemists, Inc.) [12].

25 μl of the original maximum-binding fraction is incubated with [^{125}I]-glucagon, and chromatographed on cellulose. Next, 200 μl of this eluate is applied to the identical Sephadex G-100 column, eluted with 10 mM tris-HCl, pH 7.7, and the binding profile and fluoride-stimulated adenylate cyclase activity determined by the appropriate assays [Fig. 1(b)]. If glucagon, 1×10^{-5} M, is added to the incubations prior to the addition of the [^{125}I]glucagon, no binding is observed [Fig. 1(c)]. Experiments performed in this manner show that, if solubilized adenylate cyclase is applied to the gel prior to incubation with [^{125}I]glucagon, both [^{125}I]glucagon-binding activity and fluoride-stimulated, catalytic adenylate cyclase activity are found in the void volume [Fig. 1(a)]. In contrast, if, prior to the Sephadex chromatography, the solubilized enzyme is preincubated with

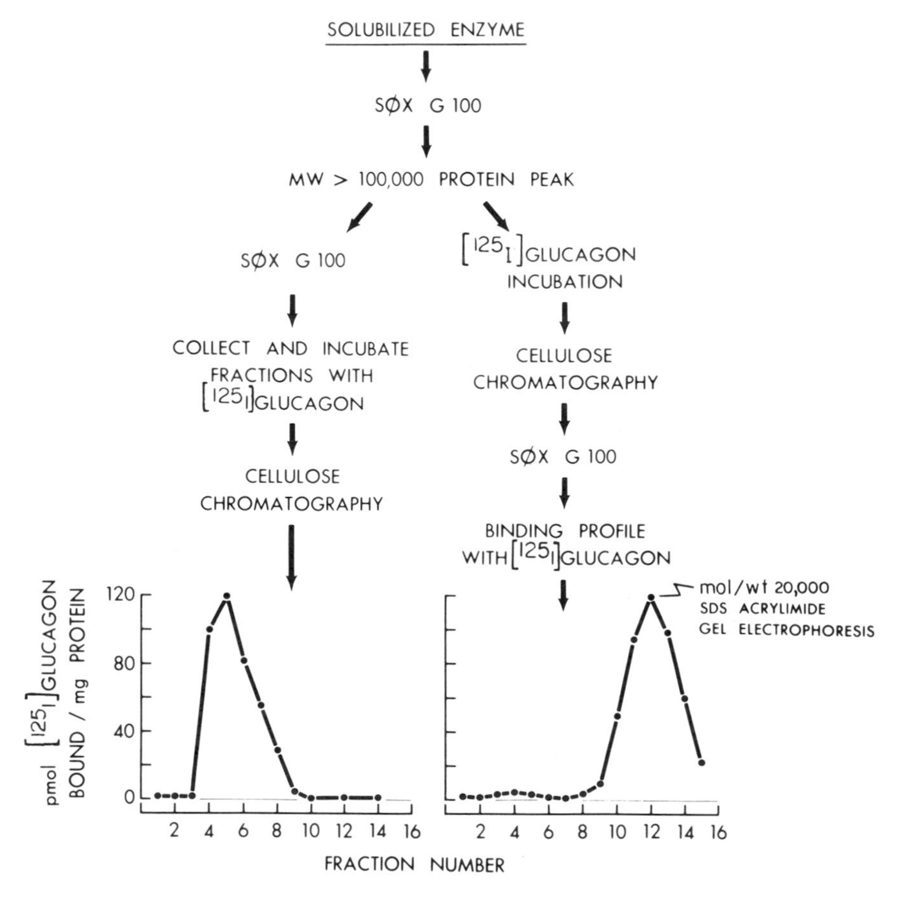

FIG. 2. Flow chart of procedure for demonstration of dissociable glucagon-binding site.

[^{125}I]glucagon and the bound glucagon rechromatographed on Sephadex G-100, the bound [^{125}I]glucagon appears in a later column eluate, consistent with a small-molecular-weight substance (see Sec. VI). Fluoride-stimulated activity remains in the void volume. This description is summarized in the form of a flow sheet depicted in Figure 2.

It became apparent, when we utilized larger (86 ml) Sephadex G-100 columns, that the bound [^{125}I]glucagon eluted [Fig. 1(b)] was located in the void volume of the column. Nevertheless, this material has the same electrophoretic mobility in the SDS polyacrylamide gels (Sec. VI) as the crude solubilized enzyme incubated with [^{125}I]glucagon. In addition, the Sephadex eluate [Fig. 1(b)] subjected to hydrodynamic flow electrophoresis in 0.1 M Veronal buffer, pH 8.6, demonstrated that the labeled material was neither free glucagon nor iodine. It is apparent that the binding site-glucagon complex interacts with the gel itself. Increasing the ionic strength of the elution buffer to 100 mM tris, pH 7.7, does not alter the elution profile, nor does elution with 200 mM acetate buffer pH 5.5. However, when elution is performed at strongly alkaline pH (0.1 or 0.025 N NaOH), a shift in the binding peak is observed toward the expected elution profile for a substance of its apparent molecular weight. These data appear to be similar to the interactions described for iodotyrosines and triiodothyronine and thyroxine on Sephadex [16, 17], indicating that a strongly positive-charged material is adsorbing to the gel. This interaction may prove to be of importance in the ultimate purification of the receptor.

VI. MOLECULAR WEIGHT ESTIMATION OF [^{125}I]GLUCAGON-BINDING FRACTION OF SOLUBILIZED CAT MYOCARDIAL ADENYLATE CYCLASE BY POLYACRYLAMIDE GEL ELECTROPHORESIS IN SODIUM DODECYL SULFATE

A. Principle

Behavior of proteins in gel electrophoresis systems can be described by the Ferguson equation [18]:

$$\text{Log } R_M = Y_o + K_R G$$

R_M is the relative mobility of the protein, calculated as the ratio of the mobility of the protein to that of the marker dye, with correction for changes in gel length after staining. Y_o is a constant representing the logarithm of the relative mobility of the protein at zero gel concentration. K_R is a constant characteristic for each protein representing the retardation of mobility by increasing gel concentration. G is the concentration of acrylamide in the gel (where the ratio of acrylamide to cross-linker is held constant). The two constants Y_o and K_R can be estimated for a given protein by determination of R_M at different acrylamide concentrations. Log R_M is plotted against gel concentration. The negative slope of the line calculated by the method of least squares is equal to K_R. Extrapolation of the line to zero percent gel yields Y_o. In gel electrophoresis without sodium dodecyl sulfate, Y_o is a function of the intrinsic net charge of a protein, and for 17 proteins K_R is directly proportional to molecular weight [19].

Recently, polyacrylamide gel electrophoresis in the anionic detergent, sodium dodecyl sulfate (SDS), has been utilized as a rapid and convenient method for the determination of the molecular weight of proteins from a plot of log molecular weight versus R_M of standard proteins at a single gel concentration [20, 21]. The accuracy of this method depends on a constancy of Y_o for most proteins when they are dissolved in SDS solutions [22, 23]. Important exceptions to this standard behavior are known. Glycoproteins with substantial amounts of carbohydrate exhibit anomalous behavior in SDS polyacrylamide gel electrophoresis because of a decreased binding of SDS and, thus, a lower value for Y_o [24-26]. When this method is applied to a substance of unknown chemical structure, as in the present application, the behavior of the substance at several gel concentrations should be compared to that of standard proteins.

B. Procedure

1. Protein Standards

Protein standards for electrophoresis are γ-, β-, and α-chains of human fibrinogen (mol wt 73,000, 60,000, and 53,000, respectively); or albumin (mol wt 46,000); human Bence-Jones protein (K type) (mol wt 22,500); and lysozyme (mol wt 14,400).

2. Polyacrylamide Gel Electrophoresis

Reagents: Gel buffer: 0.2 M sodium phosphate, pH 7.0, containing 0.2% SDS.

Electorphoresis buffer: 0.025 M Sodium phosphate, pH 7.0 containing 0.1% SDS.

Sample buffer: 0.01 M sodium phosphate, pH 7.0, containing 0.1% SDS and 0.1% mercaptoethanol.

Stock acrylamide solutions (stored at 4°C in dark): 7.5% acrylamide (16.65 g acrylamide and 0.45 g N, N'-methylenebisacrylamide (BIS) dissolved in 100 H_2O; 10% acrylamide (22.20 g acrylamide and 0.6 g BIS dissolved in 100 ml H_2O); 12.5% acrylamide (28.125 g acrylamide and 0.76 g BIS, dissolved in 100 ml H_2O).

Ammonium persulfate: 15 mg/ml in water (freshly made).

3. Preparation of gels

For a run of 12 gels, 15 ml of gel buffer (deaerated), 13.5 ml of acrylamide solution (deaerated), 1.5 ml of ammonium persulfate, and 0.045 ml of N, N, N', N'-tetramethylenediamine was prepared, rapidly mixed, and 1.2 ml added to each gel tube (7.5-cm long, 0.5-cm i.d.). A few drops of water are layered on each tube and the tubes are left at room temperature for 30 min to polymerize. Immediately before use the water is shaken out.

4. Preparation and Electrophoresis of Samples

Prior to electrophoresis, the samples are incubated at 37°C for 30 min in 50 μl of sample buffer. Three microliters of marker dye (bromphenol blue) and one drop of glycerol is added to each sample. The samples are placed on the gel columns and overlaid with electrophoresis buffer. The tubes are placed in the electrophoresis apparatus and electrophoresed at room temperature until the tracking dye is 1 cm from the bottom of the gels.

5. Processing of gels

After the electrophoresis run, the gel columns are immediately removed from the tubes, and gel length and position of marker dye determined. Those gels containing marker proteins are stained with Coomassie blue, 0.25% in 50% methanol, 9.5% acetic acid, for 18 to 24 hr and destained electrophoretically with a Canalco transverse destainer in 7.5% acetic acid. Mobilities of molecular weight marker proteins relative to mobility of marker dye (R_M) are determined after correction for changes in gel length due to staining. Those gels containing solubilized adenylate cyclase activity to which [^{125}I]glucagon had been added were sliced horizontally into 1-mm segments using a Canalco gel slicer and the radioactivity of each segment determined in a gammacounter.

C. Results

The electrophoretic behavior of the [^{125}I]glucagon-binding site in solubilized cat myocardial adenylate cyclase in 10% polyacrylamide gel with

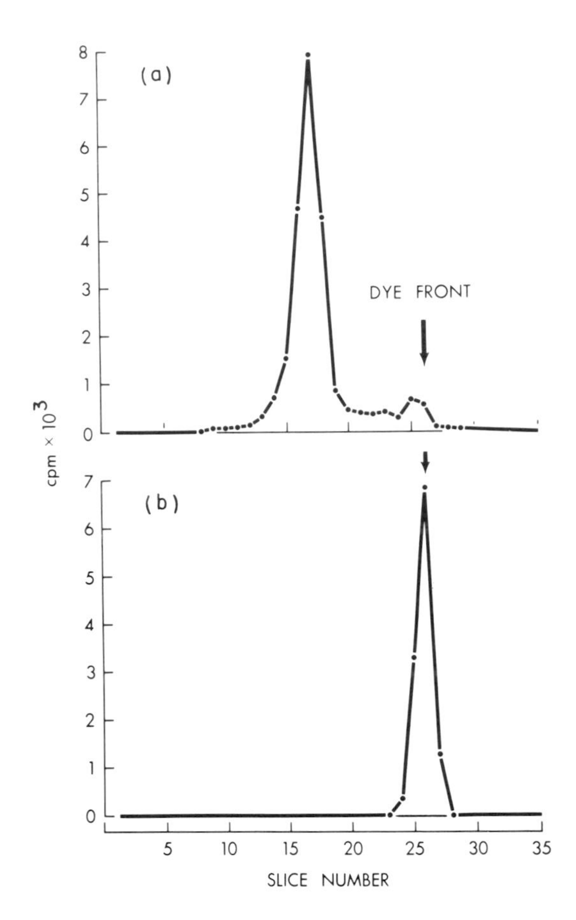

FIG. 3. SDS polyacrylamide gel electrophoresis. See discussion in the text. (Reprinted with permission of the American Society of Biological Chemists, Inc.) [13].

SDS is shown in Figure 3. The complex is represented by the large peak of slower mobility, Figure 3(a). The small peak which ran with mobility identical to that of the marker dye is $[^{125}I]$glucagon. This peak dissociated during the experiment and corresponds to the single peak seen when $[^{125}I]$glucagon alone was subjected to electrophoresis, Figure 3(b). Similar results were obtained with either crude solubilized preparations, or after passage through cellulose of Sephades. The R_M of the binding complex was also determined at 7.5% and at 12.5% acrylamide concentrations. Figure 4 shows Ferguson plots [18] of log R_M versus gel concentrations for the $[^{125}I]$glucagon-binding complex and for standard peptides.

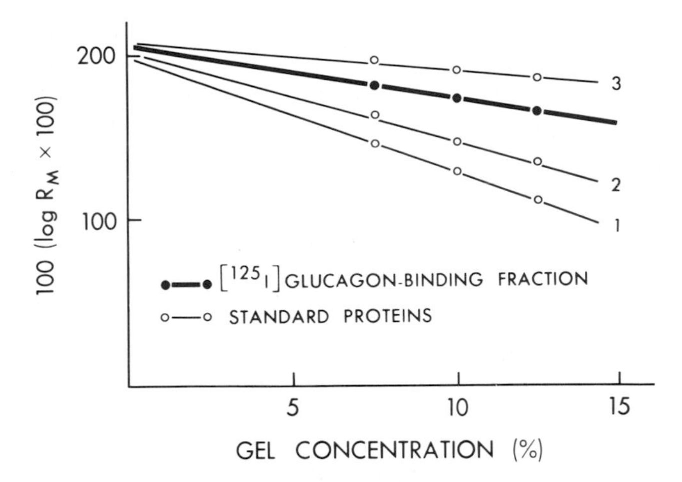

FIG. 4. Ferguson plots. (Reprinted with permission of the American Society of Biological Chemists, Inc.) [13].

Extrapolation of the lines indicates similar mobility at zero gel concentration (Y_O) for the binding complex, as compared to the standards. Retardation coefficients (K_R) were determined from the negative slopes of the Ferguson plots. K_R values for six standard peptides are plotted in Figure 5 versus log R_M at three gel concentrations, and define three nonparallel lines which intersect at the y axis. The K_R value for the complex falls on

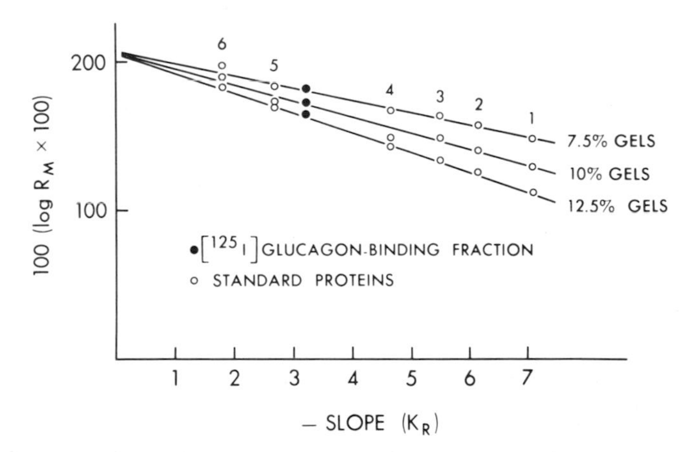

FIG. 5. Retardation coefficients. (Reprinted with permission of the American Society of Biological Chemists, Inc.) [13].

these lines. These experiments indicated that the behavior of the [^{125}I]-glucagon-binding complex did not deviate from that of the six standard peptides studied. The apparent molecular weight of the complex calculated from plots of log molecular weight versus R_M of the standard proteins was 24,000 to 28,000.

ACKNOWLEDGMENTS

These studies were supported in part by NIH grants HL-13715 and AM-16763. G.S.L. is an Investigator of the Howard Hughes Medical Institute and S.R. is Director, Research and Development, Applied Science Laboratories (Pennsylvania).

REFERENCES

1. G. A. Robison, R. W. Butcher, and E. W. Sutherland, Cyclic AMP, Academic Press, New York, 1971.
2. E. W. Sutherland, T. W. Rall, and T. Menon, J. Biol. Chem., 237:1220 (1962).
3. S. L. Pohl, H. M. J. Krans, V. Kozyreff, L. Birnbaumer, and J. Rodbell, J. Biol. Chem., 246:4447 (1971).
4. A. Rethy, V. Tomasi, A. Trevisani, and O. Barnabei, Biochim. Biophys. Acta, 290:58 (1972).
5. K. Yamashita and J. B. Field, Biochim. Biophys. Acta, 304:686 (1973).
6. G. S. Levey, Recent Progr. Hormone Res., 29:361 (1973).
7. G. S. Levey, Biochem. Biophys. Res. Commun., 38:86 (1970).
8. G. S. Levey, Ann. N.Y. Acad. Sci., 185:449 (1971).
9. G. S. Levey, J. Biol. Chem., 246:7405 (1971).
10. G. S. Levey and I. Klein, J. Clin. Invest., 51:1578 (1972).
11. G. S. Levey, in Methods in Enzymology (J. G. Hardman and B. W. O'Malley, eds.), Vol. 38, Academic Press, New York, 1974, pp. 174-180.
12. I. Klein, G. S. Levey, and M. A. Fletcher, J. Biol. Chem., 248:5552 (1973).
13. G. S. Levey, M. A. Fletcher, I. Klein, E. Ruiz, and A. Schenk, J. Biol. Chem., 249:2665 (1974).

14. G. Rouser, G. Kritchevsky, A. Yamamoto, G. Simon, C. Galli, and A. J. Bauman, in Methods in Enzymology (J. M. Lowenstein, ed.), Academic Press, New York, 1969, p. 272.

15. M. Rodbell, H. M. J. Krans, S. L. Pohl, and L. Birnbaumer, J. Biol. Chem., 246:1861 (1971).

16. R. H. Osborn and T. H. Simpson, J. Chromatog. Sci., 34:110 (1968).

17. W. L. Green, J. Chromatog., 72:83 (1972).

18. K. A. Ferguson, Metabolism, 13:985 (1964).

19. J. L. Hedrick and A. J. Smith, Arch. Biochem. Biophys., 126:155 (1968).

20. A. L. Shapiro, E. Vinuela, and J. V. Maizel, Jr., Biochem. Biophys. Res. Commun., 28:815 (1967).

21. K. Weber and M. Osborn, J. Biol. Chem., 244:4406 (1969).

22. D. M. Neville, Jr., J. Biol. Chem., 246:6328 (1971).

23. J. A. Reynolds and C. Tanford, J. Biol. Chem., 245:5161 (1970).

24. M. S. Bretscher, Nature, New Biol., 231:229 (1971).

25. J. P. Segrest, R. L. Jackson, E. P. Andrews and V. T. Marchesi, Biochem. Biophys. Res. Commun., 44:390 (1971).

26. M. A. Fletcher and B. J. Woolfolk, Biochim. Biophys. Acta, 278:163 (1972).

27. W. M. Hunter and F. C. Greenwood, Nature, 194:495 (1962).

Chapter 8

THE GLUCAGON RECEPTOR IN PLASMA MEMBRANES PREPARED FROM RAT LIVER

Stephen L. Pohl

Department of Internal Medicine
Division of Endocrinology
New England Medical Center Hospital
Boston, Massachusetts

I. INTRODUCTION

The glucagon receptor in liver is an attractive receptor system for biochemical investigation for several reasons. Glucagon has been purified and its amino acid sequence established. It is a stable, relatively simple molecule, and is available in a highly purified form. Liver can be obtained easily in large quantities and does not present many of the technical difficulties posed by more fibrous or heterogeneous organs. Plasma membrane, the organelle which contains the glucagon receptor, can easily be purified in large quantity. Finally, a very early consequence of the glucagon-receptor interaction, stimulation of 3'5'-adenosine monophosphate (cAMP) formation can be observed readily.

The purpose of this chapter is to present the basic methodologic details for working with the glucagon receptor in rat liver plasma membrane. The properties of the system have been reported in detail [1-7] and several reviews are available [8-10]. No attempt will be made to review them further here.

II. METHOD FOR PURIFICATION OF PLASMA MEMBRANES FROM RAT LIVER

The desirability of working with purified plasma membranes has been appreciated from the time that it was first proposed that receptors linked to the cAMP might reside in this organelle. However, the practical problem of preparing a membrane fraction composed only of cell surface was most difficult, and membrane preparations used for early studies of hormone receptors undoubtedly were heavily contaminated with other organelles. Although several laboratories contributed to the solution of this problem, the efforts of Neville stand out [11-13]. His procedure for isolating plasma membranes from liver [11], without modification, provides an excellent preparation for study not only of the glucagon receptor, but also other cell-surface receptors such as those for insulin and growth hormone.

A. Principles of the Method

In order to develop the method, Neville used the bile canaliculus, a tubular structure formed by the plasma membrane of two adjacent liver cells, as an unequivocal morphologic marker [12]. Thus, the procedure purifies only the "bile front," a pair of membrane sheets joined by tight junctions and including the canaliculus. The "blood front" is discarded in the original procedure. However, as will be seen, this portion of the membrane can probably be obtained as well. Gentle homogenization with a Dounce homogenizer was chosen so that membrane sheets would not be fragmented. The Dounce homogenizer consists of a cylindrical pestle and glass rod with a ball on one end, which fits the pestle. The shear force created by forcing the ball to the bottom of the pestle disrupts tissue. It is now well-established that disrupted membrane sheets form small, closed, nondescript vesicles when more vigorous cell-breaking techniques are employed. After homogenization, stirring in a chilled hypotonic medium serves to lyse unbroken cells and to wash loosely adherent protein from the membranes.

Separation of plasma membranes from other particulate structures is accomplished by three centrifugation steps. The first is a conventional low-speed differential centrifugation which sediments the plasma membranes, as well as nuclei and some mitochondria and microsomes. The second centrifugation is at a higher speed in a discontinuous sucrose density gradient. Since the plasma membranes are less dense than mitochondria and nuclei, they float to the surface with a large number of vesicular membranes of uncertain origin. The third centrifugation is a low-speed rate-zonal sedimentation in a continuous sucrose density gradient. This step is based on the fact that sedimentation rate is a function of both size and density. The large, paired membrane sheets sediment to the sucrose "cushion" faster than the smaller vesicles. The density gradient prevents disruption of the zone of sedimenting membranes by convection.

B. Procedure

The procedure given here is nearly identical to the step-by-step description in Neville's original publication [11]. The only major difference is that the amounts have been scaled up to increase the yield. In addition, a few steps have been simplified. All steps are performed at 0-4°C.

Step 1: Decapitate 16 100 to 150-g male or female rats and excise the livers. Trim each liver free from connective tissue and place immediately in an iced beaker.

Step 2: Mince two livers with a scissors and transfer to a large Dounce homogenizer. Add 25 ml of 1 mM $NaHCO_3$ and homogenize with eight vigorous strokes, using a loose-fitting pestle. Pour the homogenate into a beaker containing 500 ml of cold 1 mM $NaHCO_3$. Repeat with two more livers and add the homogenate to the same beaker. Stir for 3 min and filter, first through two layers, and then four layers of cheesecloth. Pour half of the filtered homogenate into each of two large centrifuge bottles.

Step 3: Repeat step 2 with four more livers.

Step 4: Centrifuge the four bottles in a swinging-bucket rotor for 30 min at 1500 g_{max}. Carefully decant the supernatant and place the bottles containing the gelatinous, dark-red pellets on ice.

Step 5: Repeat steps 2 through 4 using the eight remaining livers.

Step 6: Pour the pellets, scraped loose if necessary, into the cleaned Dounce homogenizer and resuspend with three strokes of the loose pestle. Pour into a graduated cylinder and adjust the volume to 70 ml with distilled water.

Step 7: Pour 90 ml of warm 69% (wt/wt) sucrose into a 500-ml Erlenmeyer flask and chill ice-cold. Pour in the resuspended pellets and swirl until thoroughly mixed. Measure the sucrose concentration in an ABBE refractometer. Mix in additional H_2O or 69% sucrose as necessary to adjust the sucrose concentration to 44.0 ± 0.1% (wt/wt).

Step 8: Distribute the sucrose-homogenate mixtures equally among six 1 × 3.5 in. cellulose centrifuge tubes. Carefully overlay each with 10 ml 42.3 ± 0.1% (wt/wt) sucrose.

Step 9: Centrifuge 120 min at 25,000 rpm in a Beckman SW 27 rotor, brake on. Make certain that caps are well-greased.

Step 10: Remove the material which has floated to the surface, using a spatula or small spoon. Add 8 ml of 1 mM $NaHCO_3$ and resuspend by drawing the mixture through a No. 22 hypodermic needle into a syringe.

Step 11: Place a 5-ml cushion of 50% sucrose in each of six 1 × 3.5 in. centrifuge tubes. Over each form a 30-ml linear sucrose gradient from 24 to 1% (wt/wt).

Step 12: Overlay the gradients with 2 ml of resuspended float and centrifuge 60 min at 2000 rpm in a Beckman SW 27 rotor, brake off.

Step 13: Aspirate and discard the supernatant to ~2 cm above the cushion, using a syringe and No. 22 needle, and transfer into two 50-ml centrifuge tubes. Fill the tubes with 1 mM $NaHCO_3$ and mix well. Pack the membranes by centrifuging for 15 min at 15,000 g.

Three points deserve emphasis. First, the preparation must be kept ice-cold throughout. All solutions and rotors must be prechilled, and all work must be done in an ice bath, cold room, or refrigerated centrifuge. Second, the sucrose concentrations in steps 7 and 8 are critical and must be measured in a refractometer. Third, the pestle must be very loose in the Dounce homogenizer. To test for looseness, pull the pestle rapidly out of the cylinder. Little or no vacuum should be drawn. The author has obtained suitable homogenizers from the Corning Glass Company. With these precautions the method is highly reproducible.

C. Morphologic and Biochemical Characterization of Membranes

Both morphologic and biochemical criteria have been used to establish that the preparation described in Sec. II. B. contains plasma membranes, and to estimate the extent of contamination by other organelles. A detailed description of these methods is beyond the scope of this chapter, and only the general approaches and conclusions are summarized here. Neville and Kahn have recently published an excellent discussion of the theoretical and practical problems involved in plasma membrane characterization [13].

Examination of the Neville membrane preparation by electron microscopy [2, 12] reveals that it is composed almost entirely of membranes, and bile canaliculi are readily identified. A small number of vesicular membranes are present, but no structures of origin definitely other than plasma membrane are seen.

Using 5'-nucleotidase activity as a plasma membrane marker, the final membrane preparation is approximately 20-fold purified with respect to the total homogenate. Using glucose 6-phosphatase and succinate-cytochrome C reductase as markers, the preparation is contaminated ~2 and 1%, respectively, with microsomes and mitochondria [2].

If the final sucrose density gradient centrifugation step is omitted from the procedure, the membrane preparation contains large amounts of vesicular membranes, the origin of which cannot be established by morphologic criteria. In addition, a few mitochondria, but no nuclei or ribosomes, are present. On the basis of the same enzymatic markers, plasma membranes are purified 16-fold with respect to the crude homogenate and are contaminated ~3 and 7%, respectively, with microsomes and mitochondria. The yield of membrane protein is increased sixfold and the specific activity of adenylate cyclase is increased by about 50% [2]. From these data, it may be inferred that most of the material removed in the final centrifugation step is blood front plasma membrane. Since the yield of glucagon receptor is markedly greater and the increased contamination relatively small, many laboratories have chosen to use this simpler preparation, and it has been designated "partially purified" [2].

D. Storage of Membranes

Plasma membranes prepared as described may be stored for long periods of time without loss of either glucagon-stimulated adenylate cyclase activity or glucagon binding. These activities are stable for at least several months at -20°C and for at least two years in liquid nitrogen.

III. METHOD FOR ASSAY OF ADENYLATE CYCLASE ACTIVITY

It is now widely accepted that the first step in glucagon action upon liver is an increase in the rate of formation of cAMP [14]. Thus, by assaying adenylate cyclase activity under various conditions, an early effect of the interaction between glucagon and its receptor can be monitored. For several years following the discovery of cAMP, studies of this sort were hampered by the difficulty of the assay methods and the very low specific activity of adenylate cyclase in mammalian tissues. A major methodologic advance was made by Krishna et al. [15], who devised an assay based on the use of high specific activity radioactive ATP as substrate and a combination of ion-exchange and adsorption procedures for purification of radioactive cAMP formed in the reaction. The method described here is based primarily on the work of Krishna et al. [15]. However, a faster, simpler purification procedure using alumina, devised by Ramachandran [16], is used.

A. Incubation Conditions

The following conditions produce maximal activity of adenylate cyclase and are appropriate for routine studies of the glucagon receptor in rat liver plasma membrane [2]. The assay medium contains 3.2 mM [α-^{32}P]-ATP (~ 1 μCi per incubation tube); 5 mM $MgCl_2$; 1 mM EDTA; 1 mM cAMP; 25 mM tris-HCl, pH 7.6; an ATP-regenerating system consisting of 20 mM phosphocreatine and 1 mg/ml of creatinine phosphokinase; and 10 to 50 μg of membrane protein, in a final volume of 50 to 100 μl. All stock solutions are adjusted to pH 7.6 prior to preparation of the medium. Hormones may be added to the system mixed in the creatine phosphokinase solution or in 1% bovine serum albumin to prevent adsorption to glass surfaces. [α-^{32}P]-ATP is used as substrate to minimize the number of labeled products which could potentially be formed by the membranes. In addition, it is the only radioactive form of ATP presently available which is suitable for the alumina purification procedure (see Sec. III. B). Unlabeled cAMP is added to prevent destruction of labeled cAMP by phosphodiesterase; 10 mmol theophylline may also be used.

The reaction is initiated by addition of the membranes. The rate of cAMP formation is linear for 10 min at 37°C. Shaking is unnecessary. The reaction is terminated by adding 100 μl of a stopping solution containing 1% sodium dodecyl sulfate and 10 mmol [^{3}H]cAMP (~20,000 cpm per tube) as a recovery marker.

B. Purification of Labeled cAMP

A variety of chromatographic methods for purification of labeled cAMP are available for adenylate cyclase assays. However, in the author's experience, none approach the efficiency and simplicity of the procedure devised by Ramachandran [16]. It is based on the fact that alumina strongly adsorbs multivalent anions such as adenosine mono-, di-, and triphosphate, and inorganic phosphate, but not cAMP, which is univalent.

Approximately 2 ml of dry alumina (neutral, activity Grade I) is poured into a small column placed directly over a liquid scintillation vial. A plastic rack to hold the columns so that they will drip directly into vials, positioned in their shipping cartons, is very convenient for this purpose. The sample is transferred from the incubation tube to the top of the column with a Pasteur pipet. After the sample is adsorbed, 5 ml of 10 mM tris-HCl, pH 7.6, is added, and the eluate is collected in the counting vial. After addition of an appropriate scintillation solution, ^{3}H and ^{32}P are determined in a dual-channel liquid scintillation spectrophotometer. Counting standards for ^{3}H and ^{32}P are required, and an unincubated sample to establish the efficiency of the column in removing [^{32}P]ATP must be prepared.

With certain precautions, this method is 99.995% efficient in removing labeled substrate, and the yield of labeled cAMP is 80 to 100%. First, the [α-^{32}P]ATP must be of adequate radiochemical purity. Some batches contain a radioactive impurity which is not adsorbed by alumina. Each batch should be tested by passing a sample through an alumina column. If the unadsorbed counts are unacceptable, another supplier should be tried. Second, if the alumina is too dry, it may adsorb cAMP. If recovery is unacceptably low, it can be increased by wetting the alumina to a lower activity grade according to the instructions provided by the manufacturer.

C. Calculation and Expression of Results

After subtraction of the counts in the unincubated sample and correction for counting overlap, the ^{32}P counts in the sample are corrected for recovery using the ^{3}H counts. The corrected ^{32}P counts are then divided by the specific activity of the [^{32}P]ATP, yielding the molar quantity of cAMP formed. Results are usually expressed as nanomoles of cAMP formed in 10 min/mg of membrane protein, but these units are not standard.

IV. METHOD FOR LABELING GLUCAGON WITH RADIOIODINE

Activation of adenylate cyclase, while a very early consequence of the hormone-receptor interaction, is still at least one step removed from the initial interaction. An alternative approach to receptor study is to label the hormone and study its interaction with the receptor directly. However, the number of receptors in a tissue is relatively small, and hormones must, therefore, be labeled to very high specific activity. This is most easily accomplished with radioiodine. The following procedure, a variation of the Hunter and Greenwood procedure [17], produces biologically active [^{125}I]glucagon with a specific activity of 2.2 Ci/μmol, an average of one atom of ^{125}I per molecule of glucagon [4].

A. Reagents

Buffer: 0.6 M NaH_2PO_4-Na_2HPO_4, pH 7.4

Albumin: 1% bovine serum albumin in 10 mmol tris-HCl, pH 7.4

Cellulose powder: 50% (vol/vol) suspension in water

$Na^{125}I$: carrier free, 2.5 to 5 mCi in 5 to 20 μl, 0.1 N NaOH

Glucagon: 3.5 mg/ml, in 0.01 N NaOH

Chloramine-T: 3.5 mg/ml in buffer

Sodium metabisulfite: 2 mg/ml in H_2O

Ethanol: 50% in 10 mM tris-HCl, pH 7.4

B. Procedure

All steps are performed at room temperature. The column of cellulose used for separating labeled glucagon from reagents, unreacted iodide, and damaged peptide is first prepared by adding 1 ml of the cellulose suspension to a Pasteur pipet plugged with glass wool. The slow rate of flow of fluid through this column can be increased by applying pressure with a syringe and rubber tubing and should be 0.5 to 1 ml/min. After the bed of cellulose has formed, it is washed with 1 ml of water and then with 1 ml of albumin solution. It may then be left with a thin film of liquid above the bed while the iodination is performed.

The stock glucagon is diluted 10-fold in buffer and 10 μl of this solution are added to a small test tube containing the ^{125}I. Chloramine-T, 10 μl of a freshly prepared solution, is then added and the reagents are mixed rapidly with a vortex mixer. Immediately (< 5 sec), 50 μl of sodium metabisulfite solution are added, and 200 μl of the mixture are applied to the cellulose column. The remainder of the reaction mixture is used for preparation of counting standards in order to estimate yield. The column is eluted successively with 2 ml of albumin solution, 1 ml of H_2O, and 0.5 ml of 50% ethanol. The biologically active [^{125}I]glucagon adsorbs strongly to cellulose and elutes in the 50% ethanol fraction, which is immediately placed on ice.

C. Chemical and Biological Characterization of [^{125}I]Glucagon

Glucagon labeled with radioiodine by the method described in Sec. IV. B has been characterized for both chromatographic behavior and for biological activity [4]. Chromatography is complicated by the strong adsorption of glucagon to surfaces, particularly chromatographic support media. However, this problem can be overcome by using organic solvent systems or by adding 6 M urea to aqueous solvent systems.

A thin layer chromatographic system has been applied successfully to [^{125}I]glucagon. Small samples of labeled and unlabeled glucagon are applied to 20 × 20-cm glass plates precoated with microcrystalline cellulose powder. The plates are then developed by ascending flow using butanol:pyridine:water:glacial acetic acid (30:20:24:6). The unlabeled material can be detected by staining with ninhydrin, and the radioactive glucagon can be detected by autoradiography. [^{125}I]Glucagon prepared as described cochromatographs with unlabeled glucagon in this system. Ion-exchange chromatography of labeled glucagon has been performed on columns of DEAE cellulose equilibrated with 0.01 M tris-HCl, pH 7.6, 6 M urea. The column is eluted with a linear gradient formed from equal volumes of 0.01 and 0.5 M tris-HCl, pH 7.6, both containing 6 M urea.

[^{125}I]Glucagon elutes as a single peak, which is biologically active. However, the ionic strength at which glucagon elutes has not been determined, nor has a comparison with the chromatographic behavior of unlabeled glucagon in this system been made. Bromer et al. [18-20] have published extensively on chromatographic and electrophoretic procedures for glucagon.

Biological activity of labeled glucagon can be established very easily by testing its ability to stimulate adenylate cyclase activity in liver plasma membranes using the methods described in Secs. II and III [4]. Using the method described in the Sec IV. B, 1 nmol of glucagon is iodinated and eluted from the cellulose column in 0.5 ml of liquid, giving a maximal theoretical concentration of 2×10^{-6} M. Glucagon stimulates adenylate cyclase activity in the liver membrane preparation at concentrations as low as 2×10^{-10} M. Half-maximal and maximal stimulation are observed at 4×10^{-9} and 10^{-7} M, respectively [2]. Consequently, this assay system has adequate sensitivity to detect biological activity of [^{125}I]glucagon directly in the column eluate. It is important to note, however, that the column eluate contains ethanol and that ethanol stimulates adenylate cyclase activity in liver plasma membrane [21]. Consequently, the [^{125}I]-glucagon solution must be diluted at least 10-fold, so that the concentration of ethanol in the adenylate cyclase assay will be less than 5%. Under these conditions, no artifact due to ethanol is observed. The concentration of glucagon may then be estimated by comparing the stimulation of adenylate cyclase by labeled glucagon with a simultaneously performed glucagon dose-response curve. Using the iodination conditions described, the recovery of glucagon-like biological activity is 70 to 100% [4]. The incorporation of radioiodine into the peptide is usually 40 to 60%. Consequently, using a molar ratio of glucagon to radioiodide of 1:2 produces a labeled peptide with an average of 1 radioiodine atom per molecule of glucagon. This estimate, however, must be considered only approximate, since Bromer et al. have established that monoiodoglucagon is significantly more biologically active than native glucagon [22, 23]. Nevertheless, [^{125}I]glucagon is by all criteria an adequate tracer for the study of binding of glucagon to liver plasma membranes.

D. Storage of Labeled Glucagon

After two weeks of storage at -20°C in 50% ethanol, glucagon loses approximately 25% of its biological activity. In addition, a new labeled peptide with relative mobility greater than glucagon on thin layer chromatography appears. This impurity can be removed and the preparation restored to its original specific radioactivity by repurification on a cellulose column as described in Sec. IV. B. The labeled glucagon is much more stable if diluted 1:100 in 1% albumin and stored frozen at -20°C.

V. MEASUREMENT OF BINDING OF LABELED GLUCAGON TO PLASMA MEMBRANES

The initial interaction between a hormone and its receptor intuitively must be a physical contact of sufficient proximity that the forces exerted by the hormone upon the receptor can initiate the next event in hormone action. If the duration of this contact or binding is sufficiently long to permit separation of bound from free hormone it should be possible to observe this initial interaction directly through the use of radiolabeled hormone. Binding of $[^{125}I]$glucagon to rat liver plasma membranes can be observed readily by very simple techniques [4]. In addition, the binding demonstrates properties of sensitivity and specificity which are appropriate for the effects of glucagon upon liver [4]. However, the time course of binding is inappropriately slow and the binding is not completely displaced by des-His-glucagon, an effective competitive inhibitor of glucagon stimulation of adenylate cyclase activity [24]. For these reasons, there is some doubt that the observed binding of glucagon to liver plasma membranes represents the initial step of glucagon action on this organ [25]. Therefore, the following methods are presented, with the caution that the reader must decide for himself whether glucagon binding represents a true interaction with the receptor, and, if not, what this binding might represent.

A. Incubation Conditions

The binding of glucagon to liver plasma membranes is critically dependent on composition of the incubation medium [5]. However, binding can be observed in a very simple medium which can then be made more complex, depending on the purpose of the binding studies. An appropriate buffer is 25 mM tris-HCl, pH 7.6. Bovine serum albumin at 0.1 to 1% is added to minimize absorption to test tube walls and other surfaces. Incubation volume may vary considerably depending upon the separation method employed. Typically a volume of 0.1 ml is used for the centrifugation method, and 4.0 ml for the filtration method described in Sec. V. B. Regardless of the incubation volume, the total amount of plasma membranes used should vary from 10 to 50 μg, since binding is usually directly proportional to the amount of membranes only within this range. The amount of $[^{125}I]$glucagon used is usually 10^6 cpm, approximately 1 pmol of monoiodoglucagon prepared as described in Sec. IV. Incubation temperature may vary from 20 to 40°C with little difference in the amount of labeled glucagon finally bound. However, incubation at 4°C results in binding to only ~25% the extent seen at 37°C. The amount of glucagon bound to the membranes increases progressively for 10 to 30 min of incubation, depending on temperature and membrane concentration. Once the maximum binding is reached, the amount of glucagon bound remains constant for a period of several hours.

B. Separation of Bound from Free Hormone

Since the plasma membrane preparation is composed of relatively large sheets and vesicles, separation of [^{125}I]glucagon bound to the membranes from free glucagon can be accomplished by either sedimentation or filtration of the membranes. However, with both these methods, the strong adsorption of glucagon to glass and plastic surfaces must be prevented. This is accomplished by selecting appropriate vessels and filters, and by incorporating an inert protein such as bovine serum albumin into all solutions.

A microfuge (Beckman Instruments Co.) is very convenient for separation by centrifugation. The tubes for this instrument are conical, plastic centrifuge tubes with a volume of ~0.4 ml. The tubes are filled with 0.3 ml of 25 mM tris-HCl, pH 7.6, with 1% bovine serum albumin. The binding incubation mixture of 0.1 ml, or less, is transferred quantitatively into the top of the tube, taking care not to mix it with the layer of albumin solution. Alternatively, an aliquot 0.1 ml or less of a larger incubation volume may be transferred into the tube. The tubes are then centrifuged for 5 min at machine speed, ~10,000 g. The membranes sediment through the albumin layer, leaving the incubation mixture containing the unbound [^{125}I]glucagon near the top of the tube. A hypodermic needle of gauge 18 to 22 attached to a vacuum line is then used to remove the supernatant from the tube. A 1-1/2-in. hypodermic needle, inserted fully into the tube, will usually remove all but the last millimeter or so of fluid. Care must be taken not to disturb the pellet on the bottom of the tube. The tubes are then filled with distilled water and aspirated again to remove the last traces of fluid from the bottom of the tube and to rinse the walls. The tip of the tube is then cut with a razor blade 1 to 2 mm above the membrane pellet, and the tube tip is placed in a test tube for counting in a gamma-scintillation counter. Controls containing no membranes or boiled membrains contain less than 0.1% of the total counts added in the tip of the centrifuge tube if this method is applied correctly.

Liver plasma membranes are retained quantitatively by 0.4-μ Millipore filters. However, ordinary cellulose nitrate filters also strongly adsorb the [^{125}I]glucagon and consequently cannot be used for separation of bound and free glucagon in studies of this type. Glass fiber filters (Whatman), soaked briefly in the tris-albumin buffer, adsorb only a very small percent of [^{125}I]glucagon filtered through them and may be used for binding studies. After incubation, the binding reaction mixture, usually in a volume of 1 ml or greater, is simply passed through the filter, which is then rinsed with a few milliliters of water. The filter is then transferred to a test tube for counting in a gamma-scintillation counter. A variety of multiple filter supports are available for this procedure.

C. Use of Glucagon Binding to Liver Plasma Membranes for Assay of the Hormone in Biological Fluids

The radioimmunoassay for glucagon, despite 15 years of development, remains plagued by several problems [26]. These include: destruction of glucagon by proteases in plasma, difficulty in generating antibodies with sufficient binding energy, and the presence in plasma of a material from the intestine which cross-reacts with most antiglucagon antisera. Because of the high degree of specificity and affinity of the binding of glucagon to liver plasma membranes [4], it appeared that this system could be developed into a competitive protein-binding assay for glucagon which might offer advantages over the convential radioimmunoassay. Using the procedures described, unlabeled glucagon competes effectively with [^{125}I]glucagon for binding beginning at a level of 10^{-10} M [4]. However, basal glucagon concentration in human peripheral blood is $\sim 10^{-11}$ M [27]. Only under unusual circumstances, such as diabetic ketoacidosis, does glucagon concentration exceed 10^{-10} M. The sensitivity of this system to unlabeled glucagon can be increased by incubating membranes with the unlabeled glucagon only for 15 min at 22°C. The labeled glucagon is then added and incubation continued for 30 min, after which bound and free glucagon are separated by filtration as described. Under these conditions, significant displacement of labeled glucagon has been seen with unlabeled glucagon concentrations as low as 10^{-12} M [28]. Consequently, the binding can be made sensitive enough for an assay of glucagon in biological fluids.

Initial studies of cross-reactivity with gut glucagon showed no cross-reactivity [28]. However, extracts from rat intestine were used for these studies. Rat is not a usual source for gut glucagon, and no attempts were made in these studies to demonstrate cross-reactivity of the extract with an antiglucagon antiserum. Bataille et al. subsequently showed that extracts of porcine intestine compete effectively with [^{125}I]glucagon for binding liver plasma membrane [29]. Consequently, it appears that the liver membrane binding system offers no increased specificity of binding over presently available antiglucagon antisera. In addition, there is a nonspecific effect of human plasma which decreases the total binding of [^{125}I]glucagon by about 50% and decreases the sensitivity to displacement of unlabeled glucagon by approximately 10-fold [30].

In summary, the liver membrane system should provide an excellent method for measurement of glucagon in incubation media in vitro. However, in its present form, it is not suitable for assay of glucagon in plasma.

VI. ESTIMATION OF INACTIVATION OF GLUCAGON BY PLASMA MEMBRANE

In addition to the glucagon-sensitive adenylate cyclase activity and glucagon-binding sites, the liver plasma membrane preparation contains at least one enzyme which alters glucagon in such a way that it neither activates adenylate cyclase nor binds to the membranes [31]. No means have yet been found for inactivating this enzyme or separating it from the plasma membrane preparation. Consequently, during the course of incubation, for either adenylate cyclase or binding, glucagon is simultaneously inactivated. In order to perform quantitative kinetic studies of binding for adenylate cyclase activation, this change in medium glucagon concentration due to inactivation must be measured and taken into account. A detailed description of various methods for estimation of hormone inactivation in receptor preparations has recently been published [32]. In general, the simplest approach is to incubate glucagon with liver plasma membranes for a period of time, remove the membranes by centrifugation, and then assay the supernatant for ability of remaining glucagon either to activate adenylate cyclase or to bind to fresh plasma membranes. Appropriate controls must be prepared and the correction must be made for the amount of glucagon bound to the membranes in the initial incubation [32].

VII. CONCLUSION

The liver plasma membrane preparation is a convenient and relatively simple system for receptor studies. Both adenylate cyclase activation and glucagon binding can be used as methods for observing the hormone-receptor interaction in this preparation. All evidence to date indicates that adenylate cyclase activation is an unequivocal expression of the hormone-receptor interaction but is at least one step removed from the receptor. Glucagon binding is potentially a much more direct way of observing the hormone-receptor interaction. However, some doubt remains that the observed binding of glucagon to plasma membranes actually represents a step in the sequence of glucagon action on liver. Further studies of the binding of glucagon to liver plasma membranes must be interpreted with caution until this doubt is resolved or methods are developed for observing the biologically significant binding only.

ACKNOWLEDGMENTS

This work was supported in part by National Institutes of Health grant AM-16826. The author is the recipient of Research Career Development Award AM-000351.

REFERENCES

1. S. L. Pohl, L. Birnbaumer, and M. Rodbell, Science, 164:566 (1969).
2. S. L. Pohl, L. Birnbaumer, and M. Rodbell, J. Biol. Chem., 246: 1849 (1971).
3. L. Birnbaumer, S. L. Pohl, and M. Rodbell, J. Biol. Chem., 246: 1857 (1971).
4. M. Rodbell, H. M. J. Krans, S. L. Pohl, and L. Birnbaumer, J. Biol. Chem., 246:1861 (1971).
5. M. Rodbell, H. M. J. Krans, S. L. Pohl, and L. Birnbaumer, J. Biol. Chem., 246:1872 (1971).
6. M. Rodbell, L. Birnbaumer, S. L. Pohl, and H. M. J. Krans, J. Biol. Chem., 246:1877 (1971).
7. S. L. Pohl, H. M. J. Krans, V. Kozyreff, L. Birnbaumer, and M. Rodbell, J. Biol. Chem., 246:4447 (1971).
8. S. L. Pohl, L. Birnbaumer, and M. Rodbell, in Annual Reports in Medicine and Chemistry, 1970 (C. K. Cain, ed.), Academic Press, New York, 1971, p. 233.
9. S. L. Pohl, in The Role of Membranes in Metabolic Regulation (M. A. Mehlman and R. W. Hanson, eds.), Academic Press, New York, 1972, p. 349.
10. J. A. Avruch and S. L. Pohl, in Biological Membranes (D. Chapman and D. F. H. Wallach, eds.), Academic Press, New York, 1973, p. 185.
11. D. M. Neville, Jr., Biochim. Biophys. Acta, 154:540 (1968).
12. D. M. Neville, Jr., J. Biochem. Biophys. Cytol., 413:000 (1960).
13. D. M. Neville, Jr. and C. R. Kahn, in Methods in Molecular Biology, Vol. 5, Dekker, New York, 1974, p. 57.
14. C. R. Park and J. H. Exton, in Glucagon (P. T. Lefebvre and R. H. Unger, eds.), Pergammon Press, New York, 1972, p. 77.
15. G. Krishna, B. Weiss, and B. B. Brodie, J. Pharmacol. Exptl. Ther., 163:379 (1968).
16. J. Ramachandran, Anal. Biochem., 43:227 (1971).
17. W. M. Hunter and F. C. Greenwood, Nature, 194:495 (1962).
18. W. W. Bromer, M. E. Boucher, and J. F. Koffenberger, Jr., J. Biol. Chem., 246:2822 (1971).

19. W. W. Bromer, M. E. Boucher, J. M. Patterson, A. H. Pekar, and B. H. Frank, J. Biol. Chem., 247:2581 (1972).

20. J. M. Patterson and W. W. Bromer, J. Biol. Chem., 248:337 (1973).

21. R. E. Gorman and M. W. Bitensky, Endocrinology, 87:1075 (1970).

22. R. W. Fuller, H. P. Anoddy, and W. W. Bromer, Mol. Pharmacol., 8:345 (1972).

23. W. W. Bromer, M. E. Boucher, and J. M. Patterson, Biochem. Biophys. Res. Commun., 53:134 (1973).

24. L. Birnbaumer and S. L. Pohl, J. Biol. Chem., 248:2056 (1973).

25. L. Birnbaumer, S. L. Pohl, and A. Kauman, in Advances in Cyclic Nucleotide Research, Vol. 4, Raven Press, New York, 1974, p. 239.

26. G. R. Faloona and R. H. Unger, in Methods of Hormone Radioimmunoassay (B. M. Jaffe and H. R. Behrman, eds.), Academic Press, New York, 1974, p. 317.

27. R. H. Unger, in Glucagon (P. J. Lefebvre and R. A. Unger, eds.), Pergammon Press, New York, 1972, p. 205.

28. S. L. Pohl and L. R. Chase, Proc. IV Int. Congr. Endocrinol., Excerpta Medica, Amsterdam, 1972.

29. D. P. Bataille, P. Freychet, P. E. Kitabgi, and G. E. Rosselin, FEBS Lett., 30:215 (1973).

30. S. L. Pohl, unpublished observations.

31. S. L. Pohl, H. M. J. Krans, L. Birnbaumer, and M. Rodbell, J. Biol. Chem., 247:2295 (1972).

32. S. L. Pohl and O. B. Crofford, in Methods in Enzymology, Vol. 37: Hormone Action, Academic Press, New York, 1975.

Chapter 9

RECEPTORS FOR GONADOTROPIC HORMONES

Kevin J. Catt
Jean-Marie Ketelslegers
Maria L. Dufau

Section on Hormonal Regulation
Reproduction Research Branch
National Institute of Child Health and Human Development
National Institutes of Health
Bethesda, Maryland

I. INTRODUCTION

The presence of high-affinity receptor sites for luteinizing hormone (LH) and follicle-stimulating hormone (FSH) in the testis and ovary has been demonstrated by localization and binding studies with gonadotropins labeled with fluorescent groups, ferritin, tritium, and radioactive iodine

[1-15]. The detailed characterization and analysis of gonadotropin receptors in target cells and subcellular fractions has been made possible by the use of tritiated or radioiodinated preparations of LH, human chorionic gonadotropin (hCG), and FSH, which retain biological activity and exhibit specific interaction with the respective receptor sites in testis and ovary [5-28]. The LH receptors of the rat testis and ovary have been shown to possess common specificity for LH and the chorionic gonadotropins of human, primate, and equine origin; and labeled hCG has been widely utilized for binding studies of the gonadal receptors for luteinizing hormone in the testis and ovary of several mammalian species. Of the radionuclides employed for labeling gonadotropin and other peptide hormones, ^{125}I has several practical advantages and has been most commonly employed for the preparation of tracer hormones for receptor-binding studies. Quantitative binding analysis of gonadotropin receptors in target tissues can only be performed if the radioactive hormone is appropriately labeled and adequately characterized. The type of gonadal tissue preparation used for gonadotropin-binding studies has ranged from tissue slices to whole cells and a variety of partially purified subcellular fractions. Due to the heterogeneous cellular composition of the gonads, relatively few binding studies have been performed with highly purified membrane fractions from the testis or ovary. The biological properties of gonadotropin receptors have been extensively defined during studies on particulate subcellular binding fractions. More recently, a number of procedures for solubilization of gonadotropin receptors has been described, and detergent-extracted binding particles have been utilized for studies on the physico-chemical properties of the LH/hCG receptor site. The complete isolation and structural analysis of a peptide hormone receptor has not yet been achieved, though such information appears likely to be derived for the cholinergic receptor within the near future. Most peptide hormone receptors are present in rather low concentrations in their target tissues, and will require extensive purification to achieve formal isolation of the receptor site.

II. PREPARATION AND CHARACTERIZATION OF LABELED GONADOTROPINS

Radioactive gonadotropins have been prepared by tritiation and iodination with ^{131}I or ^{125}I. The most satisfactory preparations for receptor-binding studies have been derived by labeling with ^{125}I to a moderately low specific activity, usually to a level of one atom of the radionuclide per molecule of gonadotropin. Selective isolation of the monoiodinated species cannot be performed after labeling the relatively large and acidic gonadotropin molecules with ^{125}I or ^{131}I, and procedures for iodination of gonadotropins have usually been designed to provide an average substitution ratio of one iodine atom per molecule of hormone. Although the chloramine-T method for radioiodination of proteins has been applied to the preparation of labeled hLH and hCG for receptor-binding studies, this procedure is less satisfactory for iodination of FSH and certain LH preparations. For this reason, the lactoperoxidase method for radioiodination has proven to be more suitable for general use in tracer preparation for receptor-binding studies. However, relatively little difference in the receptor-binding properties of the more stable glycoprotein hormones such as hCG can be detected after labeling by either procedure.

A. Radioiodination by the Chloramine-T Method

The conditions employed for preparation of tracer peptides for radioimmunoassay are usually unsuitable for radioiodination of glycoprotein hormones for receptor-binding studies. For the latter purpose, the concentration of chloramine-T should be reduced to the lowest effective level; the reaction should be performed for a short period at low temperature, and the addition of reducing agent should be avoided where possible [5]. The following conditions have been found to be suitable for radioiodination of hCG and hLH by the chloramine-T method. All solutions are prepared in 0.05 M phosphate buffer, pH 7.4, unless otherwise indicated. The iodination reaction is performed in a 5×75-mm glass tube kept in crushed ice:

1. hCG or LH solution (1 mg/ml): 25 μl
2. 0.5 M phosphate buffer, pH 7.4: 50 μl
3. Carrier-free $Na^{125}I$: 1 mCi
4. Freshly prepared chloramine-T (1 mg/ml): 25 μl

After gentle shaking for 20 to 30 sec, the reaction is stopped by addition of 25 μl of sodium metabisulfite solution (2 mg/ml). The iodination mixture is then applied to a column of Sephadex G-50 or G-100, Bio-Gel P-60, powdered cellulose [5], or agarose-Concanavalin A, for isolation of the labeled glycoprotein. To avoid the use of the reducing agent, the reaction can be terminated by transferring the iodination mixture directly to a gel column for purification of the labeled hormone. Alternatively, the reducing agent can be followed immediately by addition of a carrier protein such as BSA, to protect the labeled hormone and facilitate transfer to the column. Obviously, addition of protein carriers should not be performed unless the iodination reaction has been terminated, when the subsequent separation step is to be performed by gel filtration. However, the presence of extraneous labeled protein in the front peak is of no consequence when purification is to be performed by group-specific adsorption chromatography on Sepharose-Concanavalin A. The use of additional protein in the iodination mixture should also be avoided when purification is to be performed by adsorption to cellulose followed by subsequent elution by protein solution [5].

B. Radioiodination by the Lactoperoxidase Method

A more generally satisfactory procedure for preparation of labeled gonadotropins for binding studies is based upon the lactoperoxidase method for radioiodination of proteins [29, 30]. This labeling procedure can be applied to hFSH and prolactin, as well as to LH and hCG, and results in a higher proportion of biologically active molecules in the tracer hormone preparation:

1. hCG, LH, or FSH (1 mg/ml): 25 μl
2. 0.5 M phosphate buffer, pH 7.0: 25 μl
3. Carrier-free $Na^{125}I$: 1 mCi
4. Lactoperoxidase (100 μg/ml): 10 μl
5. H_2O_2 (1:10^5) (i.e., 1:30,000 diln. of 30% H_2O_2): 10 μl (= 100 ng)

After mixing and standing for 2 min at room temperature, the reaction is stopped by dilution and/or addition of reducing agents such as 5 mM cysteine, 5 mM mercaptoethanol, or 1 mM dithiothreitol. Because reduction of gonadotropin disulfide bonds is known to cause impaired binding to gonadal receptors [31], the use of reducing agents to terminate the iodination reaction for glycoprotein hormones should be avoided where possible. When purification is to be performed by group-specific affinity chromatography, 0.5 ml of BSA solution (1 mg/ml in 0.05 M phosphate-buffered saline, pH 7.4) can be added instead of the reducing agent, and the entire reaction mixture is transferred to a 0.5 × 14-cm column of Sepharose-Concanavalin A. In this case, the front peak of labeled material will include [^{125}I] BSA, as well as free iodide and damaged gonadotropin, and the labeled hormone is eluted later with manno- or glucopyranoside solution (see Sec. II. C).

C. Purification of Radioiodinated Glycoprotein Hormones

After labeling of peptide or glycoprotein hormones by either method of radioiodination, the radioactive hormone can be isolated from the reaction mixture by a number of purification procedures. Gel filtration of the iodination mixture on Sephadex G-50 provides an adequate separation of labeled protein from free iodine, but does not resolve "damaged" hormone from the intact labeled molecules. If gel filtration is performed, more effective fractionation of the labeled tracer is achieved by chromatography on a 100-cm column of Sephadex G-100. Adsorption and elution performed on cellulose powder has also been applied to purification of tracer hormone, with partial separation of labeled hormone from components damaged during the iodination procedure. A more selective method for purification of radioiodinated glycoprotein hormones has been based upon the use of Concanavalin A coupled to agarose beads, to achieve group-specific affinity chromatography of glycoprotein hormones [32]. During this separation technique, the labeled glycoprotein is bound to a small column of Sepharose-Concanavalin A, while free iodide, damaged components, and nonglycoproteins pass freely through the gel bed. After washing the gel with carbohydrate-free buffer, the labeled glycoprotein is eluted by a solution of 0.2 M glucopyranoside or mannopyranoside (Fig. 1). This procedure is less rapid than gel filtration or cellulose adsorption-elution, but provides tracer hormone which exhibits high binding activity and low nonspecific binding to gonadal receptor preparations. The Sepharose-Concanavalin A method is also of value to ensure the removal of desialylated molecules if present in the labeled preparation, since asialoglycoproteins are strongly adsorbed to the solid-phase lectin and are not eluted by glucose or mannose concentrations which displace the

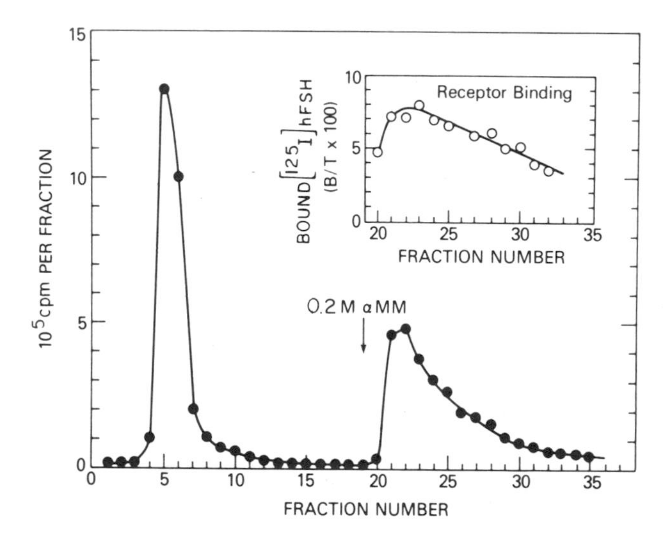

FIG. 1. Purification by group-specific affinity chromatography on agarose-Concanavalin A of [^{125}I]hFSH prepared by lactoperoxidase iodination. The labeled gonadotropin hormone is eluted as the second peak of radioactivity by buffer containing 0.2 M methyl-mannopyranoside. The insert shows the receptor binding of each fraction from the peak of labeled hormone.

fully sialylated hormone [32]. For these reasons, the combination of lactoperoxidase radioiodination and chromatographic purification on Sepharose-Concanavalin A provides a highly effective procedure for the preparation of labeled glycoprotein hormone for receptor-binding studies. Further fractionation of gonadotropin tracer purified by the Concanavalin A method, by gel filtration on Sephadex G-100, does not provide additional resolution of the labeled preparation or improvement of the properties of active hormone in terms of receptor binding (Fig. 2).

The most highly selective procedure for purification of labeled gonadotropic hormones is based upon the finding that tracer hormone bound to gonadal receptor sites displays full biological activity after elution from specific receptors by reduction of pH [33, 34]. The ability of receptors

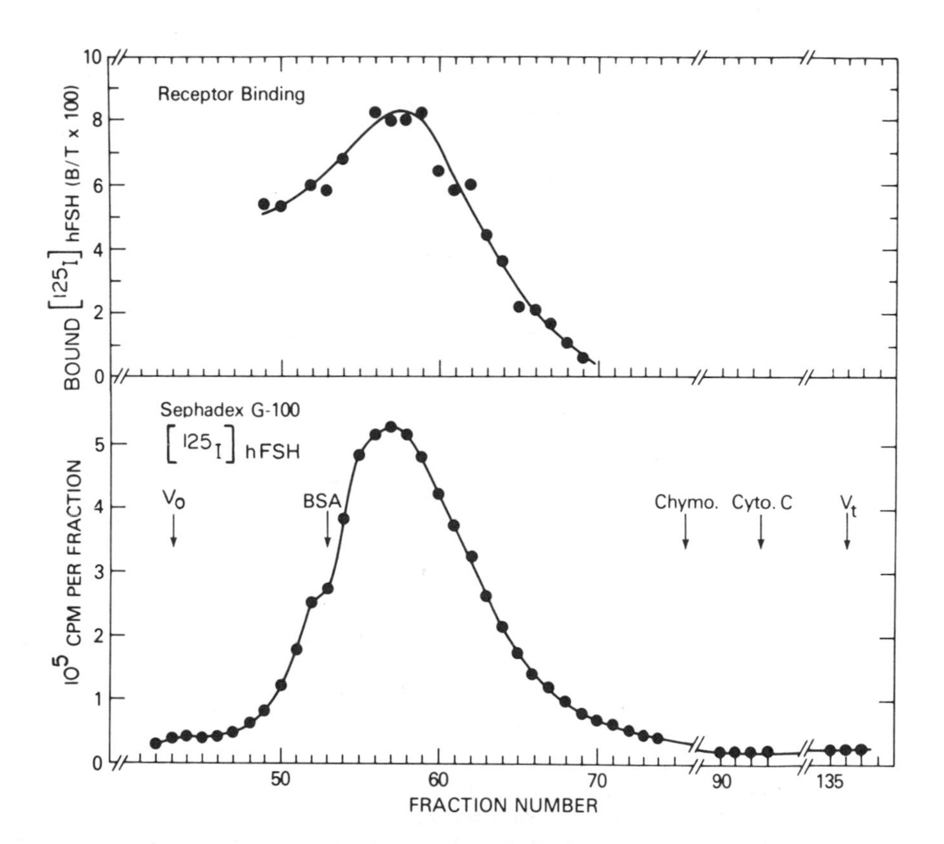

FIG. 2. Gel filtration of [^{125}I]hFSH purified by chromatography on agarose-Concanavalin A. The binding activity of the fractions eluted from the 90-cm column of Sephadex G-100 are shown in the upper panel.

to combine with the most biologically active labeled molecules can be utilized to extract the active species from radioiodinated hormone preparations containing a mixture of native and inactivated gonadotropin molecules. This method has been applied to purification of ^{125}I-labeled hCG and hFSH tracers prepared by the lactoperoxidase method, and yields radioactive hormones with considerably enhanced binding activity and biological activity (Fig. 3). The improvement of binding activity obtained

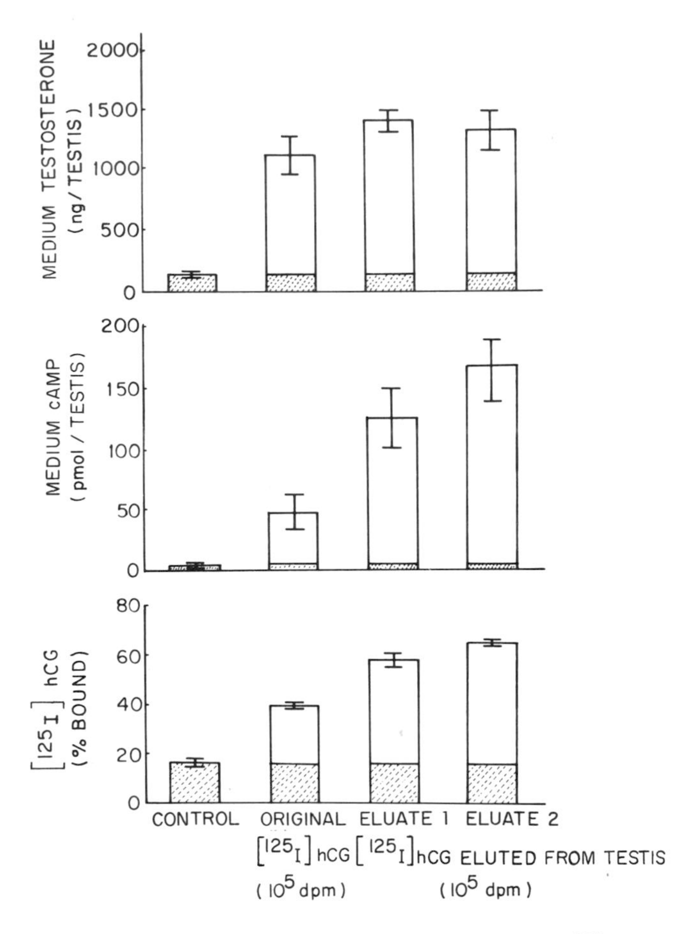

FIG. 3. Binding properties and biological activities of [^{125}I]hCG tracer before and after purification by elution from testicular binding sites at pH 3. The two acid-eluted fractions show significantly higher binding activity and enhanced ability to stimulate cAMP production by testis tissue in vitro. Production of testosterone was maximally stimulated by the original tracer, and no additional steroid response was induced by the acid-eluted purified hormone fractions.

by this procedure is particularly notable with [^{125}I]hFSH, which usually displays relatively low specific binding (~ 10-15%) to testis tubule homogenates. After purification by affinity chromatography upon gonadal receptor sites, binding activity is enhanced to the extent that up to 30% of the acid-eluted tracer is bound specifically during subsequent incubation with testis receptors (Fig. 4). In the presence of excess receptor sites, up to 40 to 50% of the acid-eluted tracer can be specifically bound to the receptor-binding sites. This purification method provides a relatively low yield of the biologically active tracer hormone (about 5% of the starting radioactivity) but the severalfold increase in binding activity is of considerable value for quantitative studies during determination of binding

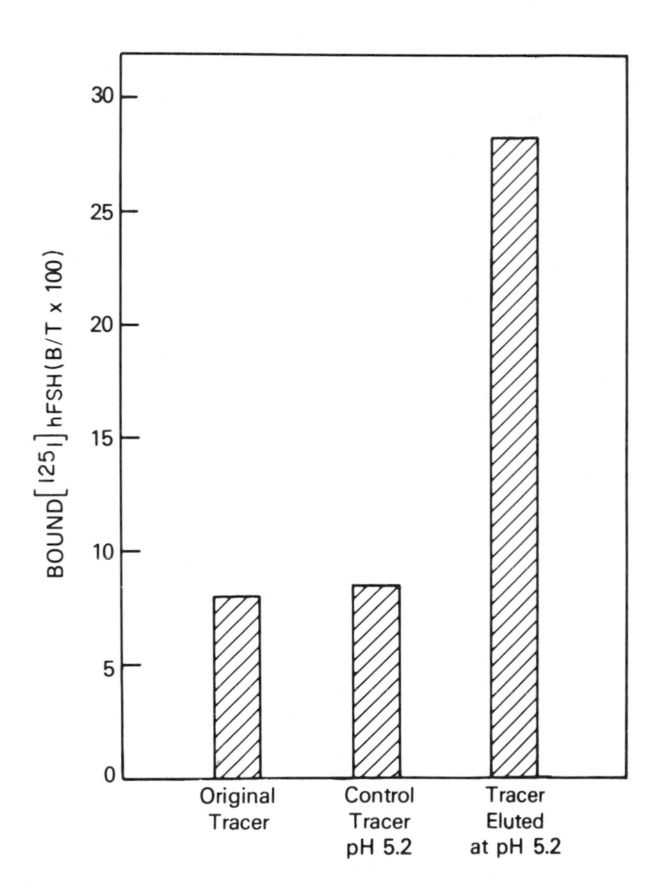

FIG. 4. Purification of [^{125}I]hFSH by elution from testicular receptor sites at pH 5.2. The eluted hormone exhibits a severalfold increase in binding activity to fresh testis homogenate, by comparison with the original tracer and control tracer treated at pH 5.2.

constants and receptor concentrations in gonadal target tissues. For routine binding studies and radioligand-receptor assay, gonadotropin tracers purified after lactoperoxidase radioiodination by gel filtration or group-specific chromatography on Sepharose-Concanavalin A are usually of satisfactory binding properties.

D. Characterization of Labeled Hormones

The majority of receptor-binding studies in the testis and ovary have been performed with radioiodinated hormones, most frequently prepared by labeling with ^{125}I using the methods described in Sec. II. A and B. The individual gonadotropins exhibit marked differences in susceptibility to damage during the radioiodination procedure, and frequently show significant loss of biological activity while retaining the capacity to react with specific antisera during radioimmunoassay. For receptor-binding experiments, it is essential that highly purified protein hormones are employed for radioiodination. In addition, the labeling procedure should not affect the biological properties of the hormone, and the radioactive hormone should be stable during binding studies and storage. Because purified gonadotropins are not usually of maximum attainable biological activity to begin with, their radioiodinated derivatives frequently contain a proportion of labeled inactive molecules, for which corrections should be made during the calculation of receptor-binding constants. In addition, the specific activity of the labeled hormone should be carefully determined in terms of the biological activity of the tracer preparation. For these reasons, the specific radioactivity and receptor-binding properties of each labeled gonadotropin preparation should be evaluated prior to use of the tracer for quantitative analysis of receptor-binding characteristics.

1. Evaluation of Physical Properties of Labeled Gonadotropins

Retention of the original physico-chemical properties of gonadotropins employed for radioiodination or tritiation should be evaluated by gel filtration and polyacrylamide gel electrophoresis (Fig. 5). Where appropriate, further characterization can be achieved by isoelectric focusing or density gradient centrifugation. An example of the analysis of radioiodinated hCG by isoelectric focusing is illustrated in Figure 6, and demonstrates that the charge of the monoiodinated hormone is not different from that of the unlabeled gonadotropin. Such analytical procedures need not be performed on every batch of the labeled hormone, but are of value during the development and initial application of labeling procedures for preparation of radioactive tracer gonadotropins.

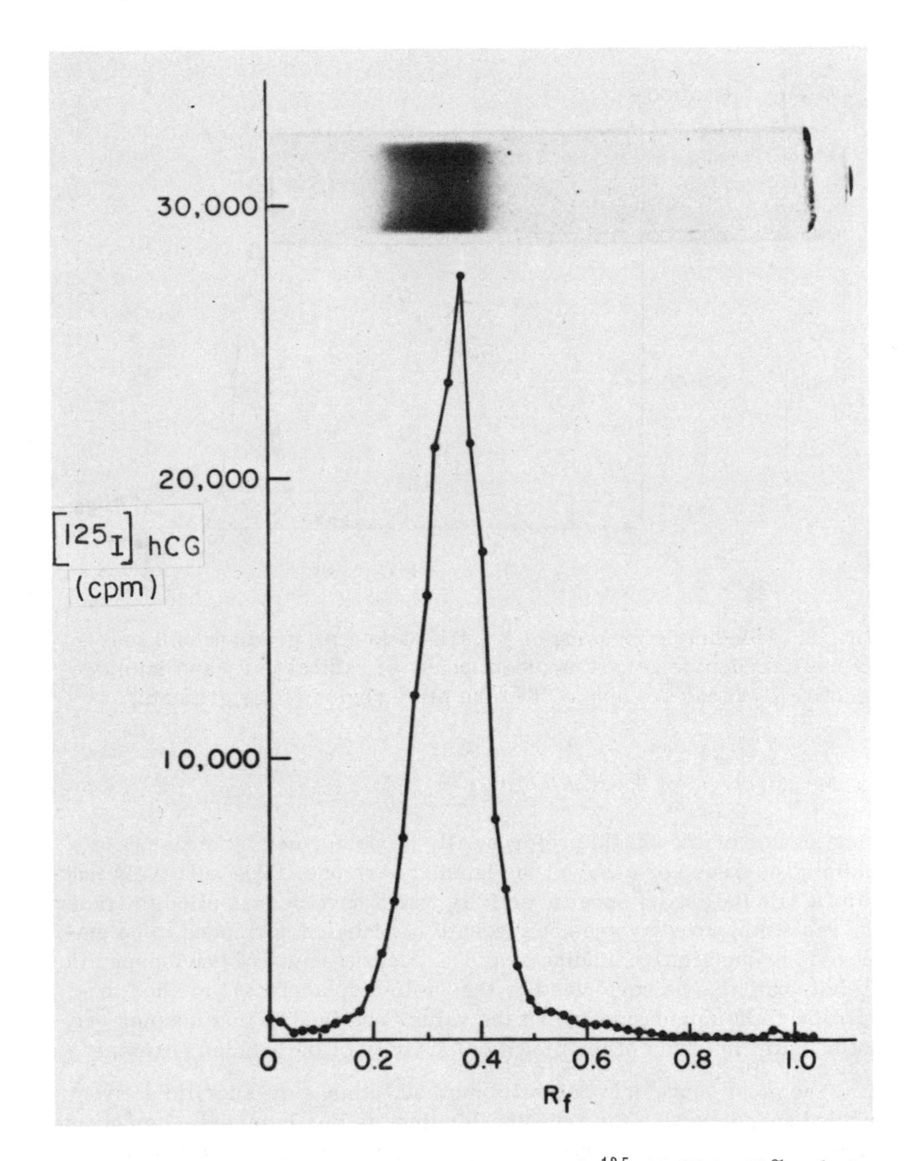

FIG. 5. Polyacrylamide gel electrophoresis of [^{125} I]hCG in 10% gel at pH 9.6. The single peak of radioactive hormone is compared with the stained pattern of the purified unlabeled gonadotropin (above). The multiple stained bands reflect the presence of charge isomers in the original preparation.

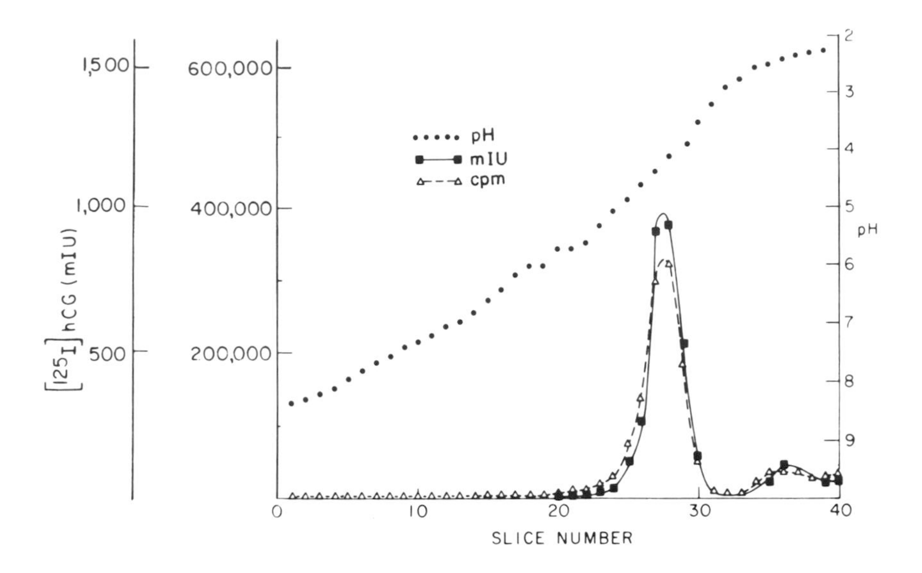

FIG. 6. Isoelectric focusing of [^{125}I]hCG in a pH gradient of 3 to 8 on 5% polyacrylamide gel. Coincident peaks of radioactivity and immunoreactive hormone are apparent in the pH-4 region of the gradient.

2. Measurement of Specific Activity

This measurement should preferentially be performed by bioassay or radioligand-receptor assay of the labeled hormone. The relatively inaccurate calculations of specific activity which have been applied to tracers for radioimmunoassay are not adequate for labeled hormones to be employed for quantitative binding studies. Measurement of tracer specific activity can also be performed by the "self-displacement" method in a specific radioimmunoassay, but the values obtained in this manner may not be valid in terms of the biological activity of the labeled hormone

The most convenient and valid method to measure specific activity of labeled gonadotropins for receptor-binding studies is by self-displacement assay of increasing concentrations of the tracer in the radioligand-receptor assay. In this way, the single isotopically labeled hormone can be employed both as "tracer" and "sample" during radioreceptor assay against standards of the original unlabeled hormone. Examples of this procedure for measurement of the specific activities of [^{125}I]hCG and [^{125}I]hFSH preparations are illustrated in Figures 7 and 8. In each case, parallel dose-response curves are obtained for binding inhibition by the labeled and unlabeled hormones, and accurate calculation of specific activity can be performed in terms of the native hormone employed for iodination.

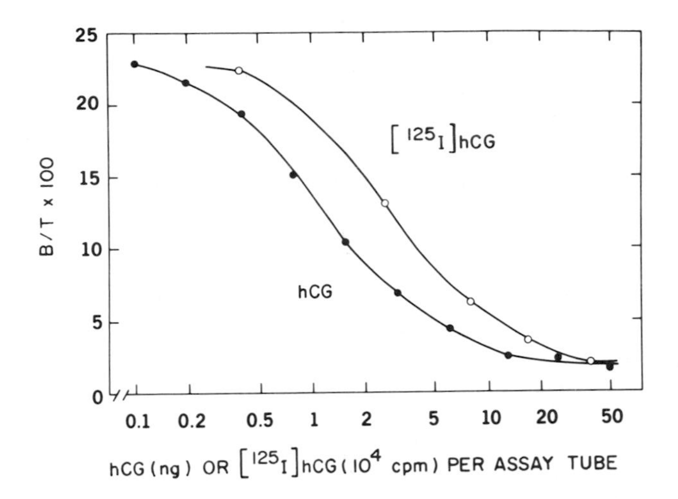

FIG. 7. Determination of specific activity of [^{125}I]hCG by "self-displacement" assay of increasing amounts of the tracer in the testis radioligand-receptor assay for LH and hCG. The mass of the labeled hormone is estimated by comparison with the binding-inhibition curve obtained with the unlabeled hCG employed for tracer preparation.

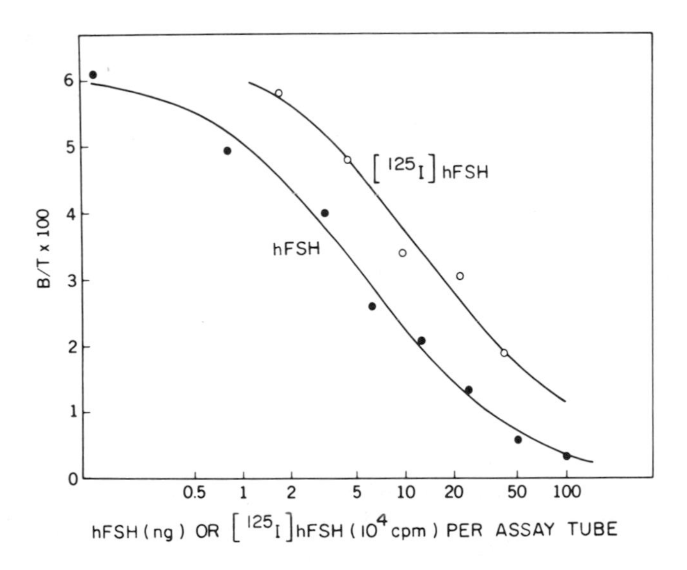

FIG. 8. Determination of specific activity of [^{125}I]hFSH, by "self-displacement" assay of tracer mass in the radioligand-receptor assay for FSH employing rat testicular homogenate.

Bioassay of radioactive tracer hormones by conventional procedures is not usually convenient for gonadotropins, due to the relatively large quantities of radioactivity which must be administered to reach the dose-response range of in vivo bioassay methods. Determination of the specific activity of labeled gonadotropins has been performed by such procedures as the ovarian ascorbic depletion assay [18], but the precision of such measurements is considerably less than that obtained during in vitro assay methods. The most convenient and precise bioassay procedures for labeled LH and hCG are provided by in vitro methods which employ the steroidogenic response of testicular or ovarian cells to gonadotropic stimulation. The most highly sensitive in vitro bioassay for LH and hCG is based upon the production of testosterone by rat testes or dispersed Leydig cells in response to low concentrations of these gonadotropins [35, 36]. Such methods are comparable to radioimmunoassay in terms of sensitivity, and are extremely useful for measurement of the biological activity of gonadotropins which stimulate steroidogenesis in gonadal target cells [36]. In similar fashion, assay of cAMP formation in dispersed target cells or adenylate cyclase activity in suitable membrane preparations could also be employed for in vitro bioassay of radioiodinated gonadotropins.

3. Determination of Binding Activity of Labeled Gonadotropins

Each labeled hormone preparation should be subjected to binding assay in the presence of excess receptor sites, to determine the proportion of radioactivity which represents "active" hormone with the capacity to interact with specific receptor sites (Fig. 9). This value represents the content of biologically active hormone in the labeled preparation, and provides an important correction factor for use during calculation of receptor-binding constants. Thus, if 60% of the tracer is bound to excess receptors, the remaining 40% of the total radioactivity will be present as inactive tracer in the "free" fraction, which should be corrected accordingly during analysis of quantitative binding data.

With certain peptide hormone-receptor systems, exposure of tracer hormone to high concentrations of cell membranes or homogenates may be accompanied by increased tracer degradation. If unrecognized, such an effect could result in a correspondingly low estimate for the proportion of biologically active hormone. However, the extent of the reduction in hormone binding is usually relatively small even at high receptor concentrations (Fig. 10), and is clearly apparent when complete binding curves are performed. If this effect is observed, the fraction of tracer with

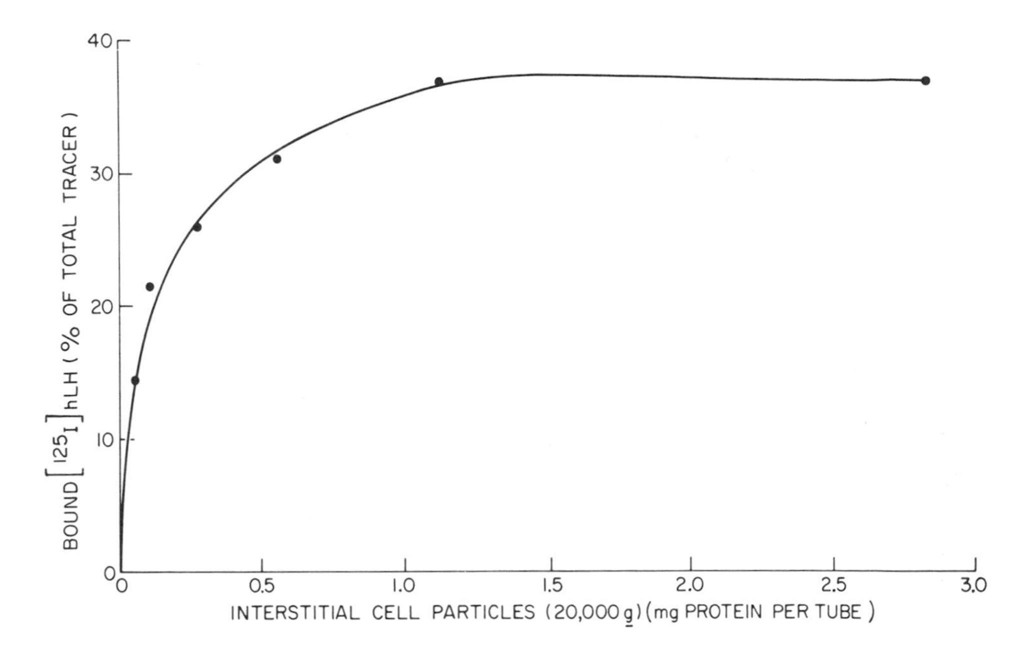

FIG. 9. Binding of [^{125}I]hLH to increasing concentrations of gonadotropin receptors, during incubation with interstitial cell particles from the rat testis. The maximum receptor-binding activity of the labeled hormone is about 37% of the total radioactivity of the tracer preparation.

binding activity is readily derived from lower receptor concentrations on the hormone-binding curve. In general, the determination of maximum binding activity of tracer hormone is a relatively simple and valuable procedure which permits more valid and precise binding constants to be derived from data obtained during binding-inhibition and saturation studies performed on particulate and soluble receptor sites. When tracer preparations contain only a relatively small proportion of active hormone, the use of a correction for maximum binding activity is particularly necessary if accurate values for binding affinity are to be derived from experiments with labeled gonadotropins. This situation is more frequently observed with FSH than with LH or hCG, as the maximum binding activity of [^{125}I]hFSH is rarely more than 10 to 15% unless selective purification by elution from gonadal receptors has been performed. An example of the maximum binding activity of [^{125}I]hFSH prepared by lactoperoxidase iodination and Sepharose-Concanavalin A chromatography is illustrated in Figure 11.

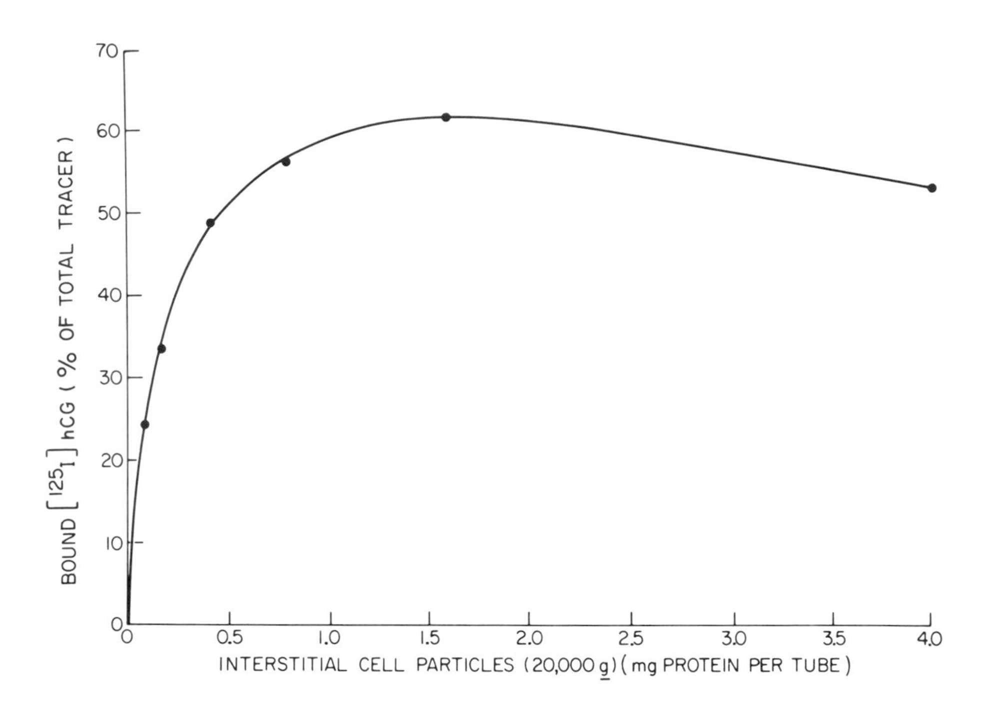

FIG. 10. Binding of [^{125}I]hCG by increasing concentrations of gonadotropin receptors, as interstitial cell particles from the rat testis. The maximum receptor-binding activity of the tracer hormone is 60% of the total radioactivity. High concentrations of testis particles show a reduction in binding attributable to increased degradation of the labeled gonadotropin.

III. PREPARATION OF PARTICULATE GONADOTROPIN RECEPTORS

A. Cellular Localization of Gonadotropin Receptors

1. LH/hCG Receptors

In the testis, specific gonadotropin receptors for LH and hCG are present only in the interstitial cell compartment, and are localized to the Leydig cells. No uptake of labeled LH or hCG by adequately isolated seminiferous tubules, tubule cells, or homogenates is detectable by in vivo and in vitro binding studies [3, 5, 37]. There is no evidence to suggest that LH receptors are present in the tubules of the adult rat testis, and homogenates of other tissues show no specific binding of radioiodinated hCG (Fig. 12).

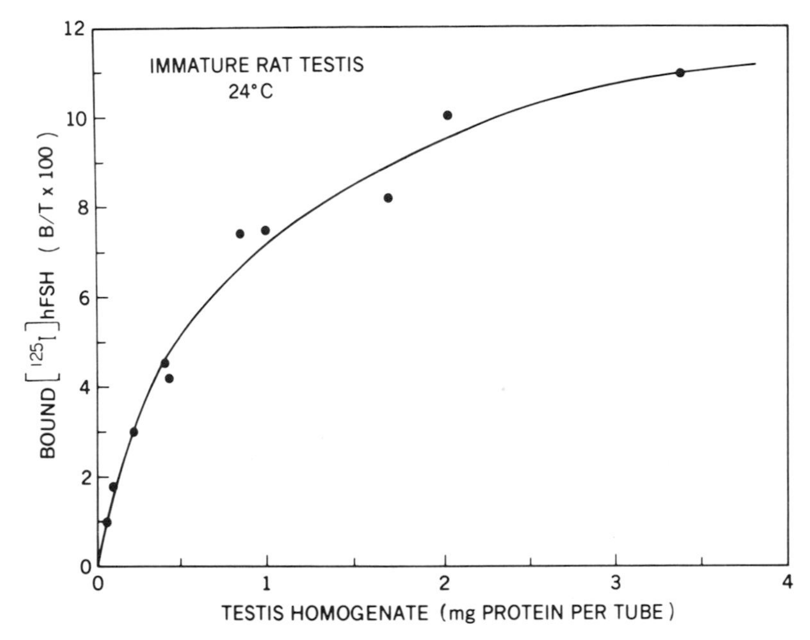

FIG. 11. Binding of [^{125}I]hFSH by increasing concentrations of testis receptors, during incubation with homogenates of the immature rat testis. The maximum receptor-binding activity of the labeled hormone is 10 to 11% of the total radioactivity of the tracer preparation.

In the ovary, the major locations of LH/hCG receptors are in the thecal and granulosa cells of the maturing follicle, and in the corpus luteum [12, 13, 19, 27, 38, 39]. Binding studies with the rat ovary have usually been performed after pretreatment of immature animals with pregnant mare serum gonadotropin (PMSG) and hCG to produce enlarged and heavily lutenized "pseudopregnant" ovaries, which exhibit considerably increased binding capacity for labeled gonadotropins [25]. Treatment of immature female rats with FSH has been shown to increase the number of LH-binding sites present in the developing ovarian follicles [40, 41].

2. FSH Receptors

Receptor sites for FSH have been more difficult to demonstrate than those for LH and hCG. In the testis, FSH receptors appear to be largely confined to the Sertoli cells of the seminiferous tubule [2, 4, 42]. Ovarian receptors for FSH have been demonstrated in the granulosa cells of

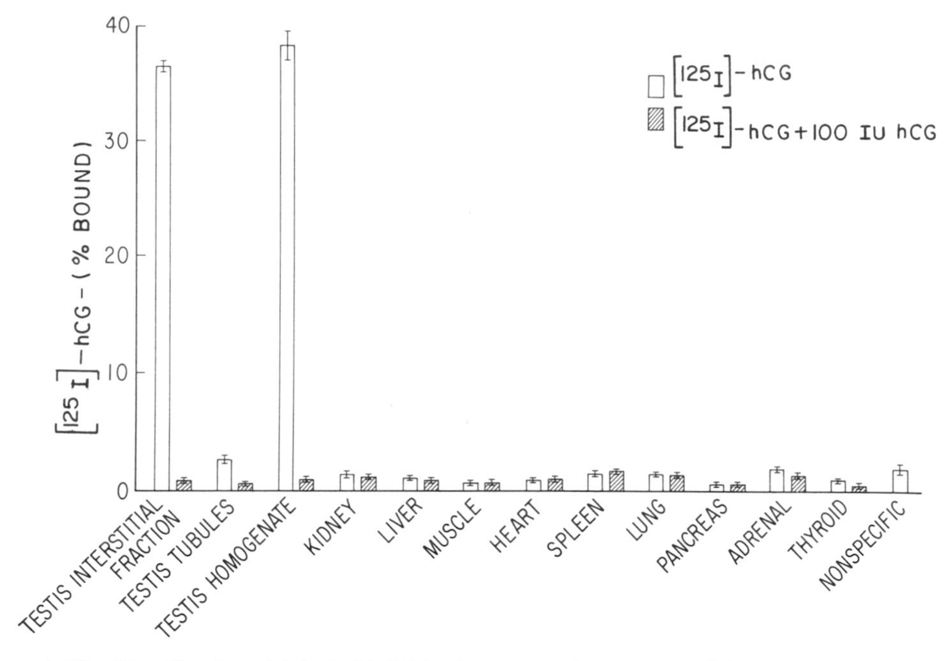

FIG. 12. Uptake of labeled hCG by homogenates prepared from several tissues of the rat. Significant levels of specific gonadotropin binding are present only in the testis, and binding within the testis is confined to the interstitial cell fraction. The small degree of binding evident in the tubule fraction is attributable to a minimum degree of contamination of tubules by Leydig cell fragments.

immature, pseudopregnant, and pregnant rats, and are present in small follicles prior to the development of LH receptors [27]. Estrogen treatment of immature female rats causes an increase in FSH binding to the ovary, probably via an effect on proliferation of the granulosa cells of the developing follicle [43].

3. Site of Subcellular Binding

Although some studies have suggested that gonadotropins may enter the cytoplasm of gonadal target cells, there is abundant evidence that the primary site of specific gonadotropin binding is located in the plasma membrane of the target cell. This has been demonstrated by autoradiography [19, 44], and by fractionation of particulate subcellular-binding sites during differential homogenization and density gradient centrifugation (Fig. 13). In homogenates of both testis [11, 37] and ovary [45],

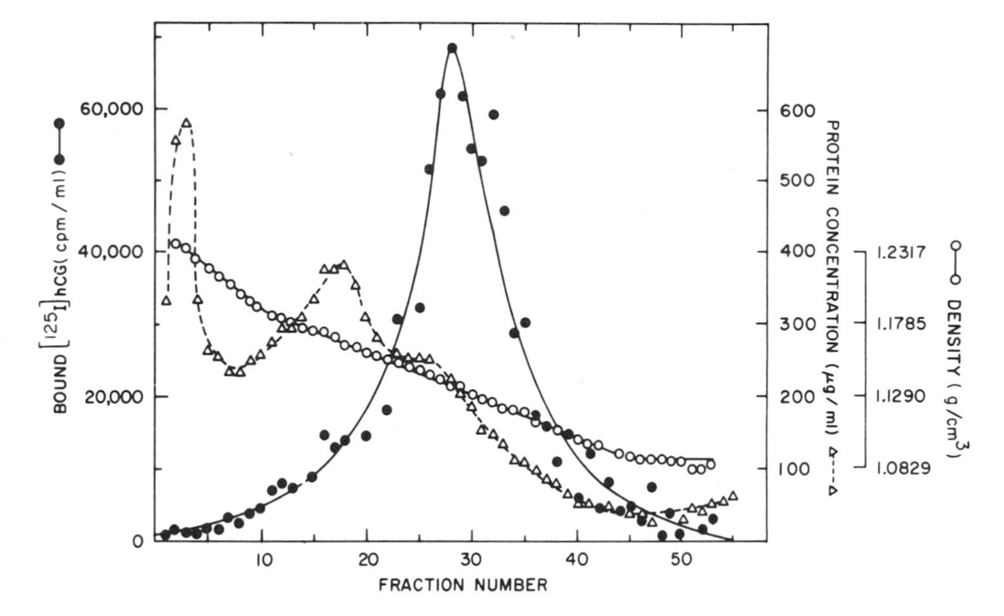

FIG. 13. Distribution of [^{125}I]hCG-binding activity of testis interstitial cell particles after isopycnic density gradient centrifugation in 10 to 50% sucrose. The single symmetrical peak of binding activity is centered at the equilibrium density value of 1.14.

such studies have shown that gonadotropin-binding sites copurify with membrane fragments, and that plasma membrane enzymes are associated with membranous subcellular particles of high gonadotropin-binding activity (Fig. 14). Indirect evidence for the plasma membrane location of gonadotropin receptors has also been provided by the retention of biological activity of agarose-coupled LH in vitro [46], and by the facility with which biologically active hCG can be eluted from testis receptors by exposure to low pH in vitro [33]. While the possibility remains that a proportion of the gonadotropin molecules bound to target cells may later appear in the cytoplasm, there is abundant evidence to indicate that the primary interaction of gonadotropins with specific receptors takes place at the cell membrane, in common with the site of action of other peptide hormones.

B. Testicular Receptors for LH and hCG

Various forms of testis particles have been employed for LH/hCG-binding studies, according to the nature and purpose of the receptor assay procedures. The interstitial cell fraction prepared by teasing apart the rat

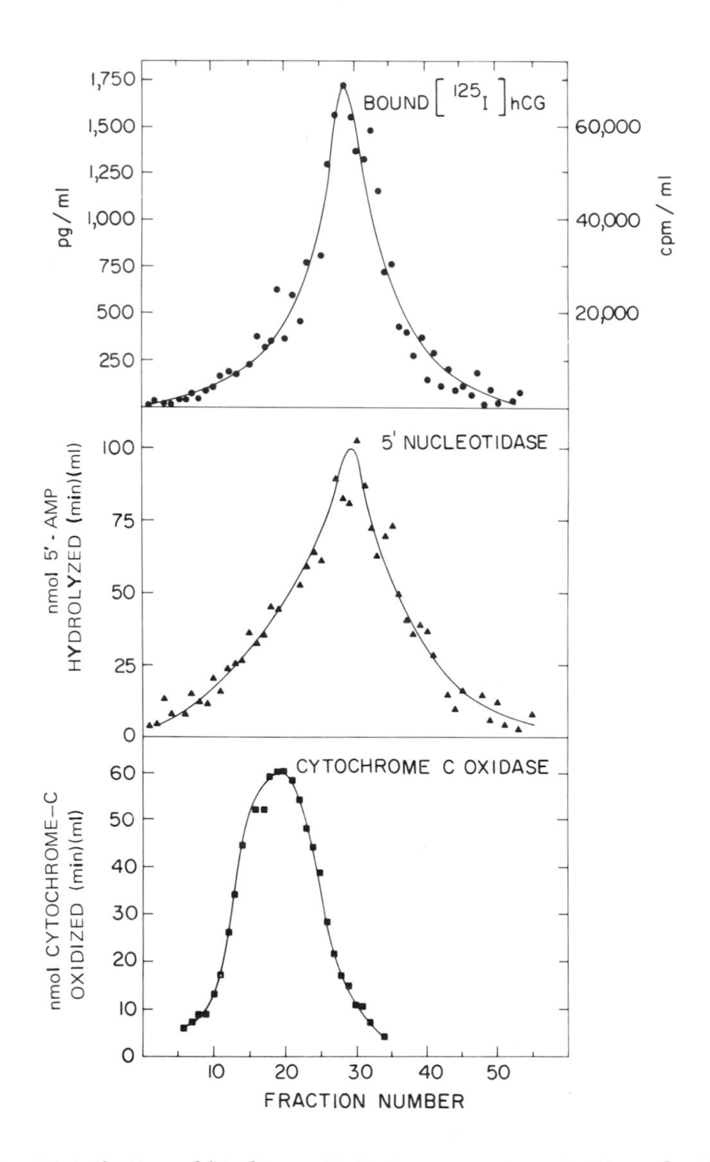

FIG. 14. Distribution of binding activity in comparison to 5'-nucleotidase and cytochrome-C oxidase activities after isopycnic density gradient centrifugation of interstitial cell particles in sucrose. Binding and 5'-nucleotidase activities are closely correlated, and well-separated from the position of mitochondrial particles indicated by cytochrome-C oxidase.

testis in phosphate-buffered saline, and filtration through cotton wool, contains a major proportion of fragmented and intact Leydig cells, and shows high binding avidity for labeled LH and hCG [5, 10]. A cell-free suspension of disrupted Leydig cell fragments can be obtained by centrifuging the teased interstitial tissue fraction at 150 g for 20 min to remove intact cells, followed by sedimentation of gonadotropin-binding particles by further centrifugation at 20,000 g for 30 min. Such particulate preparations from the rat testis interstitial-cell fraction are suitable for use in receptor-binding studies, or for further membrane purification and solubilization of gonadotropin receptors by extraction with nonionic detergents.

Less purified preparations obtained by homogenization of the whole decapsulated rat testis are satisfactory for general binding studies and radioligand-receptor assay of LH or hCG, since the presence of tubule components in the testis homogenate does not influence the binding of tracer gonadotropin by interstitial cell fragments [10]. Suitable binding fractions for radioligand assay can be prepared by teflon-glass homogenization of decapsulated testes in an equal volume of buffered saline, or by dispersion for 1 min in a Waring blendor. After homogenization, the testis suspension is centrifuged for 15 min at 1500 g to recover the majority of the binding particles. About 80% of the binding activity of rat testis homogenates is recovered in the 1500-g fraction, and the receptors remaining in the supernatant can be sedimented by further centrifugation at 10,000 g for 30 min. Testes homogenized and centrifuged in 0.3 M sucrose exhibit a broader distribution of binding particles over the sedimentation range from 1000 to 14,000 g, and 80% of the binding activity is recovered by centrifugation at 10,000 g for 30 min. After further homogenization of the 10,000 g-fraction in 0.3 M sucrose, density gradient centrifugation in 5 to 40% sucrose gives a binding fraction which is predominantly composed of membranous material on electron microscopy. As described on p. 193, isopycnic density gradient centrifugation of testis-binding particles in sucrose solutions gives a similar membranous fraction at density 1.14, with high binding activity and coincident localization with the activity of 5'-nucleotidase measured throughout the gradient (Fig. 14). The density, morphology, and enzyme content of the binding particles is consistent with their origin from the plasma membrane of the Leydig cell.

C. Testicular Receptors for FSH

In the testis, the specific binding sites for FSH are located in the tubule. This has been shown by direct binding studies of labeled FSH to isolated tubules and interstitial cells, and by autoradiographic localization of labeled hFSH in testis slices. The nature of the specific target cell for FSH in the tubular epithelium has long been debated, though cytochemical and immunohistochemical studies have indicated that the Sertoli cell is the major locus of FSH accumulation in the testis tubule [2, 4, 42]. There is now general agreement that FSH interacts primarily with the Sertoli cell. The main argument favoring this concept was provided by FSH-binding studies to "Sertoli-cell only" testis preparations of male rats irradiated in utero between day 19 and 21 of fetal life. The concentration of FSH-binding sites in the testis of the irradiated animals has been shown to be the same as in the normal rats, indicating that the Sertoli cell is probably one of the primary sites of action of FSH in the testis [47].

Particulate receptor fractions for FSH-binding studies can be prepared from homogenates of the testis of immature or adult rats. The FSH binding has generally been found to be higher in testes from immature rats than from adult animals. We have observed that on the basis of wet testis weight, the binding of FSH is maximum in gonads from animals between 10 and 15 days of age. For optimal assay conditions, and when high binding is required, the use of immature rat testis is sometimes an advantage. However, satisfactory results can also be obtained by using adult testis homogenates. This method has the advantage that less time is required to prepare relatively large amounts of homogenate, and can be used when routine assays are to be performed for derivation of potency estimates of pituitary, urinary, or plasma extracts.

The following procedure was found to provide particulate receptor preparations adequate for FSH-binding studies. After removal of the capsule, the testes are weighed and homogenized in a ground-glass tissue grinder (~ 15-20 strokes) in ice-cold PBS. The homogenate is filtered through nylon mesh and centrifuged at 20,000 g for 20 min. The supernatant is discarded and the pellets are resuspended by gentle homogenization in an appropriate volume of cold PBS.

IV. METHODS FOR RADIOLIGAND-RECEPTOR ASSAY OF GONADOTROPINS

The specific high-affinity binding sites in homogenates of rat testis and ovary provide an abundant source of material for radioligand-receptor assay of LH, hCG, and FSH. Such receptor assays combine the high precision, reproducibility, and convenience of radioligand assay with the biological specificity of conventional bioassays performed in the intact animal. The sensitivity of radioligand-receptor assay for gonadotropins is not as great as that of radioimmunoassay, but is very much higher than that of in vivo bioassays. For these reasons, radioreceptor procedures are particularly useful for assay of gonadotropin preparations, and fractions derived during purification and isolation of pituitary and chorionic gonadotropins. Such methods are also of considerable value for structure-function studies on the interactions of chemically modified gonadotropins and derivatives with the hormone-receptor sites in vitro.

A. Binding Assays for LH and hCG

Preparation of particulate receptor-sites for radioligand assay is most conveniently performed by utilizing homogenates of the adult rat testis, as described in Sec. III. The 1,500-g sediment from homogenized rat testis is diluted in phosphate-buffered saline (PBS) to a final volume appropriate for the assay conditions to be employed. For assays performed in a final incubation volume of 1 ml, the homogenate is diluted with PBS to 25 ml per testis, and 0.5 ml of the suspension is added to each assay tube [5, 11, 20]. Under these conditions, the assay is sensitive to LH and hCG levels of 0.5 to 1 ng/ml, and standard curves span the concentration range of 1 to 50 ng per assay tube (Fig. 15). For more sensitive assays, incubation volume can be reduced to 200 or 250 μl, and the quantity of homogenate per assay tube decreased to the equivalent of 1/100 testis, or about 8 mg of testis particles [37, 48]. Under these conditions, the radioligand-receptor assay is sensitive to 50 to 100 pg hCG, and the 50%-inhibition point of the binding-inhibition curve is obtained with about 1 ng of hCG (Fig. 16). The effective sensitivity of this form of the assay approaches that of radioimmunoassay, but the method cannot be applied to direct assay of low plasma gonadotropin levels, due to nonspecific interference by plasma proteins as described below [48].

To perform radioligand-receptor assay with testis particles, adult rat testes are decapsulated and homogenized in PBS by dispersion for 1 min in a Waring blendor at room temperature. The homogenates are centrifuged at 1500 g for 15 min and the sediment resuspended in 10 ml PBS per testis. The reagents for assay are added to 12 × 75-mm glass or polyethylene tubes in the following order:

1. 100 μl of testis homogenate
2. 100 μl of gonadotropin standards (e.g., 0.1-20 ng/hCG) or samples to be assayed, in PBS-BSA (1 mg/ml)
3. 50 μl of [^{125}I]hCG tracer (~ 10^5 dpm/ng, 50,000 dpm) in PBS-BSA containing 0.05% Neomycin sulfate

Nonspecific binding is determined in the presence of 100 IU (~ 10 μg) hCG per assay tube. After mixing and incubation at room temperature for 16 hr, 3 ml of ice-cold PBS is added to each tube and the diluted receptor suspension is either filtered or centrifuged at 1500 g for 20 min to isolate the receptor-bound radioactive hormone. For small numbers of assay tubes, and particularly if kinetic studies are to be performed, filtration through albumin-soaked 0.45-μ Millipore cellulose filters can be employed.

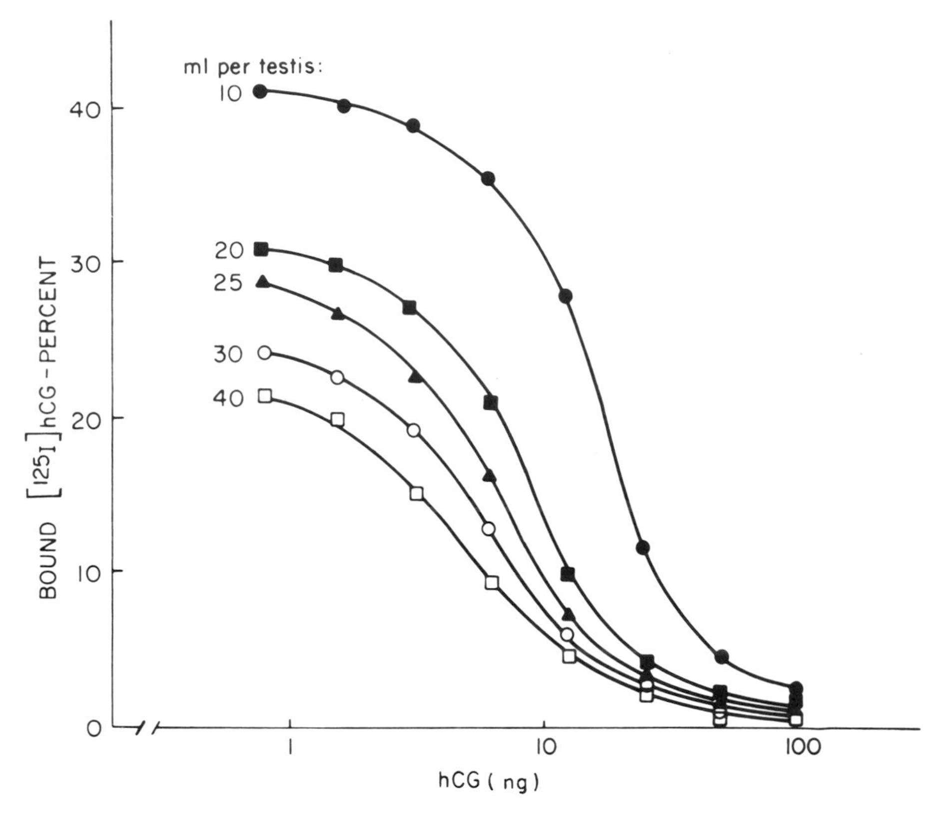

FIG. 15. Standard curves obtained from radioligand-receptor assay with [^{125}I]hCG and decreasing concentrations of rat testis homogenate. Lower binding and increased sensitivity are evident when less concentrated homogenates are employed for the binding assay. These dose-response curves were obtained with 0.5 ml of homogenate in a final assay volume of 1 ml.

For larger numbers of assay tubes, centrifugation followed by aspiration of the supernatant solutions is a considerably more rapid and convenient procedure for isolation of the particle-bound radioactive hormone. The filtration technique is more laborious and gives less uniform results, and requires the additional step of transferring the filters to counting vials prior to measurement of the bound radioactivity. Current preference is for the centrifugation method unless filtration is required for specific applications of the assay, e.g., to kinetic studies of hormone association or dissociation.

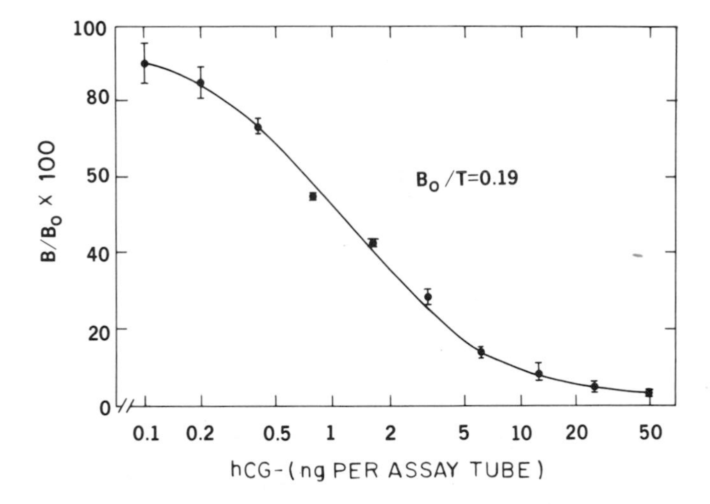

FIG. 16. Radioligand-receptor assay for LH/hCG employing 1/100 of a rat testis per assay tube, performed in an incubation volume of 0.25 ml. Each point is the mean ±SD of four determinations.

Although radioligand-receptor assay of LH and hCG can also be performed with homogenates of ovarian tissue from PMS/hCG-treated rats [18, 49] or of bovine corpora lutea [50], the method based upon rat testis homogenate is the simplest form of the assay for general use. No pretreatment of the animals is required, and the binding particles obtained from one adult rat are adequate for about 200 assay tubes. The specificity and sensitivity of binding assays performed with testis or ovarian homogenates are similar, and the methods are applicable to measurement of a wide variety of LH and chorionic gonadotropin preparations. Satisfactory binding of ^{125}I-labeled LH or hCG can also be obtained with homogenates of testes from other species, and the use of boar testes for radioligand-receptor assay is illustrated in Figure 17. However, the rat testis provides the most convenient and abundant source of LH/hCG receptors for radioligand assay, and the isolated testis or prepared testis homogenate can be stored frozen with adequate retention of binding activity. The extent of hormone binding is highest when fresh testis homogenates are employed, but satisfactory assays can be performed with homogenates of rat testes obtained from bulk suppliers and stored frozen at -60°C. Storage of frozen testes at -15°C is accompanied by a more significant loss of binding activity over a period of several months. Once prepared, testis homogenates can be stored on ice for several days with little change in binding activity.

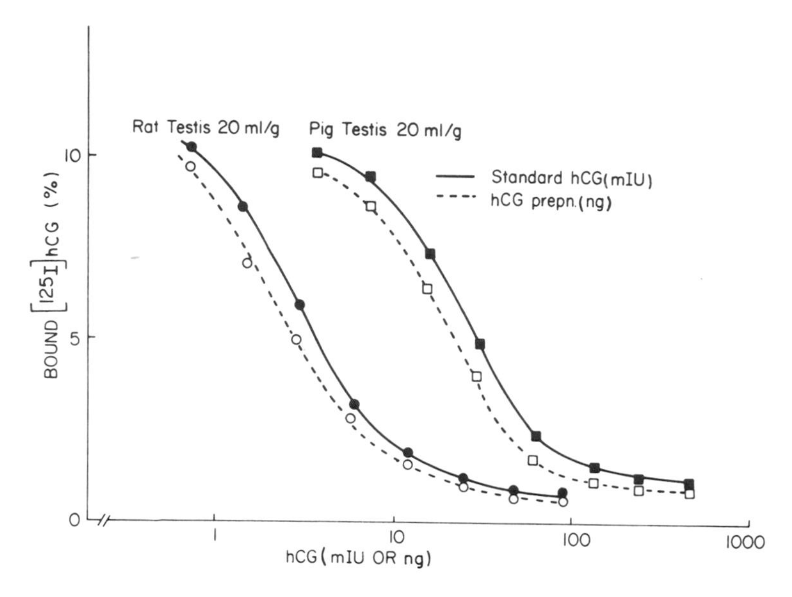

FIG. 17. Radioligand-receptor assay of hCG, comparing the binding-inhibition curves obtained with homogenates of rat and pig testis. Although the Leydig cell content of the testis is relatively higher in the pig, the binding of hCG tracer is usually higher per unit weight of rat testis homogenate. In these assays, performed in an incubation volume of 1 ml, the pig testis homogenate gave less sensitive standard curves than those obtained with rat testis homogenate.

B. Binding Assay for FSH

Radioligand-receptor assays for FSH can be performed using testis homogenates prepared from gonads of immature or mature rats as the source of FSH receptors [34, 51, 52]. For routine assays, the homogenate is diluted to contain the equivalent of 100 to 200 mg of wet tissue per ml of PBS.

To perform the assay, the following reagents are added sequentially to 12 × 75-mm glass tubes:

1. 100 μl of testis homogenate
2. 100 μl of gonadotropin standards (e.g., 0.25-100 ng FSH) or the samples to be assayed, in PBS-BSA (1 mg/ml)
3. 50 μl of [^{125}I]hFSH (40,000-50,000 dpm) in PBS-BSA containing 0.05% Neomycin sulfate

Nonspecific binding is determined in triplicate tubes containing 1 to 2 IU of Pergonal. After brief application to a vortex mixer, the assay tubes are incubated at 24 or 37°C for 18 hr with continuous shaking. Bound and free hormone are separated by filtration through Millipore filters (0.45 μ), previously soaked in 3% PBS-BSA to reduce nonspecific binding of tracer FSH to the cellulose filter. Filtrations are performed immediately after addition of 3 ml of ice-cold PBS to each incubation tube, and the filters are washed once with the same buffer. Each filter is then transferred to a plastic scintillation counting vial and its radioactivity determined in a γ-spectrometer.

A typical binding-inhibition curve obtained in the FSH radioligand-receptor assay is illustrated in Figure 18. The sensitivity of the assay is ~ 0.5 ng of hFSH per assay tube, and the 50% binding-inhibition dose lies between 5 and 8 ng of hFSH. The specificity of the assay is illustrated by the low cross-reaction of hLH and hCG in this system, at doses, respectively, 100 and 750 times higher than hFSH. Parallel dose-response curves are obtained with highly purified hFSH and the Second International

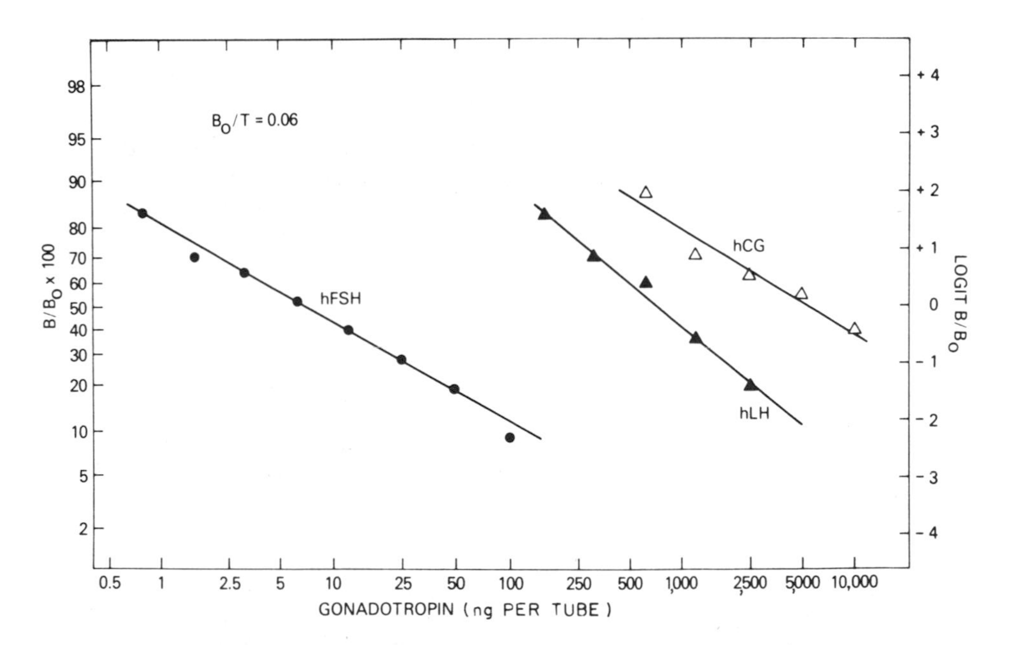

FIG. 18. Radioligand-receptor assay for hFSH employing immature rat testis homogenate. The assay is performed at 24°C. hLH and hCG reacted in the systems only at doses respectively 100 and 750 times greater than hFSH.

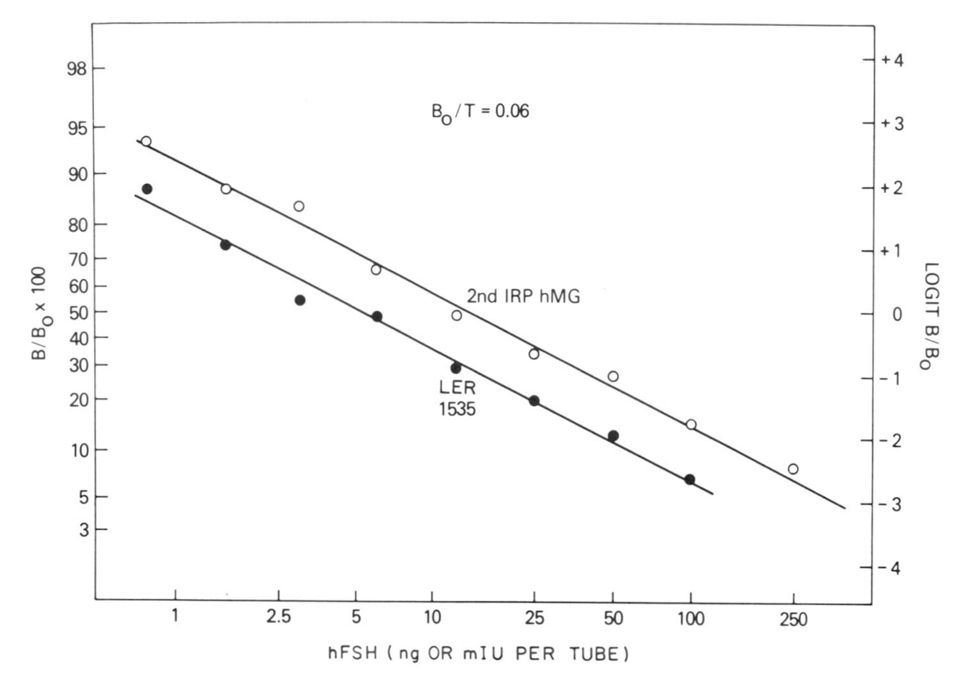

FIG. 19. Comparison in the radioligand-receptor assay for hFSH of the binding-inhibition curves obtained with the Second International Reference Preparation of Human Menopausal Gonadotropin (2nd IRP-hMG) and highly purified hFSH (LER 1535). The slopes are parallel, and the potency of hFSH interms of 2nd IRP-hMG is of 2,690 IU/mg (95% confidence limits: 2,130-3,420).

Reference Preparation for Human Menopausal Gonadotropin (2nd IRP hMG), as shown in Figure 19. In the same assay, purified hFSH, ovine FSH, and PMSG reacted with parallel slopes. When 15-day-old rat testis homogenates are used, the binding of [^{125}I]hFSH is significantly higher at 37°C than at 24°C; the binding-inhibition curve is somewhat more sensitive, and its slope is steeper (Fig. 20). In contrast, when adult testis homogenate is used, the binding is lower at 37°C. This could be attributable to a higher degradative activity in adult testis homogenates than in those prepared from immature animals. Therefore, if adult testis homogenates are employed for the binding assays, the incubation should be performed at room temperature. Satisfactory assays can be performed with homogenates kept on ice for several days, and the particulate receptors can be stored at -60°C for several weeks without significant loss of binding activity.

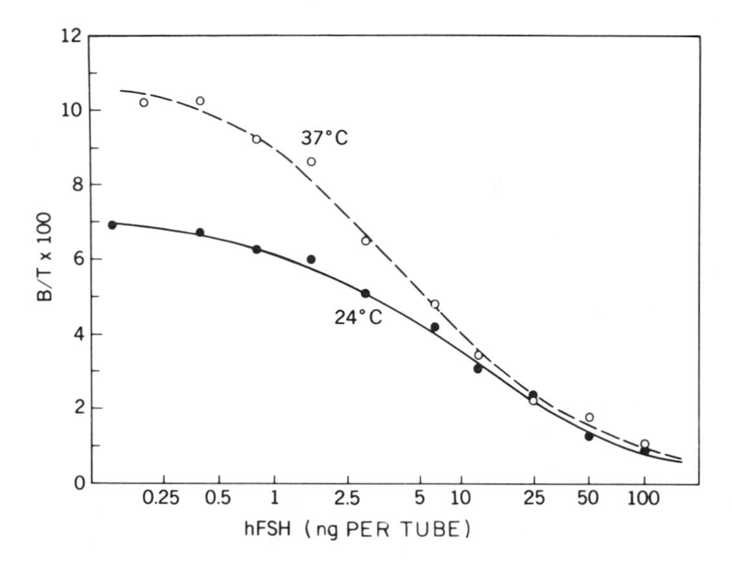

FIG. 20. Effect of the temperature on the radioligand-receptor assay for hFSH, using immature rat testis. The initial binding is higher at 37°C and the binding-inhibition curves are significantly steeper than at 24°C.

V. APPLICATIONS OF RADIOLIGAND-RECEPTOR ASSAYS FOR GONADOTROPINS

A. Measurement of LH and hCG Activity

The use of radioligand-receptor assay procedures for measurement of LH and chorionic gonadotropin has been described in detail elsewhere [5, 10, 20]. An example of the binding-inhibition curves given by a number of standard pituitary hormone preparations during radioligand-receptor assay is shown in Figure 21. The potency estimates for gonadotropin preparations derived by receptor assay are usually commensurate with, or somewhat higher than, those obtained by conventional bioassay. The only notable exception to this general finding has occurred with desialylated gonadotropin prepared by treatment with neuraminidase [7, 18, 53]. Such modified gonadotropins are characterized by high binding affinity for gonadotropin receptors in vitro, but display low biological activity during in vivo bioassay due to rapid hepatic clearance and removal from the circulation [53]. This characteristic of desialylated gonadotropins has been demonstrated with enzymically modified hCG and LH, but probably does not occur to a significant degree in pituitary extracts containing LH, or in purified urinary hCG preparations. The occurence of variably

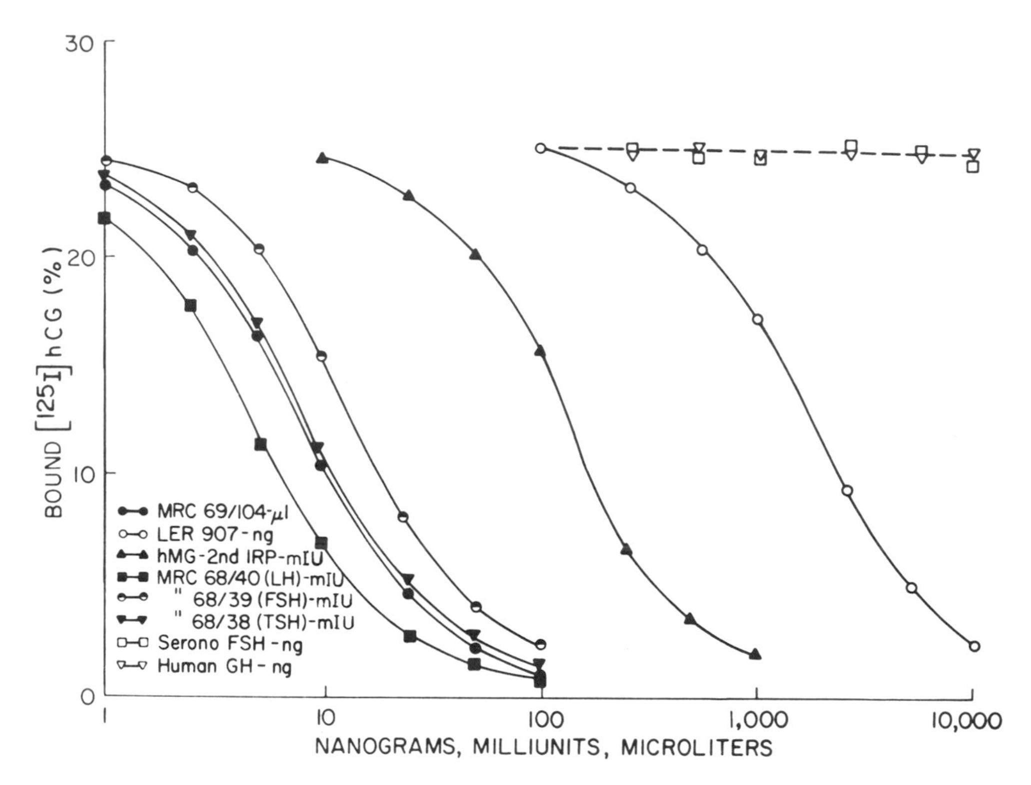

FIG. 21. Radioligand-receptor assay of standard gonadotropin preparations. The activity detected in FSH and TSH preparations is attributable to the known contamination of these standards with residual LH.

desialylated hCG molecules may occur during urinary excretion of the hormone in pregnancy, though the extent of such a change has not been determined. Due to the rapid hepatic clearance of asialoglycoproteins in vivo, it is unlikely that a significant quantity of desialylated gonadotropin molecules will be present in circulating blood and plasma samples. For these reasons, the application of receptor assay to measurement of naturally occurring gonadotropins can be regarded as a valid method for in vitro estimation of the biological activity of gonadotropic hormones. However, gonadotropins which exhibit a short plasma half-life in vivo will commonly exhibit higher biological activity by any in vitro assay method, including radioligand-receptor assay. This is a notable feature of ovine LH, which contains no sialic acid and is characterized by rapid clearance from plasma during in vivo bioassays, with consequent difficulty in comparison with standard hormone preparations of more prolonged half-life, e.g., the 2nd IRP hMG. For such hormones, in vitro bioassay or receptor assay methods provide a more accurate measure of the intrinsic biological activity of the molecule at the target cell level.

For measurement of plasma gonadotropin levels, radioligand-receptor procedures are most satisfactory when applied to determination of the relatively high levels of chorionic gonadotropin which occur during pregnancy in human and equine plasma (Fig. 22). Measurement of gonadotropin levels in plasma of nonpregnant subjects by receptor assay is compromised by significant nonspecific interference by plasma proteins [37], and by the low basal levels of circulating gonadotropins in comparison to the detection limit of the assay system. Such interference prevents the accurate measurement of LH or hCG in plasma below levels of 5 to 10 ng/ml, and can be minimized by careful equalization of the serum or plasma content of all assay tubes, by addition of hormone-free serum. In addition, extraction of gonadotropins from plasma samples by the agarose-Concanavalin A method [32] can be employed to avoid interference and to concentrate the hormone prior to radioligand assay. An example of the effect of serum protein upon the standard curve for the LH/hCG-receptor assay

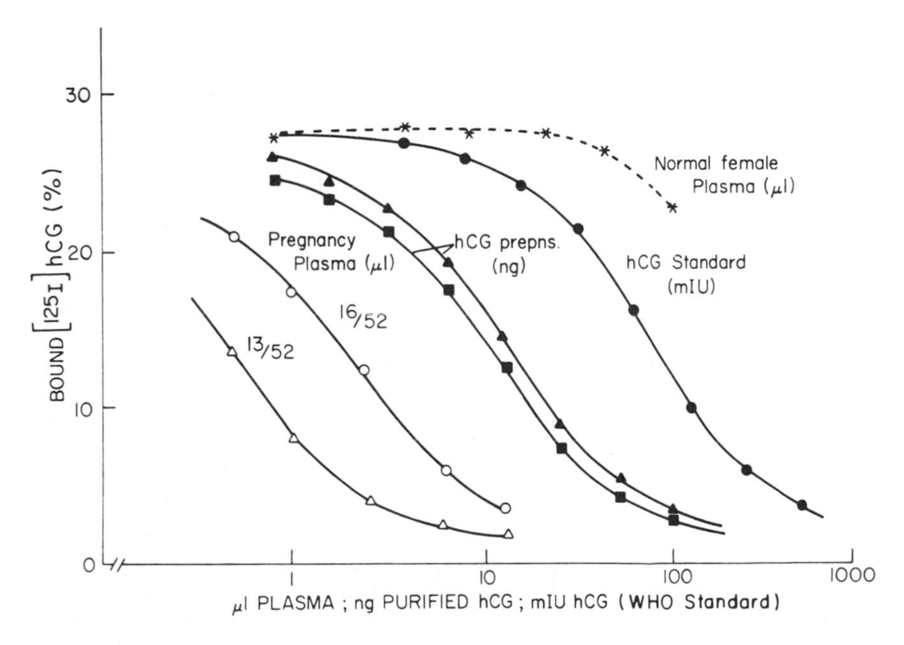

FIG. 22. Radioligand-receptor assay of hCG in pregnancy plasma, with reference standards of purified hCG (nanograms) and the Second International Standard preparation of hCG (milliunits).

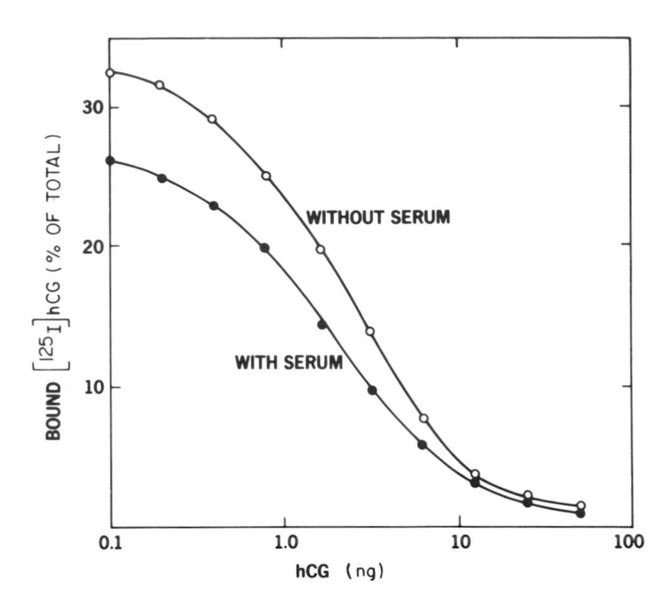

FIG. 23. Radioligand-receptor assay of LH/hCG performed with rat testis homogenates in the presence and absence of 20% hormone-free serum. The interfering effect of serum protein upon [^{125}I]hCG binding by testis receptors is apparent throughout the standard curve.

performed with testis particles is illustrated in Figure 23. The nonspecific interfering effect of proteins upon receptor binding of tracer gonadotropin is probably responsible for the apparent presence of a "gonadotropin" in human plasma within 4 to 6 days after fertilization [50], as well as in the blastocyst fluid of early rabbit embryos [54]. Measurement of plasma gonadotropin during the first two weeks of pregnancy by radioligand-receptor assay performed with addition of constant serum concentrations to standard and sample tubes has given results in general agreement with radioimmunoassay. However, such assays performed in the absence of serum from the standard-curve tubes gives spuriously high values for plasma "gonadotropin" levels, due to the nonspecific effect of serum proteins in the sample tubes. An example of the artifact produced by nonspecific serum protein interference during assay of hCG levels in plasma from early pregnancy is shown in Figure 24. The correlation between immunoreactive and biologically active hCG during early pregnancy has been recently demonstrated by specific radioimmunoassay and bioassay of hCG levels after fertilization, demonstrating that no rise of endogenous hCG occurred until 9 to 13 days after ovulation, at the presumed time of implantation of the blastocyst [48].

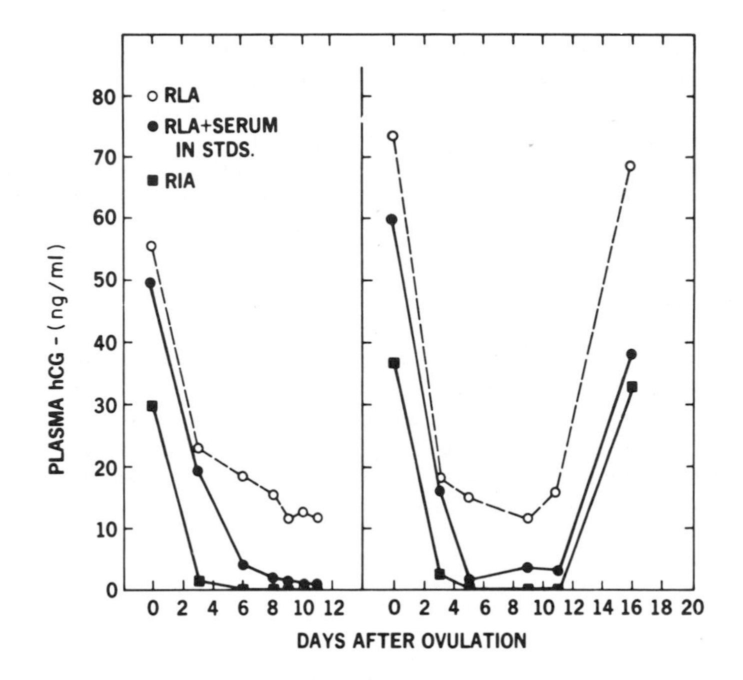

FIG. 24. Measurement of hCG in plasma after induction of ovulation with Pergonal and hCG, and subsequent fertilization. The decline of exogenous hCG levels in plasma after an unsuccessful attempt to achieve conception is shown in the left panel. Plasma gonadotropin levels measured by radioimmunoassay (RIA) and radioligand-receptor assay (RLA) with serum in the standard tubes had declined to zero when high apparent hCG values were still detected by RLA performed without addition of serum. In the right panel, a similar pattern is observed after a further induction of ovulation, followed on this occasion by conception and a subsequent rise of plasma hCG levels after 12 days, when implantation has commenced.

B. Hormone Structure and Receptor Binding

In addition to the assay of LH and hCG concentrations in hormone preparations and body fluid or tissue extracts, the radioligand-receptor assay method is of particular value for experiments upon the structure-function relation of gonadotropins. The in vitro assay method permits the evaluation of receptor-binding activity in the absence of complicating effects caused by changes in metabolic clearance and plasma half-life of native or chemically modified hormones. In addition to the studies upon desialylated hormones described, the receptor assay method has been of value during analysis of the activities of hormone subunits and chemically modified gonadotropins. The α- and β-subunits of LH and hCG have been shown to be completely devoid of receptor-binding activity in vitro [10, 55], and the reassociation kinetics of isolated subunits have been determined by receptor-binding assay [56]. In addition, the effects of a number of chemical modifications of the gonadotropin molecule have been shown to reduce receptor-binding activity in proportion to the impairment of biological activity in vitro (Fig. 25).

Quantitation of changes in receptor-binding activity of gonadotropins after derivative formation by nitration, succinylation, and maleylation has also been utilized to evaluate the charge and conformational requirements for interaction of LH with the testicular receptor sites [57]. The use of receptor assay procedures for these types of studies provides a marked advantage over conventional bioassay procedures, and illustrates one of the more important applications of the receptor-binding assay to the analysis of hormone-receptor interactions.

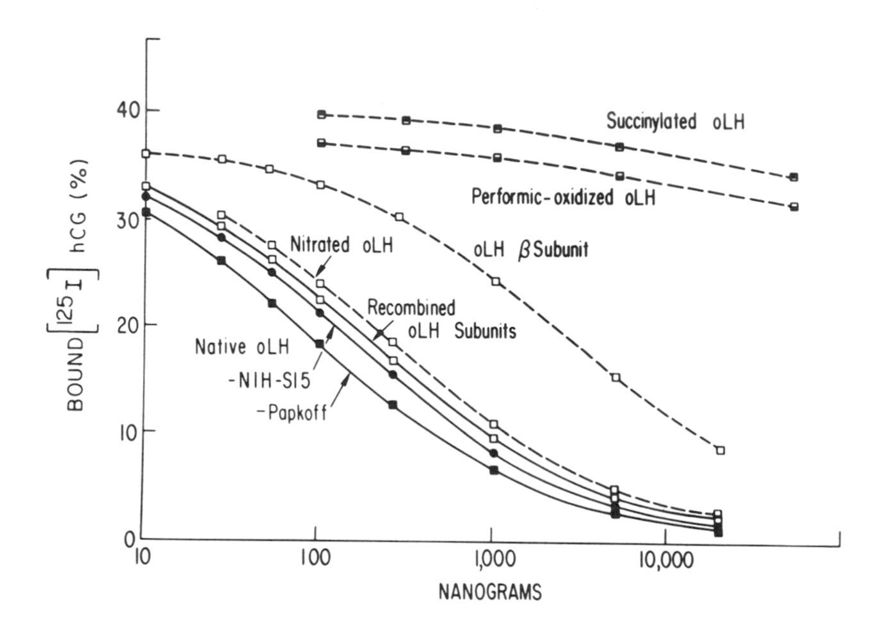

FIG. 25. Radioligand-receptor assay of derivatives of ovine LH, prepared by Dr. H. Papkoff. The activity of each modified hormone preparation measured by radioligand assay was similar to the potency estimate determined by bioassay.

VI. DETERMINATION OF BINDING CONSTANTS AND CONCENTRATION OF RECEPTORS

A. Analysis of Binding Data

For optimal determination of the binding constants of the hormone-receptor interaction, the tracer preparations should be extensively characterized, as described in Sec. II. D., both in terms of specific activity and maximum binding activity. When the specific activity of labeled gonadotropin is measured in a bioassay or in a radioligand-receptor assay, it expresses the amount of radioactivity corresponding to a given amount of intact biologically active hormone. This value can be used to calculate the total concentrations of tracer hormone in the incubation media. If the maximum binding activity of the tracer is less than 100%, only a fraction of the total radioactivity will correspond to intact hormone. The radioactivity specifically bound to the receptor preparation represents only intact hCG. Therefore, the values of the specific activities used to compute the bound hormone concentrations have to be corrected by a factor expressing the fraction of the tracer preparation corresponding to the inactive labeled material. This correction factor is determined by the maximum-bindability experiments outlined in Sec. II. D. Failure to apply this correction will result in an underestimate of bound-hormone concentrations; an overestimation of the free-hormone concentrations will also result from this, if the latter parameter is calculated by subtracting bound from total hormone concentrations. For instance, if 100,000 cpm of labeled gonadotropin corresponds to 1 ng of unlabeled hormone in the radioligand-receptor assay, and if the maximum binding activity of the labeled preparation is 60%, 60,000 cpm will be bound when 1 ng of hormone has reacted specifically with the receptor. If the maximum binding activity of the labeled hormone preparation is not taken into account, it would have been erroneously concluded that only 0.6 ng of hormone are bound and that 0.4 ng are free. Even though they are highly purified, gonadotropins used for binding studies are not always of maximum attainable bioactivity. When the specific activities of the labeled gonadotropins are expressed in terms of such preparations, the calculations of the molar concentrations of the hormone should be corrected to their expected theoretical maximum bioactivity.

1. Model for the Analysis of Hormone-receptor Interaction

The simplest model for the hormone-receptor interaction is represented by a single, bimolecular, reversible reaction:

$$P + Q \underset{k_2}{\overset{k_1}{\rightleftarrows}} PQ \tag{1}$$

where

P = free-hormone concentration

Q = free-receptor concentration

PQ = hormone-receptor complex concentration

= B, the bound-hormone concentration

k_1 = association-rate constant ($M^{-1}\ min^{-1}$)

k_2 = dissociation-rate constant (min^{-1})

The association equilibrium constant, K_a (M^{-1}), is defined by the law of mass action as

$$K_a = \frac{k_1}{k_2} = \frac{PQ}{P.Q} = \frac{B}{P.Q} \tag{2}$$

The kinetics of this reaction are expressed by the second-order chemical kinetics differential equation:

$$\frac{dB(t)}{dt} = k_1[P_0 - B(t)]\,[Q_0 - B(t)] - k_2 B(t) \tag{3}$$

where B(t) is concentration of hormone-receptor complex at time t, with $B(0) = 0$; and P_0 and Q_0 are total hormone and receptor concentrations.

According to this model, the dissociation process of the hormone-receptor complex follows first-order kinetics:

$$\frac{dB(t)}{dt} = -k_2 B(t) \tag{4}$$

where B(t) is the concentration of hormone-receptor complex at time t, during the dissociation experiments, with $B(0) = B_0$; and B_0 represents the hormone-receptor complex concentration at the time at which the dissociation experiment is started (time zero).

After differentiation, Eq. (4) becomes

$$B(t) = B_0 e^{-k_2 t} \tag{5}$$

2. Analysis of Equilibrium Data

The value of K_a characterizing a hormone-receptor system can be derived from saturation curves at equilibrium. Such data can be generated by incubating increasing concentrations of labeled hormone with a constant amount of receptor preparation and measuring, after equilibrium has been reached, the specifically bound hormone concentrations. The same type of data can be obtained by incubating the receptor preparation with increasing concentrations of unlabeled gonadotropin and a constant amount of labeled hormone, to give the usual binding-inhibition curve. The total hormone concentrations are then the sum of labeled and unlabeled ligand; the bound-hormone concentrations are calculated from the specifically bound radioactivity, using for the specific activity at each data point the ratio between total radioactivity (corrected for maximum bindability) and the sum of labeled and unlabeled hormone present in the incubation medium. When binding-inhibition studies are to be used for derivation of receptor-binding constants, it is most important to ensure that two practical requirements are observed during the experimental procedure. These are (1) that the tracer and unlabeled hormones are added first and mixed before addition of the receptor preparation, and (2) that incubations are performed for sufficient time to achieve equilibrium for the lowest hormone concentration employed.

Equilibrium binding data are usually analyzed by applying the Scatchard model [58], which relates the bound hormone concentrations (B) to the bound-over-free hormone ratios (R) at equilibrium; Q_0 represents the total concentration of binding sites.

$$R = K_a [Q_0 - B] \tag{6}$$

When a single class of independent binding sites is present, the plot of R versus B will be linear; the slope of the line will be $-K_a$ and its intercept with the horizontal axis, Q_0 [Fig. 26(b)].

Several assumptions are implicit in the use of the Scatchard model:

1. The hormone-receptor interaction is a reversible reaction following second-order chemical kinetics, and the system is at equilibrium when bound and bound-over-free are measured.
2. There are no site-site interactions (positive or negative cooperativity).
3. Labeled and unlabeled hormones behave identically and are present in an homogeneous form.
4. Bound and free can be separated perfectly, and the separation procedure does not perturb the equilibrium.
5. No reactant degradation occurs during incubation.

Departures from linearity in the Scatchard plots can be observed under certain circumstances. If more than one class of independent sites are present, concave Scatchard plots will be observed. The mathematical theory of such complex ligand-binding systems at equilibrium has been extensively developed [59-72]. In particular, the relation between bound

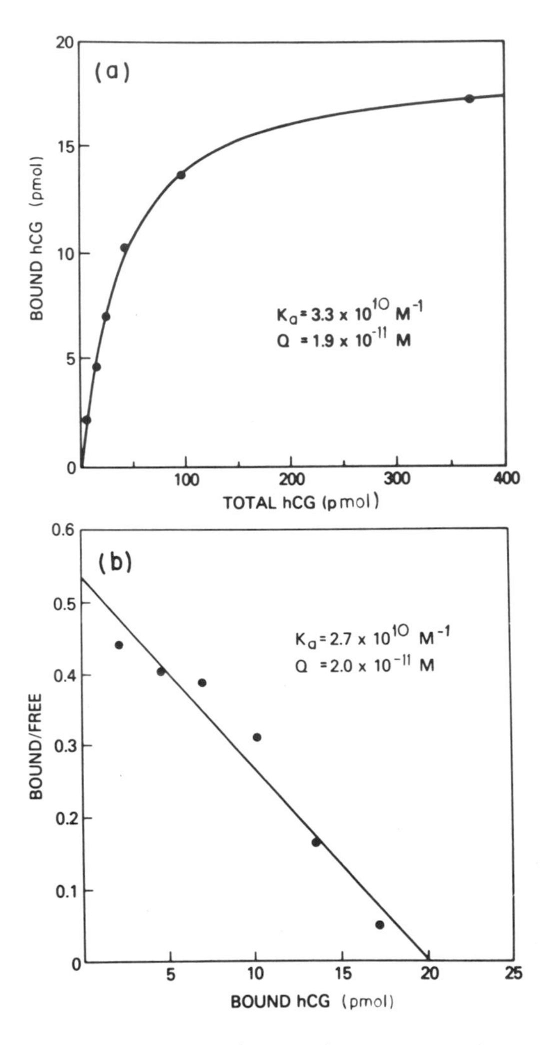

FIG. 26. (a) Saturation curve of testis homogenate obtained, at 24°C, by incubating constant amounts of homogenate (0.5 ml) with increasing concentrations of hCG (constant amount of [^{125}I]hCG with increasing amounts of unlabeled hCG). The final incubation volume was 1 ml. The solid line represents the best fit of Eq. (8) with the data. (b) The same data expressed as a Scatchard plot. The solid line represents the fit of the Scatchard model [Eq. (6)] with the transformed data.

and bound-over-free for one ligand and two classes of binding been explicitly expressed [69, 70]. The properties of one-ligand group Scatchard curves have also been analyzed, and graphical methods have been described to determine the binding constants and concentration of each class of sites from these plots [60, 66, 70]. It is important to note that, as emphasized by Klotz and Hunston [66], "the graphical parameters for multiple classes of sites are generally not what one might have expected from 'intuitive' generalization from a single class." More accurate estimations of these parameters can be obtained by analysis of the data with nonlinear curve-fitting programs such as MLAB [73, 74], MODELAID [75], or SAAM [76], using as models the expressions referred to above. Curvilinear Scatchard plots can also suggest cooperative interactions between binding sites [77]. Many other factors can affect the slope and/or the shape of Scatchard plots: systematic errors in tracer mass, error in the measurement of "nonspecifically" bound radioactivity, misclassification error or perturbation of equilibrium during separation of bound and free tracer [78], differences in the affinities of labeled and unlabeled hormone [70], heterogeneity of hormone preparations, or failure to reach equilibrium. A case of particular interest is represented by the effects of degradation of free hormone and unoccupied binding sites upon the shape of saturation curves and Scatchard plots. To study this problem, a mathematical model has been developed for the analysis of the binding curves for reversible bimolecular reactions in which the free reactants undergo degradation during the incubation process [79]. Computer simulations of binding curves, based upon this model, have shown that degradation of the free reactants had relatively little influence upon the linearity of the Scatchard plots or the apparent binding capacities of the receptor preparations. In contrast, degradation of free receptors or free hormone caused a significant and marked reduction of the apparent values of the equilibrium constant, K_a (Fig. 27).

In the expressions usually employed to analyze equilibrium binding data, the specifically bound hormone concentrations are related to the free-hormone concentrations. This is the case for the Langmuir isotherm and the Scatchard expressions. In most experimental binding systems, the free hormone concentrations are not directly measured, but are calculated by subtracting the bound- from the total-hormone concentrations over the range of the experimental data points. This practice cumulates the errors of measurement of both the bound- and total-hormone concentrations. The total-hormone concentrations can be measured independently and with high precision, in contrast to the inherently greater error of the determination of the bound hormone. Therefore, this parameter could be considered as an "error-free" independent variable which would permit weighted nonlinear regression analysis of the saturation curves. For these reasons, the relations between bound- and total-gonadotropin

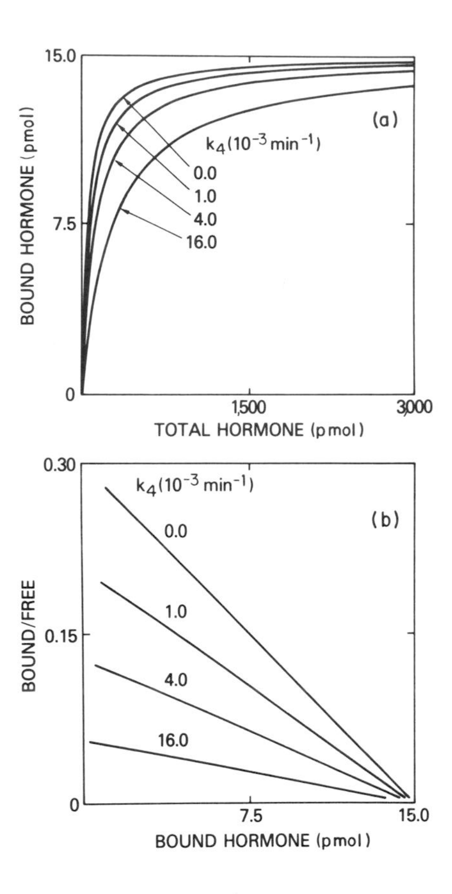

FIG. 27. (a) Effect of increasing rates of free-receptor degradation on the shape of computer-simulated saturation curves. The rate of receptor degradation was treated as a first- or pseudo-first-order kinetic process, with rate constant k_4. No hormone degradation was expected to occur. The values for the association (k_1) and dissociation (k_2) rate constants were set respectively to 10^8 M^{-1} min^{-1} and 5×10^{-3} min^{-1}; the binding capacity (Q_0) was fixed at 15 pM. The concentrations of bound hormone were computed for each total hormone concentration at a time long enough to ensure equilibrium or steady-state at the lowest concentration of total hormone (5 pM). (b) The same data, expressed as Scatchard plots. The free-hormone concentrations were computed by subtracting the bound from the total hormone concentrations.

concentrations have been established. By defining equilibrium as the stage at which the rate of change of formation of the hormone-receptor complex reaches zero, the second-order kinetics differential equation becomes

$$k_1 [P_0 - B(t_e)] [Q_0 - B(t_e)] - k_2 B(t_e) = 0 \qquad (7)$$

where t_e is the time required to reach equilibrium. Solving for $B(t_e)$, the following reaction is obtained:

$$B(t_e) = 0.5\{[P_0 + Q_0 + 1/K_a] - ([P_0 + Q_0 + 1/K_a]^2 - 4P_0Q_0)^{1/2}\} \qquad (8)$$

An example of the computer fit of a saturation curve according to this model is presented in Figure 26 (a).

3. Analysis of Kinetic Data

An important aspect of the characterization of the interaction between a hormone and its specific receptors in the target organ is represented by the kinetics of the binding process. The time course of the association reaction is determined by measuring the concentrations of the hormone-receptor complex at various time intervals following the initiation of the binding reaction. The dissociation rate of the hormone-receptor complex can be monitored after addition of a large excess of unlabeled hormone to a receptor preparation previously incubated with labeled gonadotropin. This procedure prevents any detectable binding or rebinding of free labeled hormone to the receptors; under these conditions, the decay of the radioactive hormone-receptor complex concentrations represents its rate of dissociation. The same effect can be obtained by performing "infinite" dilution of the preincubated mixture of hormone-receptor complex, free hormone, and unoccupied receptors. The concentrations of free hormone and free binding sites are then reduced almost to zero and the forward reaction becomes infinitely slow, since the rate of the second-order association reaction is proportional to the product of these two parameters. By contrast, the first-order dissociation process follows a single exponential decay and will not be perturbed by the dilution factor. In some receptor systems, discrepancies between these two methods for studying the dissociation process have been observed and attributed to negative cooperative interaction between binding sites [77].

The association rate constant, k_1, can be determined by graphical methods from the initial binding velocities. When the reaction is considered to be irreversible, the second term of Eq. (3) is neglected, and after differentiation, the following reaction is established [80]:

$$\frac{2.303}{[P_0 - Q_0]} \log \frac{Q_0 [P_0 - B(t)]}{P_0 [Q_0 - B(t)]} = k_1 t \tag{9}$$

Plotting the left term of this expression as a function of time, a straight line will be obtained, with slope k_1.

This method has the disadvantage of assuming irreversibility of the reaction. However, when the contribution of dissociation becomes significant, departure from linearity will be observed in experimental data plotted according to this expression. In addition, Eq. (9) cannot be used when $P_0 = Q_0$, since it then reduces to $k_1 t = 0/0$, which is indeterminate [80]. In this case, again assuming irreversibility, Eq. (3) can be integrated to give the expression

$$\frac{B(t)}{P_0[P_0 - B(t)]} = k_1 t \tag{10}$$

Values of k_1 can also be calculated from the initial reaction velocity derived from the slope of the tangent drawn at zero time to association curves plotted as hormone-receptor complex concentrations versus time. However, the determination of this tangent by graphical methods is relatively inaccurate [81].

All the methods presented have in common the disadvantage of using only part of the information contained in a complete association time course performed until equilibrium. The use of nonlinear curve-fitting computer programs overcomes this problem, and allows a more precise and valid analysis of the association time course. An analytical solution of the second-order kinetics differential equation [3] is available and can be used as the model with such programs [82, 83]. With more sophisticated systems, models defined by the differential equations of the process can be used directly. This feature is particularly convenient for problems involving modeling of kinetic data derived during studies on hormone-receptor interactions, as we will describe.

It is important to notice that all methods of derivation of association-rate constants require an exact determination of both P_0 and Q_0; P_0 can be measured directly, and Q_0 is to be derived from equilibrium studies analyzed as described in Sec. VI. A. 2. Another requirement for the determination of association-rate constants is to establish that k_1 is independent from reactant concentrations. Any systematic trend of k_1 to follow hormone or receptor concentrations invalidates the application of the second-order kinetics model to the experimental system. This parameter has, thus, to be determined on a wide range of hormone and receptor concentrations.

The binding kinetics of hCG with its receptors in the rat testis have been extensively studied, using as the basic model for the analysis of the data the differential equation [3] describing second-order chemical kinetics [79, 84]. The quantitative analysis of the data was performed by using the MLAB system [73, 74]. This interactive program for mathematical modeling runs on a PDP-10 computer, and contains algorithms for nonlinear curve-fitting procedures which adjust the parameters of a model function to minimize the squared errors with respect to a set of data points, which can be weighted according to their individual variances. An important advantage of this system is the ability to directly utilize differential equations during the construction of mathematical models of kinetic processes. In addition, graphic routines permit rapid examination of the results of the fitting process by visual display, and provide high quality graphic outputs. As an example, Figure 28 shows the fit of a complete

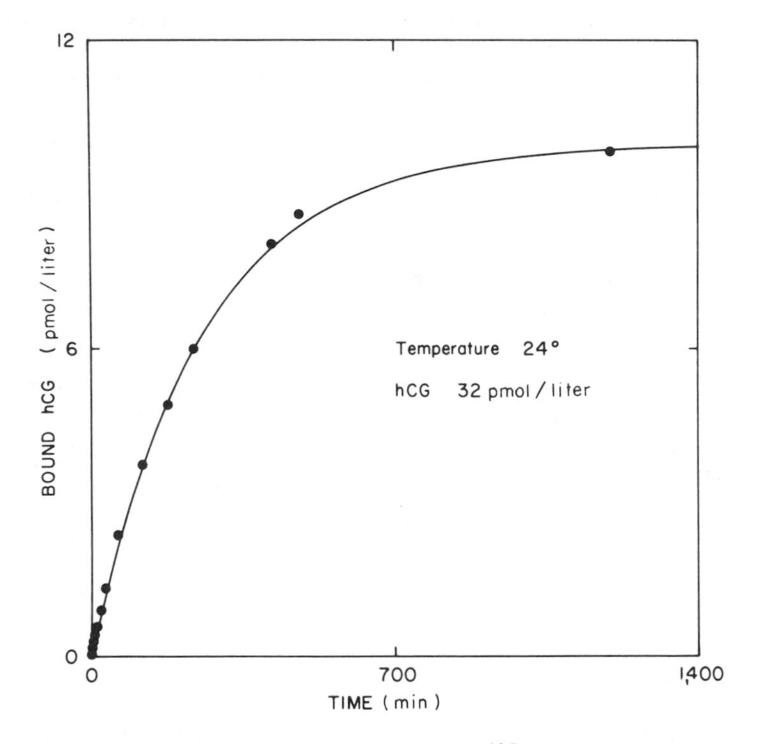

FIG. 28. Time course of the association of [^{125}I]hCG with its receptors in the rat testis at 24°C; the experiment was continued until equilibrium was reached. The solid line represents the best fit of the data with the differential equation for second-order chemical kinetics [Eq. (3)].

association time course for the binding of hCG to its receptors in the rat testis, performed according to the second-order model represented by Eq. (3). The results of this study are summarized in Sec. VI. B.

Dissociation curves can be analyzed graphically by plotting the logarithm of bound-hormone concentrations as a function of time. If the dissociation process follows first-order kinetics, with a single exponential decay function, a straight line will be observed. The value of k_2 is obtained by the following relation:

$$k_2 = \frac{-\ln 2}{t_{1/2}} \tag{11}$$

where $t_{1/2}$ is the half-life of the hormone-receptor complex. The dissociation studies should be performed for sufficient lengths of time to test whether the reaction is completely reversible. Only then can it be established that the dissociation rate proceeds according to simple first-order kinetics (single exponential decay) since in that case, the horizontal asymptote of the curve is zero. Because of the very slow dissociation of hCG and LH from their binding sites, such tests are often practically impossible. Departures from linearity in the semilogarithmic plots, or the presence of nonzero asymptotes require the use of more complex models for the characterization of the dissociation-time curves. Multiexponential models with or without a nonspecific irreversible component should then be evaluated [81]. This kind of analysis is also facilitated by the use of nonlinear curve-fitting methods.

A common problem in the quantitative analysis of hormone-receptor interaction is represented by the degradation of both reactants during incubation. Such degradation has been observed in many hormone-receptor systems [79, 85-89], and its effects on the derivation of the binding constants of the system should always be evaluated. Therefore, the kinetics of hormone and receptor degradation have to be determined. To measure the time course of free-hormone inactivation, labeled gonadotropin is preincubated with the receptor preparation; at various periods of time, aliquots of the incubation mixture are centrifuged or filtered, and the radioactive material from the supernatants or the filtrates is tested for its ability to react specifically with fresh receptors. The rate of gonadotropin inactivation is derived from the loss of binding activity of these fractions. For these calculations, it is important to take into account the amount of inactive labeled material initially present in the fresh tracer, and the amount of labeled hormone bound to the receptor during the first incubation. Ideally, the conditions of the second incubations should be selected so as to minimize the effects of degradation, i.e., low temperature or short-term incubations. Similarly, the kinetics of free-receptor degradation are measured by the effect of preincubation upon the ability of the preparation to bind labeled gonadotropin. In order to determine if the hormone and receptor degradation follow first- or second-order kinetics, the time courses of these processes can be determined for various reactant concentrations.

In order to take free-reactant degradation into account during the computation of binding constants, the terms $[P_0 - B(t)]$ and $[Q_0 - B(t)]$ of Eq. (3), representing the concentrations of free hormone and binding sites at time t, have to be replaced by more complex expressions. These terms should express the fact that during incubation, the rate of change of free hormone and free receptor are the results of two phenomena: (1) degradation of the free hormone and free receptor; and, (2) the formation and dissociation of the hormone-receptor complex.

Based on the finding that the bound hormone is protected from degradation and retains full biological activity [33], the assumption can be made that only free hormone and free receptor sites are subject to the degradative process.

Such a model has been described in detail, extensively evaluated by computer simulation, and applied to the binding of hCG to the rat testis receptors [79]. In this experimental system, the effects of degradation are particularly evident at 37°C. After an initial steady-state has been reached, a progressive drop of hormone-receptor complex was observed. This type of association time course could be fitted entirely with the model which takes reactant degradation into account (Fig. 29). Computer-simulated association curves obtained with the same model showed that the hormone-receptor complex concentrations of steady-state were more affected by degradation than the initial binding velocities (Fig. 30).

B. Binding Constants of Gonadotropin Receptors

Kinetic and equilibrium studies have shown that the equilibrium association constant (K_a) of the testicular and ovarian receptors is $2\text{-}6 \times 10^{10}\ M^{-1}$. Such high affinity of the gonadal receptor sites is concordant with the low plasma concentration of gonadotropins, and favors maximum sensitivity of the target cells to circulating trophic hormones. In addition, each target cell contains a large proportion of excess or "spare" receptors, about 100 times more than the number which must be occupied to evoke a maximum steroidogenic response [20]. The presence of such spare receptors possibly provides a further mechanism for maximizing the sensitivity of the gonads to trophic hormone.

The magnitude of the equilibrium association constant of the testis receptors for chorionic gonadotropin is comparable to that observed in homogenates of the pseudopregnant rat ovary ($1.5 \times 10^{10}\ M^{-1}$) [18]. Somewhat lower values were reported for rat ovarian receptors ($1.2 \times 10^9\ M^{-1}$)

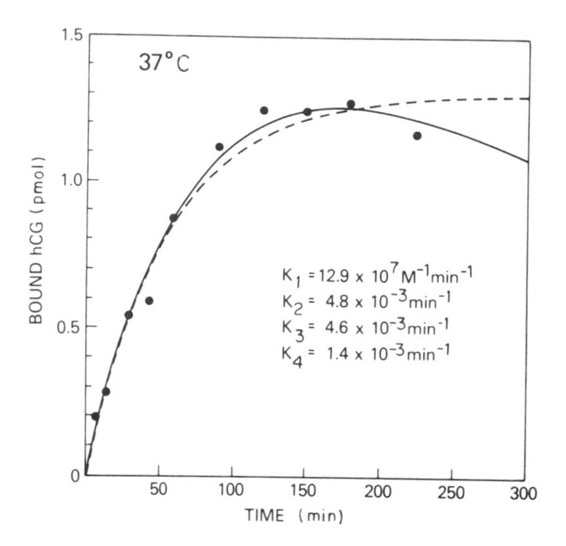

FIG. 29. Time course of the binding of [^{125}I]hCG by its receptors in the rat testis at 37°C, fitted with (——) and without (- - - -) corrections for hormone and receptor degradation. k_3 and k_4 represent the rate constants of hormone and receptor degradation treated as first- or pseudo-first-order kinetic processes.

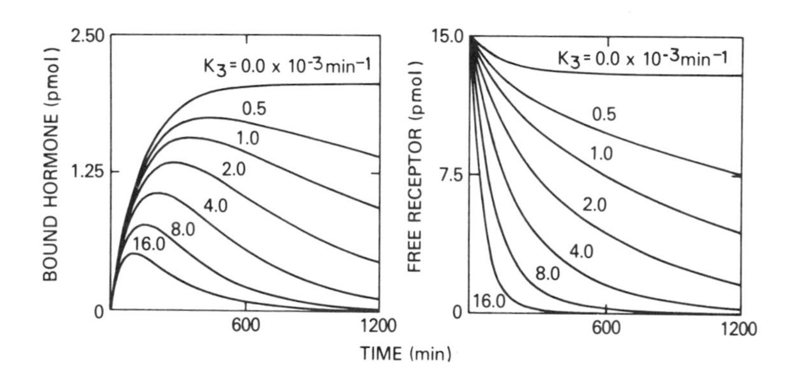

FIG. 30. Computer simulation of the effects of various rates of free receptor degradation on association curves and on the rate of change of free receptor. The meaning and values of the different parameters are as in Figure 27.

[15] and bovine corpus luteum receptors (0.3×10^9 M^{-1}) [24]. Most of these estimates have been based upon binding-inhibition data obtained with increasing concentrations of unlabeled hormone in the presence of a constant amount of ^{125}I-labeled gonadotropin, either LH or hCG. In binding-inhibition systems employing ^{125}I-labeled human gonadotropin tracer (either hCG or LH), significant differences in the slopes of displacement curves of human and ovine LH have been consistently observed [10]. This probably reflects differing affinity of the receptor sites for tracer human hormone and the displacing ovine hormone. As already noted, binding-inhibition studies with ovine LH in homologous binding systems of corpus luteum membranes and ^{125}I-labeled ovine LH [24] also give an association constant (0.3×10^9 M^{-1}), which is significantly less than those obtained in binding systems employing rat gonadal preparations with human gonadotropin tracers, and human hormone for displacement.

Kinetic studies of gonadotropin binding by particulate receptors of the testis and ovary have shown that the rates of association and dissociation of gonadotropins and receptors are markedly influenced by temperature, with higher rates at 37°C than at lower temperatures (Fig. 31). At 0 to 4°C, the association rate is extremely slow, and equilibrium is not reached for two to three days in the presence of physiological concentrations of labeled tracer hormone. The association rate is increased

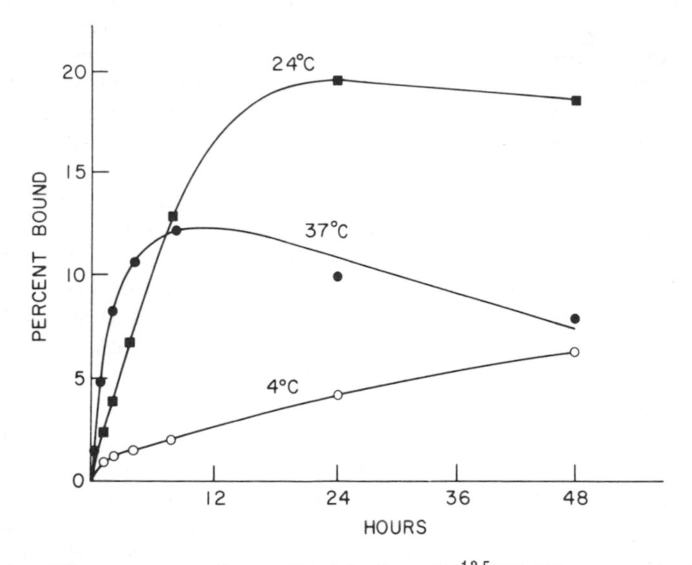

FIG. 31. Time course of specific binding of [^{125}I]hCG by particulate receptors in rat testis homogenates (40 mg) during incubation at 4, 24, and 37°C for 48 hr.

substantially at 24°C, and further at 37°C. However, degradation of both receptors and labeled ligand are much more rapid at the higher temperature, and the high initial association rate at 37°C is accompanied by lower steady-state binding than that observed at 24°C. For this reason, the most satisfactory studies of gonadotropin association are performed at lower temperatures, such as 24°C. The results of such studies indicate that the interaction between testis receptors and [^{125}I]hCG follows second-order kinetics at the concentrations employed for binding assay, with association-rate constant at 24°C of 1.3×10^7 M^{-1} min^{-1} [10]. Dissociation-rate studies are similarly influenced by temperature. At 0 to 4°C, preformed receptor-hormone complexes dissociate extremely slowly, with a half-time of about 72 hr. The dissociation rate is also temperature dependent, being higher at 24°C and at 37°C. The first-order dissociation-rate constant determined at 24°C was 2.1×10^{-4} min^{-1}, and the equilibrium constant derived from the ratio between the association and dissociation rate constant was 6.2×10^{10} M^{-1}.

Additional studies of the kinetics of the binding reaction between particular testis receptors and [^{125}I]hCG have been analyzed by computer fitting, and have given kinetic constants comparable to those above [79]. Such methodology has the advantage that allowance can be made for the measured extent of degradation of receptor sites and trophic hormone during incubation, and provides more precise rate constants than those derived without corrections for degradation. The mean values for k_1 at 24 and 37°C were, respectively, 4.7×10^7 M^{-1} min^{-1} and 11.0×10^7 M^{-1} min^{-1}. These values were derived from studies performed over wide ranges of hormone concentration (Fig. 32) and were independent of these reactant concentrations. The first-order dissociation-rate constant at 24°C has been calculated to be 5.3×10^{-4} min^{-1}. More extensive studies based on short preincubations followed by subsequent observations of the dissociation process have been performed. When fitted with a single negative exponential model, the initial dissociation velocities were found to be of the order 1.7×10^{-3} min^{-1} at 24°C and 4.6×10^{-3} min^{-1} at 37°C (Fig. 33). When the dissociation process was observed for long periods, the best fit of the data was obtained with a multi-exponential model. At 24°C, the dissociation curve exhibits an initial rapid component with mean rate constant of 2.1×10^{-3} min^{-1}, and a very slow component with rate constant of the order of 10^{-12} min^{-1}. At 37°C, the dissociation-rate constant of the rapid component was increased to 3 to 5×10^{-3} min^{-1}. The relative proportions of the rapid and slow components of various receptor preparations and their dissociation-rate constants were found to be influenced by the preincubation time. The dissociation of hLH and hCG from rat ovarian receptor sites has also been shown to be a multi-exponential process [26]. Whether the presence of the slow components is due to a secondary binding process of the labeled ligand with the particular

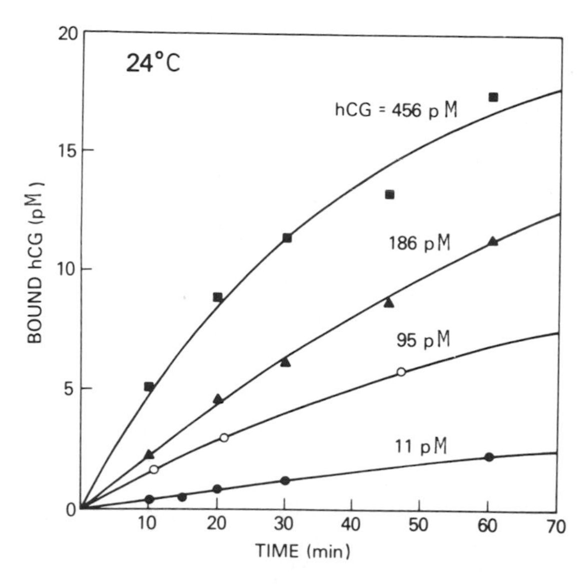

FIG. 32. Time course of gonadotropin binding by testis receptor at 24°C in the presence of various concentrations of hCG. The solid lines are the fit of the data obtained by computer analysis using the second-order chemical kinetics differential equation as the model [Eq. (3)].

receptor preparation, and/or to an effect of degradation, has still to be determined. The equilibrium association constant calculated from the ratio of the corrected rate constants at 24°C is 2.8×10^{10} M^{-1}, in good agreement with that previously derived from the rate constants determined without correction for degradation.

VII. EXTRACTION AND CHARACTERIZATION OF SOLUBLE GONADOTROPIN RECEPTORS

The particulate high-affinity gonadotropin receptors of the rat testis and ovary are readily solubilized by treatment with nonionic detergents such as Triton X-100 and Lubrol. Extraction of gonadal particles with nonionic detergents provides soluble gonadotropin receptors with uniform and reproducible binding properties. Such soluble preparations have been employed to determine several of the physico-chemical characteristics of the gonadotropin-receptor sites [90-97].

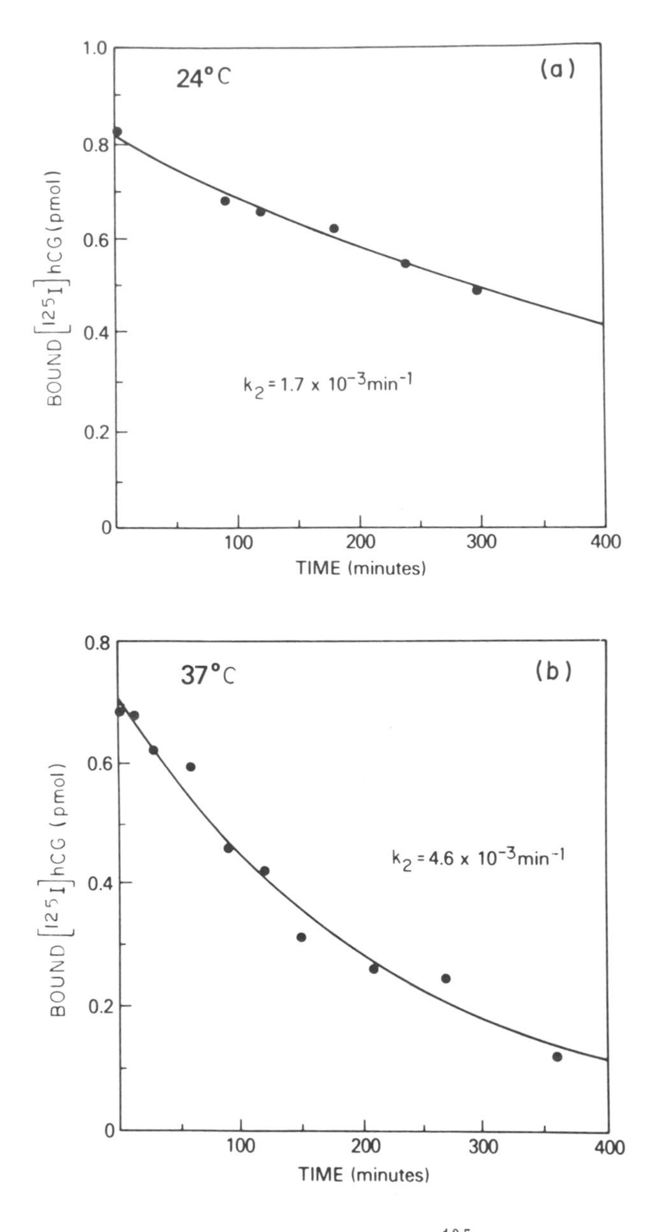

FIG. 33. Time course of the dissociation of [^{125}I]hCG from testis receptor sites, at 24°C (a) and 37°C (b). The preincubation period was 3 hr. The solid lines represent the fit of the data with a single negative exponential model.

A. Solubilization of Testis Particles

Decapsulated testes from adult male Sprague-Dawley rats are teased apart in phosphate-buffered saline (PBS), pH 7.4, to release interstitial cells and particles with high binding affinity for LH and hCG [5, 10, 91]. The highest yield of interstitial cell particles can be obtained by mixing the dispersed tubules for 5 to 10 min with a magnetic stirrer, followed by filtration through nylon mesh. The filtered particulate suspension is centrifuged at 120 g for 20 min to remove intact cells and tissue fragments, and the supernatant solution is then centrifuged at 27,000 g for 30 min to sediment particles with gonadotropin-binding activity. The pellets are resuspended in 1% Triton X-100 in PBS or 50 mM tris-HCl buffer, pH 7.4, for 30 min at 2 to 4°C, then centrifuged at 27,000 g for 20 min to remove undissolved material. Treatment of the particulate binding fraction from 10 testes with 0.5 ml of 1% Triton X-100 extracts the majority of the surviving gonadotropin-binding sites into solution, with total protein content of about 4 mg. Such solubilized binding sites are not sedimented by further centrifugation at 360,000 g for 3 hr [90, 91, 93].

When testicular binding sites are to be labeled with [^{125}I]hCG prior to extraction with detergent, interstitial cell particles (40 mg) are incubated at 4°C for 16 hr with 10^6 cpm (25 ng) of ^{125}I-labeled hCG, then washed with PBS to remove free hCG, and recovered by centrifugation at 20,000 g. Under these conditions, up to 50% of the labeled hCG is taken up by the particulate fraction and remains bound during storage of the particles at 4°C for several days. Such labeled particles are particularly useful for determination of the rate and efficacy of procedures for receptor solubilization. The receptor-hormone complex extracted from such prelabeled particles is considerably more stable than the free soluble receptors extracted from testis particles not previously equilibrated with hCG.

B. Assay of Receptor Binding

The relatively large size of the radioactive hormonal ligand (mol wt 38,000) restricts the range of separation procedures available for isolation of receptor-bound hormone after binding studies with ^{125}I-labeled hCG. Adsorbent procedures for removal of the free tracer are not readily applicable, and precipitation methods to isolate the bound complex give relatively high nonspecific values, due to partial coprecipitation of free hCG. However, a satisfactory method for isolation of the bound complex has been based upon repeated precipitation with polyethylene glycol (Carbowax 6000), employed at a final concentration of 12% [90]. Separation of antibody- and receptor-bound ligands of smaller molecular weight has been previously performed by a single precipitation with polyethylene glycol [98, 99]. For the hCG-receptor complex, precipitation at 12%

polyethylene glycol, followed by immediate redissolving in 0.1% Triton X-100, and a second precipitation with 12% polyethylene glycol, is necessary to achieve a satisfactory blank value. For binding assays, 0.5-ml aliquots of soluble receptor are mixed with 0.1 ml of PBS containing no hCG, or known quantities of unlabeled hCG, followed by 0.1 ml of PBS containing 50,000 cpm of ^{125}I-labeled hCG. Nonspecific binding is determined from tubes in which labeled hormone is incubated with receptor in the presence of 20 μg unlabeled hCG, and also in the absence of soluble receptor. The assay tubes containing 0.7 ml are kept at 4°C for 16 hr; 1 mg bovine γ-globulin (0.2 ml of 5 mg/ml solution) is then added as protein carrier, followed by 0.5 ml of 30% polyethylene glycol (wt/vol) in PBS. After standing for 15 min at 4°C, the tubes are centrifuged at 2000 *g* for 10 min at 4°C, and the supernatants aspirated. The precipitates are immediately redissolved in 0.9 ml of 0.1% Triton X-100 in PBS; after standing for 10 min at 4°C, the precipitation step is repeated with 0.5 ml of 30% polyethylene glycol. After further centrifugation and aspiration of the supernatants, the bound hormone present in the precipitates is determined by measuring the radioactivity remaining in the assay tubes. By this procedure, the nonspecific binding is less than 1% of the added radioactivity, and the specific binding ranges from 20 to 40%, depending on the concentration of receptor [90, 91].

C. Determination of Physical Characteristics of Soluble Gonadotropin Receptors

1. Gel Filtration

Analytical gel chromatography of the soluble gonadotropin receptors can be performed at 4°C on 100-cm columns of Sepharose 6B in 50 mmol tris-HCl buffer, pH 7.4, containing 0.1% Triton X-100. Adsorption of the soluble receptors to agarose during gel filtration studies with Sepharose 6B is minimized by the inclusion of 0.01% bovine serum albumin in the buffer solution employed for column chromatography. Columns are calibrated with Blue Dextran to define the void volume (V_o), and tritiated water to define the total or bed volume (V_t). Standard marker proteins employed for column calibration, either unlabeled or labeled with ^{14}C or ^{125}I, include thyroglobulin, apoferritin, human γ-globulin, albumin, and myoglobin. Blue Dextran should not be used to determin V_o during gel filtration of soluble receptors, as it complexes with the receptors and causes virtually all of the bound radioactivity to appear in the void volume. A small peak of radioactive microaggregates with detectable absorbance at 280 nm is consistently present during gel filtration of receptor preparations previously equilibrated with [^{125}I]hCG, and can be used to define the void volume of the columns during gel filtration of soluble receptors in the absence of Blue Dextran.

2. Density Gradient Centrifugation

Solubilized gonadotropin receptors can be analyzed by density gradient centrifugation in 5 to 20% (wt/vol) sucrose in 50 mM tris-HCl buffer, pH 7.4, containing 0.1% Triton X-100, in a Beckman Model L2-65B ultracentrifuge employing the SW40 rotor. Sample solutions are mixed with standard protein markers, usually BSA and 7S human γ-globulin; other useful marker proteins include apoferritin, thyroglobulin, and free [^{125}I]hCG.

After centrifugation at 38,000 rpm for 18 hr at 4°C, a total of 40 to 45 0.3-ml fractions is usually collected from each tub. The position of protein markers is determined by measurement of optical density at 280 nm, and labeled peaks are located by counting the radioactivity present in each fraction.

3. Gel Electrophoresis

a. Polyacrylamide Gel Electrophoresis in Detergent-containing Buffer

The solubilized hormone-receptor complex or free receptors can be subjected to analytical gel electrophoresis at pH 10 in 5% polyacrylamide gels containing 0.1% Triton X-100. The migration of the preformed complex is determined by measuring the radioactivity in individual gel slices, or by autoradiography if slab gels are used. To monitor the migration of free receptors in cylindrical gels, the individual 1-mm gel slices are homogenized and dispersed in 0.1% Triton X-100, then incubated overnight with [^{125}I]hCG at 4°C. The bound radioactivity is then determined, after sedimenting and washing the gel particles. Staining procedures for proteins and glycoproteins after electrophoresis in gels containing Triton X-100 are characterized by excessive background retention of dye, caused by binding to the detergent. The most satisfactory staining procedure for proteins separated by gel electrophoresis in Triton X-100 was found to be amidoblack in methanol:acetic acid, followed by electrolytic destaining to remove the background of dye-detergent complex.

b. SDS Gel Electrophoresis

Gel electrophoresis in SDS is mainly of value for estimation of the homogeneity and molecular weight of the receptor protein or subunits after extraction with detergents and purification by affinity chromatography. The binding activity of receptors is destroyed by exposure to SDS, and there is no virtue in searching for receptor binding after electrophoresis of soluble extracts in SDS gels. Similarly, preformed soluble hormone-receptor complexes are dissociated during SDS electrophoresis, and only free [^{125}I]hCG can be detected in gels after this procedure. A possible approach to the resolution and localization of hormone-receptor complexes by SDS electrophoresis is provided by the use of cross-linking agents such as glutaraldehyde, carbodiimide, or suberimidate to form covalent bonds within the complex and maintain its molecular association during electrophoretic fractionation in SDS.

VIII. PROPERTIES OF TESTICULAR RECEPTORS SOLUBILIZED WITH NONIONIC DETERGENTS

After extraction of particulate binding fractions with 1% Triton X-100, the undissolved material recovered by centrifugation at 27,000 g for 30 min contains little or no residual binding activity. The small pellet resulting from further centrifugation of the detergent extract for 60 min at 360,000 g is also devoid of significant binding activity. Detergent solubilization of testis particles previously labeled with [^{125}I]hCG has demonstrated that more than 90% of the labeled sites are extracted with 1% Triton X-100, and that 80% of the sites are extracted by 0.1% Triton. Such experiments have indicated that Triton X-100 is an effective agent for extraction of both testicular and ovarian receptors for LH/hCG. However, measurement of the recovery of unlabeled receptor sites during Triton extraction, as measured by quantitative binding assays of particulate and soluble binding sites with [^{125}I]hCG, reveals significant loss of binding activity during detergent extraction. In general, less than 10% of the original particulate receptors can be recovered after extraction with Triton X-100, and the affinity of the solubilized receptors for hCG is reduced by about 50% [91, 97].

A. Stability of Soluble Receptors

In addition to the difference in recovery of binding sites observed with free or occupied gonadotropin receptors, the binding sites solubilized by Triton X-100 are significantly less stable than the original particulate binding fractions. The detergent-extracted receptors are reduced by 50% during storage for 24 hr at 0 to 4°C, and incubation of the preparation for 20 min at 34°C causes a similar loss of binding capacity. The rapid loss of binding activity during incubation at 34°C is not significantly altered by the presence of trypsin inhibitor or Trasylol [91]. The relative lability of the uncharged soluble receptor is in marked contrast to the much higher stability of the solubilized hormone-receptor complex and the original particulate receptors. The loss of binding capacity of the soluble receptor during incubation at 34°C in 0.1% Triton is probably attributable to enzymatic degradation, since treatment of the solubilized preparation with 20 mM NEM causes a significant increase in [^{125}I]hCG binding to soluble receptors [31]. Such reduced receptor degradation could arise from blocking of thiol groups of degradative endopeptidases present in soluble preparations. In addition, the LH/hCG receptor isolated by affinity chromatography of solubilized gonadal particles appears to be quite stable [94, 100], and retains binding activity for several days during storage at 4°C.

B. Effects of Enzyme Treatment and Reducing Agents

The uptake of ^{125}I-labeled hCG by the particulate binding fraction is slightly enhanced by pretreatment with neuraminidase, and is significantly reduced by exposure to trypsin and Phospholipase A. Testis particles treated with Phospholipase C show unaltered or slightly reduced uptake of [^{125}I]hCG. Treatment of the solubilized receptors with trypsin causes almost complete loss of binding activity, and reduced binding is also apparent after exposure to Phospholipase A [91, 93]. By contrast, neuraminidase increases the gonadotropin binding by soluble receptors, and Phospholipase C has a similar effect as determined by polyethylene glycol precipitation. The enhanced binding after neuraminidase treatment of the soluble receptor is attributable to desialylation of the labeled gonadotropin, since asialo-hCG shows higher affinity for gonadotropin receptors than the native hormone [7, 18, 53].

Incubation of soluble receptors with 1 mmol dithiothreitol at 34°C for 20 min reduces the subsequent binding of [^{125}I]hCG to less than 10% of the control value. Similar effects upon hormone binding are also produced by incubation of soluble receptors with 12 mM mercaptoethanol or 42 mM cysteine, indicating that the integrity of disulfide bonds is an important factor in hormone-receptor interaction [31].

C. Binding Characteristics of Soluble Receptors

Specific binding of ^{125}I-labeled hCG by soluble gonadotropin receptors increases serially with rising concentration of the solubilized testis preparation. At high protein concentrations, decreasing binding is sometimes observed due to increased degradation of the tracer hormone. Saturation of receptor sites by labeled or unlabeled hCG is readily demonstrable, with an approximate binding capacity of 10^{-13} mole/mg of soluble protein. The equilibrium binding of [^{125}I]hCG by solubilized testis receptors shows a relatively sharp pH optimum, with maximum association at pH 7.4. The rate and extent of hormone binding are markedly influenced by temperature, but not by variations in calcium ion concentration. Binding is progressively inhibited by increasing concentrations of LH and hCG (Fig. 34),

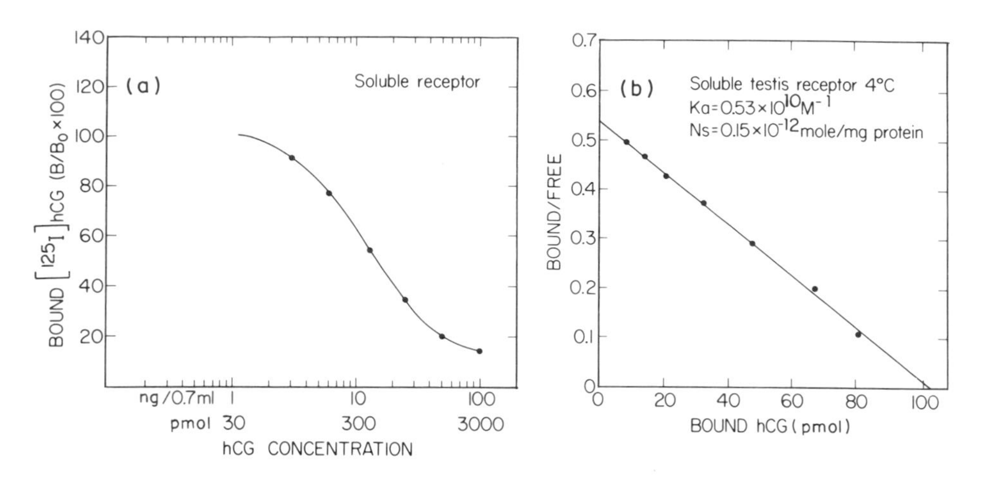

FIG. 34. (a) [^{125}I]hCG binding-inhibition curve obtained with soluble receptor. (b) Scatchard plot of receptor-hCG binding data.

but not by other peptide hormones, including follicle-stimulating hormone, prolactin, growth hormone, thyrotropin, and corticotropin [91].

1. Kinetics of Association and Dissociation

The rate of gonadotropin binding by the solubilized receptors is temperature-dependent, with higher initial association rate at 34°C than at 24 and 4°C. Due to the more rapid degradation of receptors and tracer hormone during incubation at 34°C, greater binding occurs at the lower temperatures. At 24°C, uptake of [^{125}I]hCG continues to rise until a maximum is reached at 6 hr; at 4°C, comparable levels of binding are attained after 24 hr. The second-order association-rate constant calculated from the binding velocity and the estimated binding capacity of the soluble testis receptor is 6.1×10^5 M^{-1} min^{-1} at 4°C. Dissociation of ^{125}I-labeled hCG from soluble receptor-hormone complexes occurs extremely slowly, with a first-order dissociation-rate constant of 1.2×10^{-4} min^{-1}. The equilibrium constant determined from the association and dissociation-rate constants at 4°C is 0.5×10^{10} M^{-1}.

2. Equilibrium Binding Constant

The rapid degradation of soluble receptors and tracer hormones at 34°C prevents accurate determination of equilibrium constants at higher temperatures. At 4 and 24°C, valid studies on the binding of [^{125}I]hCG at equilibrium can be performed during incubation with increasing concentrations of unlabeled hCG, and association constants determined from Scatchard plots or direct analysis of the binding-inhibition data (Fig. 34). The soluble testis receptors behave as a single order of binding sites, with equilibrium association constants of 0.6 to 1×10^{10} M^{-1} at 4 and 24°C, and 0.2 to 0.4×10^{10} M^{-1} at 34°C. The binding capacity of the soluble receptor preparation is significantly reduced when incubations are performed at temperatures above 4°C [91, 97]. Binding-inhibition curves obtained with the soluble receptor are similar to those observed during incubation of [^{125}I]hCG with rat testicular homogenate and with interstitial cell fractions from which the soluble receptor was extracted by treatment with Triton X-100. However, the association constant of the soluble receptor is 30 to 50% of that of the original particulate preparation, and the binding capacity is also significantly reduced.

D. Gel Filtration and Density Gradient Analysis of Soluble Receptors

1. Solubilized Receptors Equilibrated with [^{125}I]hCG

The elution profile obtained by gel filtration of the [^{125}I]hCG-receptor complex on Sephadex G-200 in 0.1% Triton X-100 shows a prominent peak of radioactivity coincident with the void volume indicated by Blue Dextran, and a retarded peak of free hCG with $K_{av} = 0.33$. The elution pattern obtained by gel filtration on Sepharose 6B also exhibits a major radioactive component coincident with the front peak of Blue Dextran (Fig. 35). This finding suggested adsorption of the receptor-hormone complex to the front marker during gel filtration, and this phenomenon was confirmed by gel filtration performed in the absence of Blue Dextran [91]. Dissociation of the hormone-receptor complex and free receptor from Blue Dextran can be achieved by passing the Blue Dextran-receptor complex through a small column of DEAE-Sephadex A-50 previously equilibrated with 5% ammonium sulfate in 0.1 M phosphate buffer. Elution of the receptor from the column can be performed with the same buffer, the Blue Dextran remaining adsorbed to the column. The association between protein ligands and Blue Dextran is believed to depend upon ionic interaction with the negatively charged chromophore, and suggests the presence of a net positive charge on the receptor molecule at pH 7.4 [94]. Gel filtration performed on Sephadex G-200 in the absence of Blue Dextran shows a slightly retarded peak of receptor-bound radioactivity ($K_{av} = 0.08$), which can be completely abolished by preceding incubation with excess unlabeled hCG (Fig. 36).

Gel filtration on Sepharose 6B in the absence of Blue Dextran shows a small front peak of aggregated material, followed by two closely adjacent peaks of receptor-bound hormone and free hCG. The receptor-hormone complex appears as a shoulder on the larger peak of free hCG in Figure

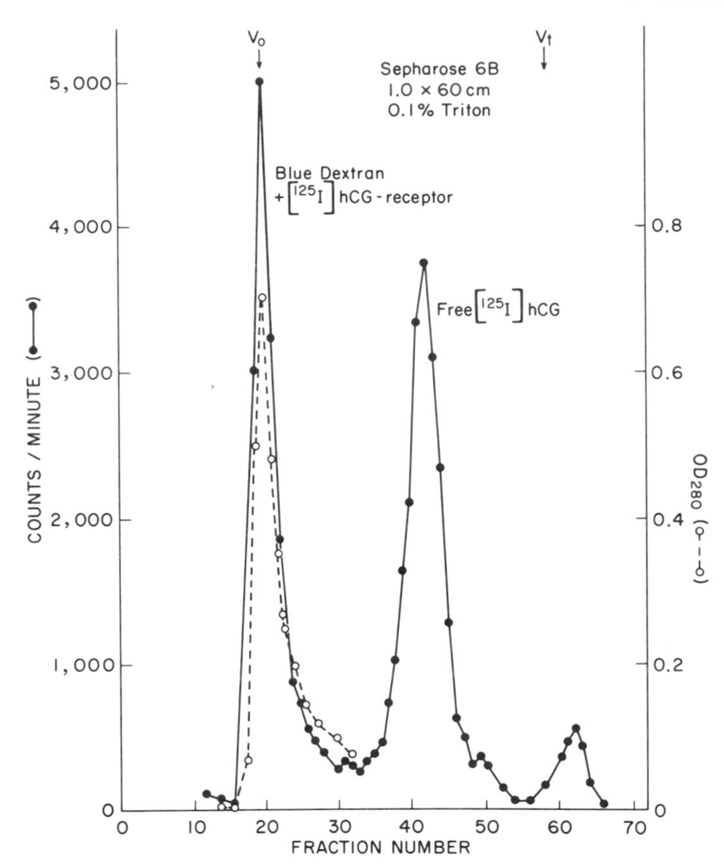

FIG. 35. Gel filtration of [^{125}I]hCG testicular receptor equilibration mixture on Sepharose 6B. A major radioactive of hormone-receptor complex was eluted coincident with the void volume shown by the Blue Dextran marker, and the free hormone is eluted with K_{av} of 0.56.

37(a) and could be abolished by preincubation with an excess of unlabeled hCG. The receptor-hormone complex is more clearly resolved after concentration and refiltration, as shown in Figure 37 (b), and is eluted with K_{av} of 0.32, in contrast to the K_{av} for free hCG of 0.56. Comparison of the partition coefficient of the receptor-hormone complex with those of reference proteins by the method of Laurent and Killander [101] gave a value of 64 Å for the hydrodynamic radius of the complex.

Density gradient centrifugation of the equilibrium hormone-receptor complex in 5 to 20% sucrose for 16 hr at 190,000 g gives the sedimentation pattern shown in Figure 38. The mean sedimentation coefficient of

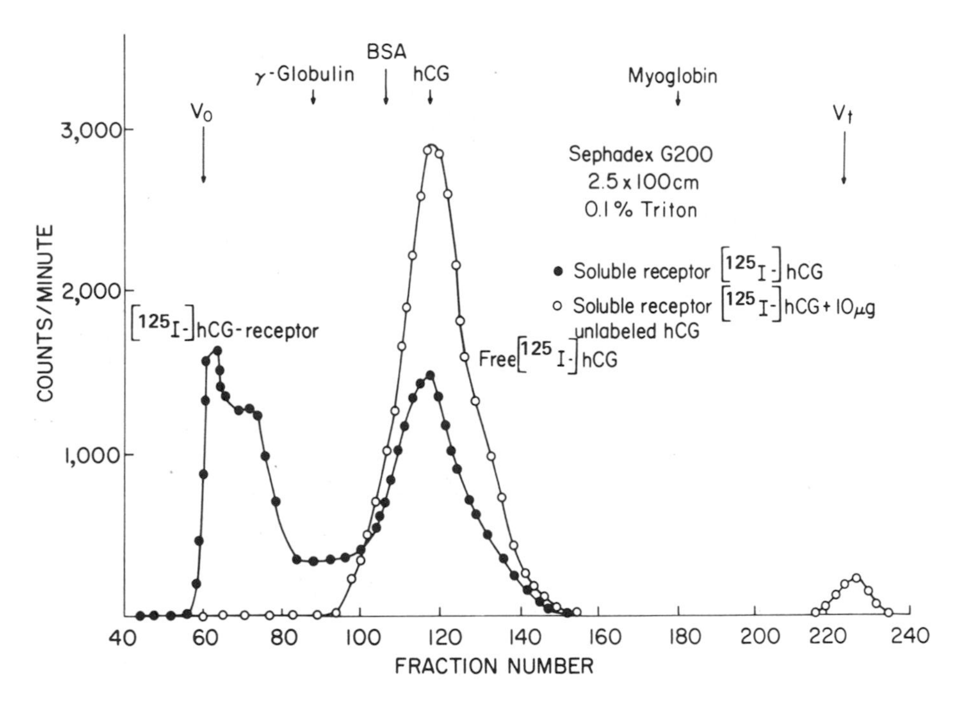

FIG. 36. Elution profile of [^{125}I]hCG testicular receptor complex during gel filtration on Sephadex G-200 in the absence of Blue Dextran. The hormone-receptor complex is eluted as a broad peak immediately behind the void volume. A second radioactive peak of free hCG is eluted with K_{av} of 0.32.

the soluble receptor-hormone complex calculated by comparison with reference proteins is 7.5 ± 0.35 (SD), and the complex is clearly separated from the 2.9S peak of [^{125}I]hCG.

2. Solubilized Particles Prelabeled with [^{125}I]hCG

In addition to greater stability than the free receptors, the soluble hormone-receptor complex extracted with Triton X-100 from testis particles previously labeled with [^{125}I]hCG exhibits a higher sedimentation constant (8.8S) than that of the 7.5S complex formed by equilibrating the free receptors with [^{125}I]hCG [93]. Despite this consistent difference in sedimentation constants, the two forms of the hormone-receptor complex behave identically during gel filtration on Sepharose 6B columns, with K_{av} values of 0.32 and calculated Stokes radii of 64 Å. When the detergent

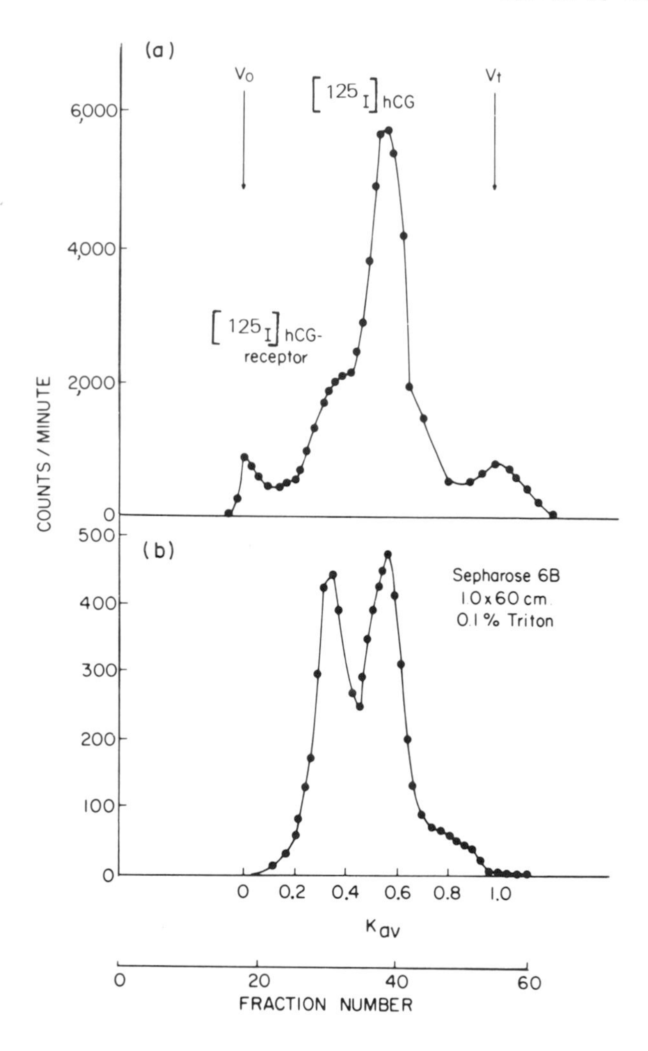

FIG. 37. (a) Elution profile of hCG testicular-receptor complex during gel filtration on Sepharose 6B, in the absence of Blue Dextran. A small peak of radioactivity is present at the void volume, coincident with a minor and constant peak of aggregated protein. The hCG-receptor complex is eluted as a shoulder preceding the major peak of free hCG. (b) Fractions corresponding to the hCG-receptor peak (above) are pooled with concentrated fivefold dry Sephadex G-50. Upon refiltration of an aliquot of the pooled fraction, the elution profile shows two clearly separated radioactive peaks, corresponding to free hCG and the hormone-receptor complex.

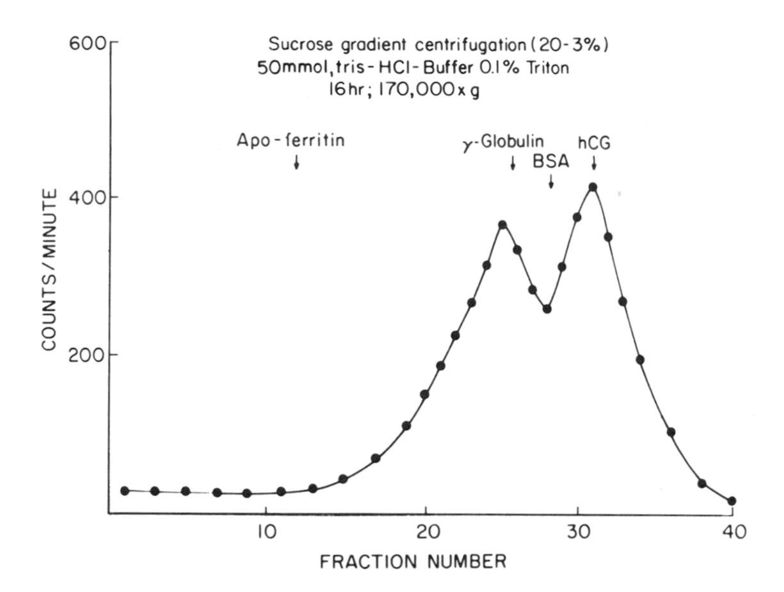

FIG. 38. Sucrose density gradient centrifugation of the soluble hormone-receptor complex obtained by gel filtration on Sepharose 6B [Fig. 37 (b)]. The two discrete peaks of radioactivity are free hCG (2.9S) and the hormone-receptor complex (7.5S), which is precipitable with polyethylene glycol.

concentration is reduced by dialysis, the preformed 8.8S hormone-receptor complex was converted to a broad and rapidly sedimenting peak (Fig. 39). This pattern is consistent with the formation of aggregates, and such dialyzed preparations are eluted in the void volume on Sepharose 6B. Such aggregation is reversible upon addition of Triton X-100 to a concentration of 0.1%, with conversion of the rapidly sedimenting complexes to the original 8.8S form. Dissociation of the hormone-receptor complex is favored by low pH and denaturing agents. Exposure of the complex to pH 3.5 for 1 to 5 min causes extensive dissociation of the labeled hormone from the receptor sites. Subsequent reassociation in 0.1% Triton after neutralization to pH 7.4 gives the 7.5S form of the complex. Exposure to 2 M urea or guanidine-HCl leads to irreversible dissociation of the hormone from the receptor sites, and no binding activity is recovered after removal of the denaturing agents by dialysis.

Extraction of prelabeled testis particles with Lubrol-PX or Lubrol-WX gave a 7S species on density gradient centrifugation [93, 94]. This form of the receptor also showed aggregation during dialysis to reduce the detergent concentration, with reversion to the 7S form when equilibrated

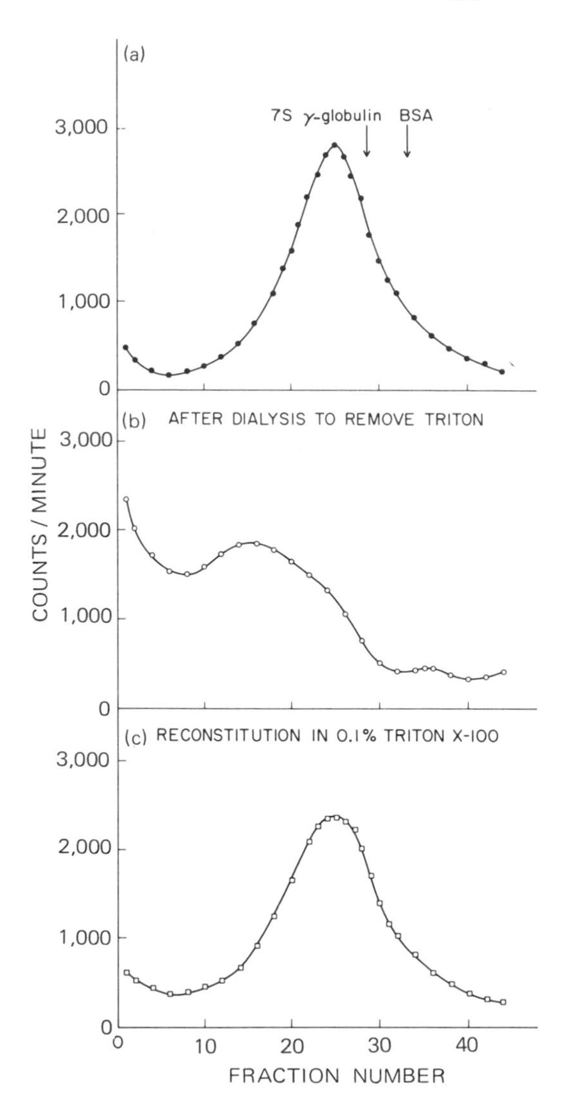

FIG. 39. Sucrose density gradient centrifugation of the hormone-receptor complex extracted from [^{125}I]hCG-labeled testis particles with Triton X-100. The complex sediments as an 8.8S species, (a), and aggregates reversibly when the detergent concentration is reduced by dialysis, (b) and (c).

with higher concentrations of Lubrol. The concentration of Lubrol necessary to extract the 7S species was higher than the concentration of Triton X-100 required to solubilize the corresponding species from the rat testis. Extraction should be performed with 1% Lubrol to obtain the 7S form of the receptor-hormone complex, and the minimum concentration of detergent required to avoid aggregation of the complex is 0.3% for Lubrol-PX and Lubrol-WX [100].

3. Free or Uncharged Soluble Receptors

The free gonadotropin receptor can also be characterized by gel filtration on Sepharose 6B in 50 mM tris-HCl buffer, pH 7.4, containing 0.1% Triton X-100 and 0.01% BSA. The receptor content of each eluate fraction is determined by binding assay performed by incubation of 0.5-ml aliquots with [^{125}I]hCG tracer at 4°C for 16 hr, in the presence and absence of excess unlabeled hCG, to determine specific binding. Separation of receptor-bound and free-tracer hCG by precipitation with polyethylene glycol reveals a single peak of binding activity, with K_{av} of 0.31, as shown in Figure 40. Sucrose density gradient centrifugation of the free gonadotropin receptor gives a single peak of binding activity with sedimentation coefficient of 6.5S.

E. Molecular Weight of Testicular Gonadotropin Receptor

The values for the Stokes radius and sedimentation coefficients of the receptor and hormone-receptor complex have been employed to calculate the molecular weight and frictional ratio of the components from standard equations [102, 103]. For the gonadotropin receptor-hormone complex, the partial specific volume derived from the density of the molecule in cesium chloride gradients is 0.776 cm^3/g. The relative viscosity of the solvent containing 0.1% Triton X-100, determined in an Oswald viscometer, is 0.95.

The calculated molecular weight of the free gonadotropin receptors is 194,300, and that of the receptor-hormone complex is 224,200. The difference between these values (30,100) is in reasonable agreement with the molecular weight of 37,000 for hCG, determined by structural analysis, and with the value of 39,000 calculated from the sedimentation coefficient (2.9S) and Stokes radius (31 Å) of the labeled hCG molecule. The frictional ratios of the receptor and the receptor-hormone complex are 1.65 and 1.56, corresponding to axial ratios (prolate) of 12.0 and 10.2, respectively.

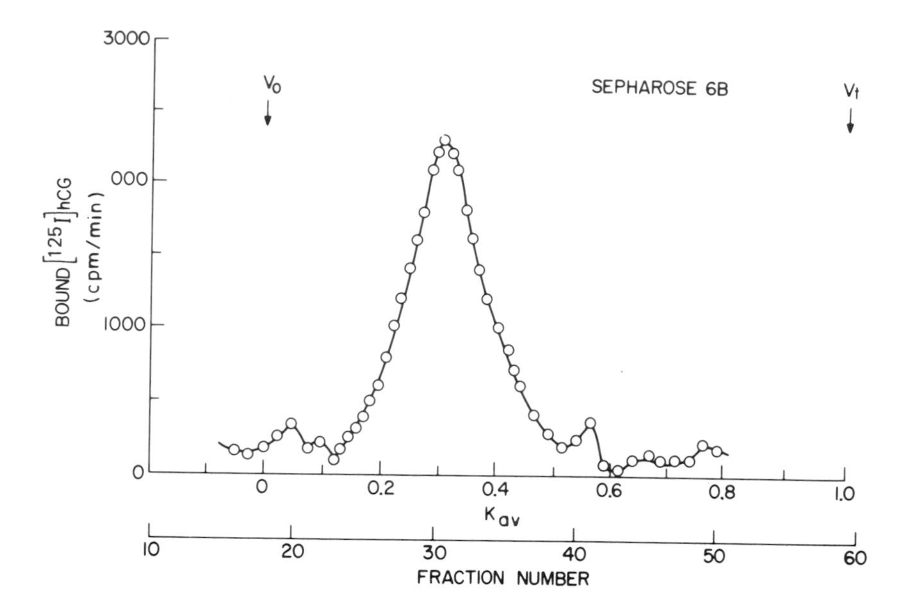

FIG. 40. Gel filtration of free testicular gonadotropin receptors on Sepharose 6B. After elution, an aliquot of each fraction is incubated with 15,000 cpm of [^{125}I]hCG in the presence or absence of 10^{-7} M hCG for 16 hr at 4°C, and the bound radioactivity isolated by polyethylene glycol precipitation. Nonspecific values are determined in tubes containing excess hCG and subtracted from the total precipitable radioactivity.

F. Gel Electrophoresis of Hormone-Receptor Complex

Analytical electrophoresis of the 8.8S hormone receptor complex in 5% polyacrylamide containing 0.1% Triton X-100 at pH 10.0 showed the presence of a single peak of radioactivity with R_f of 0.30. Under the same conditions of electrophoresis, free [^{125}I]hCG migrated more rapidly than the complex, with R_f of 0.60. This method provides satisfactory resolution of the hormone-receptor complex and free hormone during electrophoretic analysis of equilibrium mixtures of the two species (Fig. 41).

IX. SOLUBILIZATION AND CHARACTERIZATION OF OVARIAN GONADOTROPIN RECEPTORS

To prepare ovarian binding particles for solubilization studies, the luteinized ovaries from PMS/hCG-treated immature female rats are homogenized in all-glass homogenizer with two volumes of phosphate-

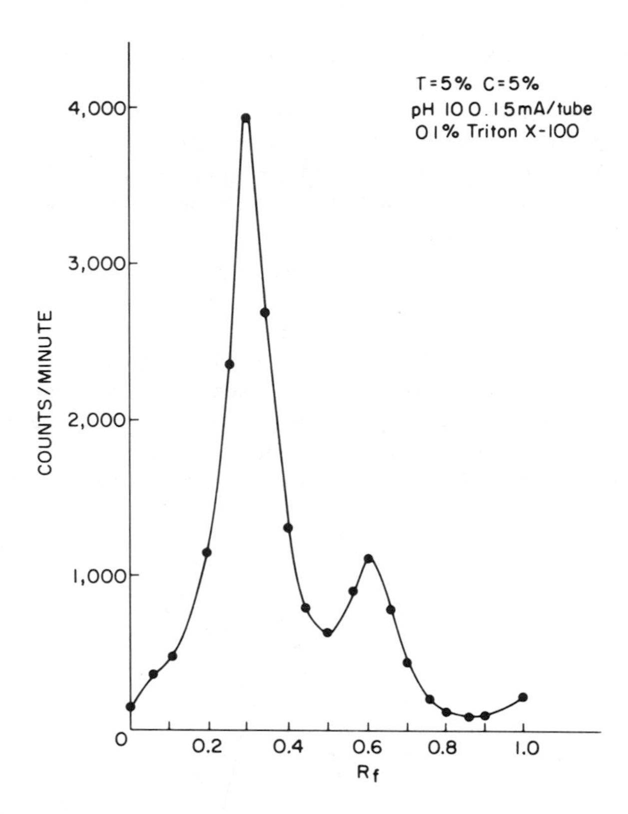

FIG. 41. Polyacrylamide gel electrophoresis of 8.8S hormone-receptor complex extracted from prelabeled testis particles with Triton X-100.

buffered saline or 50 mM tris-HCl buffer [92, 96]. After centrifugation at 120 g for 30 min at 4° C, the supernatant solution is discarded and the pellets are either (1) solubilized in 1% Triton X-100 in PBS for 30 min at 4°C, then diluted to 0.1% Triton for binding studies and physico-chemical characterization; or (2) resuspended in buffer and equilibrated with ^{125}I-labeled hCG for 16 hr at 4°C, then washed to free the particulate preparation from unbound hCG. The labeled pellet is then resuspended in 1% Triton X-100, Lubrol-PX, or Lubrol-WX for 30 min at 4°C to solubilize the preformed hormone-receptor complex.

The soluble receptors extracted from unlabeled ovarian particles are diluted to a final volume equivalent to 25 ml per ovary with 0.1% Triton X-100. Binding-inhibition curves are performed with 50,000 cpm (1ng) of ^{125}I-labeled hCG and increasing concentrations of unlabeled hCG, from 1 to 100 ng per tube. After incubation at 4°C for 16 hr, separation of receptor-bound and free tracer hCG is performed by precipitation with polyethylene glycol (PEG) as previously described. Ovarian receptors solubilized with Triton X-100 appear to be more stable than similarly prepared testis receptors, are of comparable binding affinity, and exhibit closely similar physical properties on gel filtration and density gradient centrifugation. The binding capacity of receptors solubilized from ovarian particles is 0.15×10^{-12} mole/mg protein, and the equilibrium constant (K_a) of the soluble preparation at 4°C is 0.66×10^{10} M^{-1}.

The three forms of soluble receptors with characteristic sedimentation constants extracted from testis particles with Triton X-100 are also demonstrable in the solubilized ovarian receptors. Thus, extraction and sucrose density gradient centrifugation of the free receptors gives a 6.5S peak of binding activity, whereas the free receptors equilibrated with [^{125}I]hCG after extraction, and those extracted from free labeled ovarian particles, show hormone-receptor peaks which sediment as 7.5S and 8.8S, respectively (Fig. 42). During gel filtration on Sepharose 6B, the 6.5S

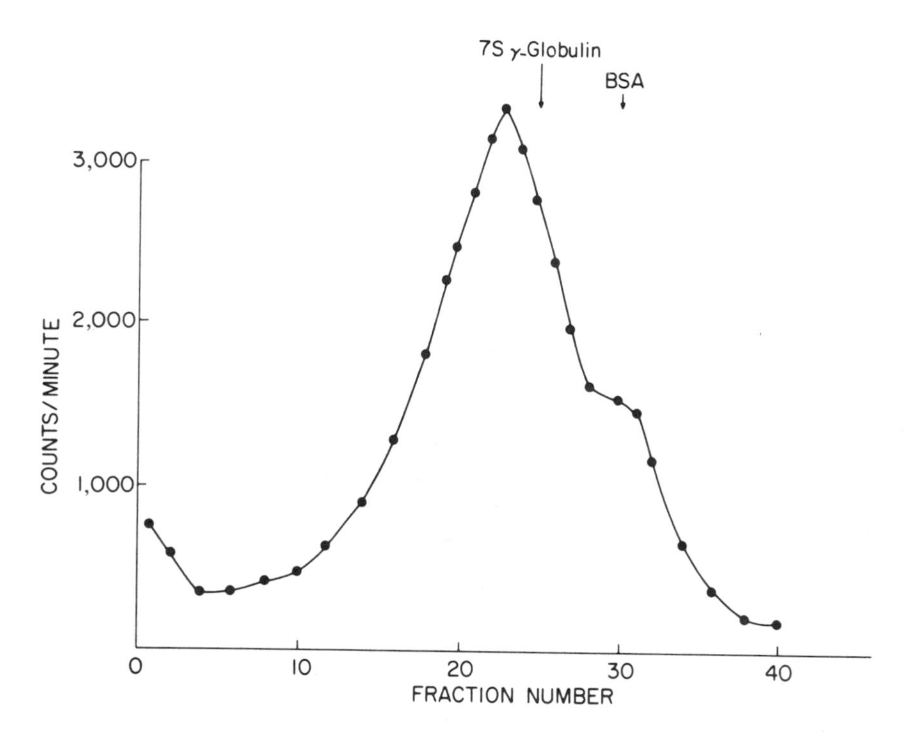

FIG. 42. Sucrose density gradient centrifugation of the prelabeled ovarian receptors extracted with Triton X-100 from particles equilibrated with [^{125}I]hCG. The hormone-receptor complex corresponds to the major 8.8S peak of radioactivity.

and 7.5S forms of the receptor are eluted with K_{av} of 0.36, equivalent to a hydrodynamic radius of 60 Å (Fig. 43). Under the same conditions, the K_{av} of the 8.8S hormone receptor complex is 0.30 [92]. Thus, the properties of the detergent-solubilized ovarian receptors are closely similar to those of the sites extracted from testis particles with Triton X-100 and Lubrol.

X. PURIFICATION OF SOLUBILIZED GONADOTROPIN RECEPTORS BY AFFINITY CHROMATOGRAPHY

The use of affinity chromatography for purification of hormone receptors is an obvious and attractive application of this isolation technique, which has been used with considerable success for isolation of cholinergic receptors from tissues such as the electric tissue of *Electrophorus electricus* [104]. Peptide hormone receptors which have been purified by affinity chromatography include insulin [105], LH/hCG [94, 100, 106], and prolactin [107]. The application of this technique to ligand pairs with extremely high affinity (i.e., $K_a > 10^{10}\ M^{-1}$) presents special problems during the elution procedure, which cannot be accompanied by competing ligands and requires the use of pH changes, chaotropic ions, or denaturing agents. Such conditions are not always favorable for isolation of biologically active receptors in satisfactory yield, but can be employed with success under appropriately controlled conditions.

For isolation of LH/hCG receptors of the rat testis, detergent-solubilized preparations of particulate binding sites were prepared by Triton X-100 extraction of the fragmented interstitial cell fraction obtained by physically teasing apart the tubules of the rat testis as described. The 200-20,000-*g* fraction containing interstitial cell particles was extracted with 1% Triton X-100 for 30 min at 4°C, then diluted 1:5 with PBS, and centrifuged at 360,000 g_{av} for 1 hr at 4°C. The clear supernatant solution contained more than 90% of the gonadotropin-binding sites extracted from the original particulate fraction, as determined by quantitative binding studies with [^{125}I]hCG.

Affinity chromatography media for isolation of the LH/hCG receptors were prepared by coupling partially purified hCG (Pregnyl; Organon) to agarose beads by a variety of conjugation procedures. Preliminary experiments were performed with hCG coupled directly to cyanogen bromide-activated Sepharose, and hCG coupled by glutaraldehyde treatment after previous binding of the gonadotropin to Sepharose-Concanavalin A [32]. In each case, the uptake of gonadotropin receptors from solution by the hCG-substituted gel was almost complete, but the recovery of receptor-binding activity was relatively low during subsequent elution under a variety of dissociating conditions, including low pH and agents such as

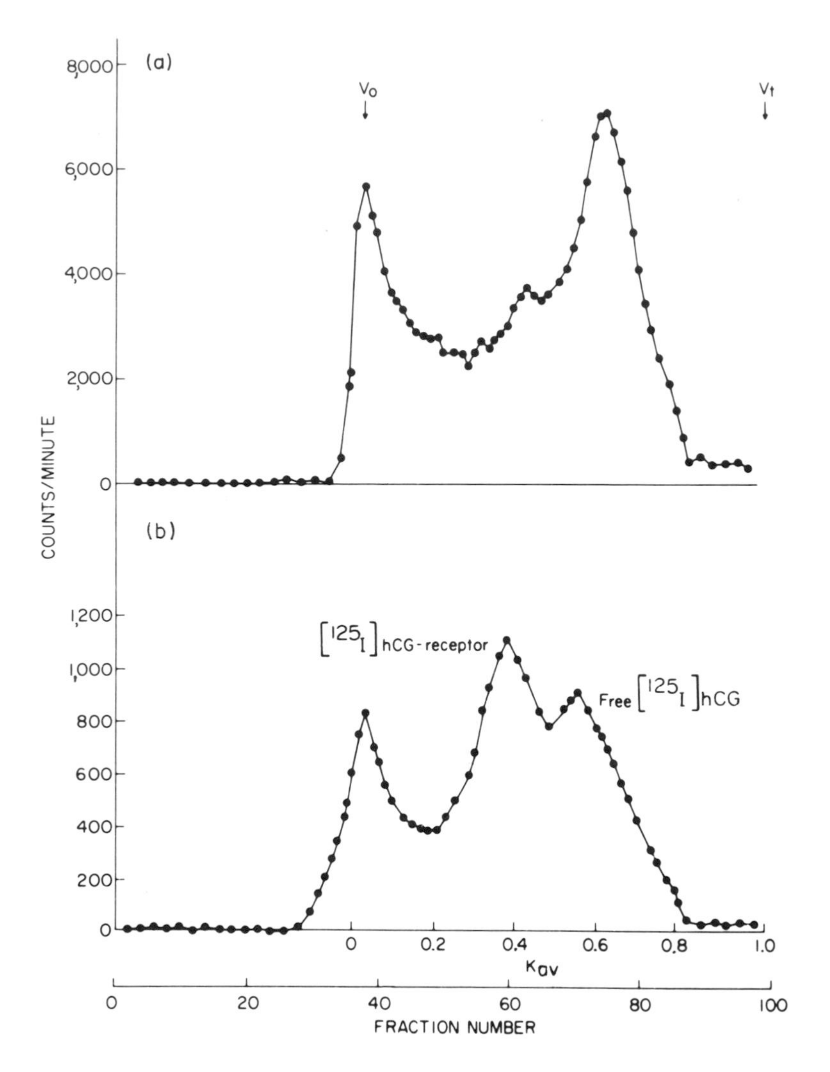

FIG. 43. (a) Gel filtration on Sepharose 6B (1 × 60 cm) of Triton-extracted ovarian gonadotropin receptors after equilibration with [^{125}I]hCG at 4°C for 16 hr. A peak of aggregated hormone-receptor complex is present at the void volume, and the hCG-receptor complex is eluted as a small shoulder preceding the major peak of free hCG. (b) Gel filtration of the pooled and concentrated fractions corresponding to the hormone-receptor complex in (a). The elution profile shows an aggregation peak at the void volume, followed by two clearly separated radioactive peaks which correspond to free hCG (K_{av} 0.56) and the hormone-receptor complex (K_{av} 0.36).

urea and guanidine. The most satisfactory gel-gonadotropin complex for affinity chromatography was prepared by conjugation of hCG to agarose beads bearing a 10-Å aliphatic chain terminating in N-hydroxy-succinimide ester (Affigel-10; Biorad). The gel-hCG complex was prepared by shaking 2 g of Affigel-10 beads with 100,000 IU hCG, equivalent to about 10 mg of purified hormone. The coupling reaction was performed in 20 ml PBS at 4°C for 16 hr, followed by washing of the gel with 5 liters of PBS, then 1 liter of 0.025 M acetic acid, and a further 2 liters of PBS. The uptake of hCG by activated gel particles was calculated from the binding of $[^{125}I]$hCG tracer during the coupling procedure to be 1,500 μg of pure hCG/gm of dry gel. Of the elution procedures tested, dissociation of the hormone-receptor complex at low pH [93], as previously described for the release of gonadotropin bound to particulate testis receptors [33], was found to give the highest and most consistent yield of soluble gonadotropin receptor sites.

For selective adsorption of gonadotropin receptors, the agarose-hCG derivative was shaken with solubilized testicular particles (1 g of wet gel per extract of 20 testes) for 16 hr at 4°C. The gel suspension was then allowed to settle in an 0.7 × 4-cm polypropylene column, and washed successively by shaking and elution with 4-ml aliquots of PBS, followed by 4-ml aliquots of 0.025 M acetic acid, pH 3.2. The acid eluates were immediately adjusted to pH 7.0 with 0.1 M ammonium hydroxide, and aliquots of 0.5-ml were equilibrated with $[^{125}I]$hCG by overnight incubation at 4°C for assay of receptor-binding activity. The remainder of the neutralized aliquots were frozen and lyophilized for quantitative binding studies and protein determination. The elution profile of gonadotropin binding activity from the affinity column is shown in Figure 44.

The hCG binding capacity of the original Triton-solubilized material was 0.15 pmol/mg protein, while the binding capacity recovered in fractions 1 through 4 of the eluted receptor was 225 pmol/mg protein, i.e., a 1,500-fold purification factor [94, 100]. The expected binding capacity for the completely pure receptor, based on the approximate molecular weight of 200,000 for the free receptor is ~5,000 pmol/mg protein. It was observed that the initial acid-eluted fractions contained two protein components demonstrable after SDS electrophoresis in 10% polyacrylamide gel electrophoresis as discrete bands with R_f's of 0.44 and 0.57. These components were coincident with minor and major bands observed in the crude soluble receptor preparation, and the more rapidly moving protein could be eliminated by discarding the first two acid eluates from the affinity column. The degree of purification of the receptor obtained by pooling the acid-eluted fractions 3 through 6 was calculated to be 15,000-fold, corresponding to an estimated binding capacity of 2,500 pmol/mg protein. This figure is equivalent to ~50% purity of the receptor isolated by affinity chromatography, and indicates the value of this technique for

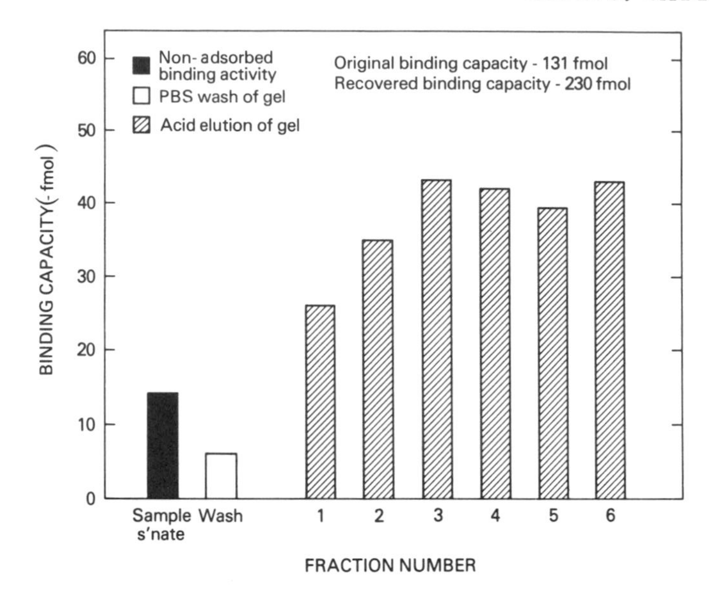

FIG. 44. Adsorption of soluble gonadotropin receptors by affinity column of agarose-hCG, and elution by 0.025 M acetic acid. The solid bar (left) indicates the binding activity remaining in the original receptor solution, and the hatched bars indicate the quantities of receptor recovered during each acid-elution step.

purification of high-affinity gonadotropin receptors with retention of biological activity [106].

A valuable feature of the present acid-elution procedure is the ability to recover specific receptor sites in the free form, rather than as the combined complex obtained by ligand elution, or the denatured form eluted with dissociating agents such as urea or guanidine. Although the purification technique is relatively simple, each detail of the method has been found to be necessary to achieve high recovery of the active receptor, and minor variations in technique can result in a considerable loss of binding activity. It is anticipated that application of this procedure to large-scale purification of LH/hCG receptors could provide sufficient quantities of the gonadotropin receptor for physico-chemical and immunological studies, and ultimately for structural analysis.

REFERENCES

1. B. Lunenfeld and A. Eshkol, in Vitamin and Hormones (I. G. Wood and J. A. Loraine, eds.), Vol. 25, Academic Press, New York, 1967, p. 137.

2. R. E. Mancini, A. Castro, and A. C. Seigeur, J. Histochem. Cytochem., 15:516 (1967).

3. D. M. DeKretser, K. J. Catt, H. G. Burger, and G. C. Smith, J. Endocrinol., 43:105 (1969).

4. A. E. Castro, A. D. Seiguer, and R. E. Mancini, Proc. Soc. Exptl. Med., 133:582 (1970).

5. K. J. Catt, M. L. Dufau, and T. Tsuruhara, J. Clin. Endocrinol. Metab., 32:860 (1971).

6. D. M. DeKretser, K. J. Catt, and C. A. Paulsen, Endocrinology, 88:332 (1971).

7. M. L. Dufau, K. J. Catt, and T. Tsuruhara, Biochem. Biophys. Res. Commun., 44:1022 (1971).

8. C. Y. Lee and R. J. Ryan, Endocrinology, 89:1515 (1971).

9. J. L. Vaitukaitis, J. Hammond, and G. Ross, J. Clin. Endocrinol. Metab., 32:290 (1971).

10. K. J. Catt, T. Tsuruhara, and M. L. Dufau, Biochim. Biophys. Acta, 279:194 (1972).

11. K. J. Catt, M. L. Dufau, and T. Tsuruhara, J. Clin. Endocrinol. Metab., 34:123 (1972).

12. H. Rajaniemi and T. Vanha-Perttula, Endocrinology, 90:1 (1972).

13. D. H. Espeland, F. Naftolin, and C. A. Paulsen, in Gonadotropins (E. Rosemberg, ed.), Geron X, Los Angeles, 1968, p. 177.

14. S. Kammerman and R. E. Canfield, Endocrinology, 90:384 (1972).

15. C. Y. Lee and R. J. Ryan, Proc. Natl. Acad. Sci. U.S.A., 69:3520 (1972).

16. F. Leidenberger and L. E. Reichert, J. Endocrinol., 91:901 (1972).

17. A. R. Means and J. L. Vaitukaitis, Endocrinology, 90:39 (1972).

18. T. Tsuruhara, E. V. Van Hall, M. L. Dufau, and K. J. Catt, Endocrinology, 91:463 (1972).

19. Y. Ashitaka, Y. Y. Tsong, and S. S. Koide, Proc. Soc. Exptl. Biol. Med., 142:395 (1973).

20. K. J. Catt and M. L. Dufau, Advan. Exptl. Med. Biol., 36:379 (1973).

21. C. P. Channing and S. Kammerman, Endocrinology, 92:531 (1973).

22. B. Danzo, Biochim. Biophys. Acta, 304:560 (1973).

23. J. Frowein, W. Engel, and H.-C. Weise, Nature, New Biol., 246: 148 (1973).

24. D. Gospodarowicz, J. Biol. Chem., 248:5042 (1973).

25. Ch. V. Rao and B. B. Saxena, Biochim. Biophys. Acta, 313:372 (1973).

26. C. Y. Lee and R. J. Ryan, Biochemistry, 12:4609 (1973).

27. A. R. Midgley, Advan. Exptl. Med. Biol., 36:365 (1973).

28. M. Rajaniemi and T. Vanha-Pertulla, J. Endocrinol., 57:199, (1973).

29. J. J. Marchalonis, Biochem. J., 113:299 (1969).

30. J. I. Thorell and B. G. Johansson, Biochim. Biophys. Acta, 251:363 (1971).

31. M. L. Dufau, D. Ryan, and K. J. Catt, Biochim. Biophys. Acta, 343:417 (1974).

32. M. L. Dufau, T. Tsuruhara, and K. J. Catt, Biochim. Biophys. Acta, 278:281 (1972).

33. M. L. Dufau, K. J. Catt, and T. Tsuruhara, Proc. Natl. Acad. Sci. U.S.A., 69:2414 (1972).

34. J.-M. Ketelslegers and K. J. Catt, J. Clin. Endocrinol. Metab., 39:1159 (1974).

35. M. L. Dufau, K. J. Catt, and T. Tsuruhara, Endocrinology, 90:1032 (1972).

36. M. L. Dufau, C. Mendelson, and K. J. Catt, J. Clin. Endocrinol. Metab., 39:610 (1974).

37. K. J. Catt, T. Tsuruhara, C. Mendelson, J.-M. Ketelslegers, and M. L. Dufau, in Hormone Binding and Target Cell Activation in the Testis (M. L. Dufau and A. R. Means, eds.), Plenum, New York, 1974, p. 1.

38. J. Presl, J. Pospisil, V. Figarova, and V. Wagner, J. Endocrinol., 52:585 (1972).

39. C. P. Channing and S. Kammerman, Biol. Reprod., 10:179 (1974).

40. R. L. Goldenberg, E. O. Reiter, J. L. Vaitukatis, and G. T. Ross, Endocrinology, 92:1565 (1973).

41. A. J. Zeleznik, A. R. Midgley, and L. E. Reichert, Endocrinology, 95:818 (1974).

42. A. E. Castro, A. Alonso, and J. Mancini, J. Endocrinol., 52:129 (1972).

43. R. L. Goldenberg, J. L. Vaitukaitis, and G. T. Ross, Endocrinology, 90:1492 (1972).

44. S. S. Han, H. J. Rajaneimi, I. C. Moon, A. N. Hirshfield, and A. R. Midgley, Endocrinology, 95:589 (1974).

45. H. J. Rajaneimi, A. N. Hirshfield, and A. R. Midgley, Endocrinology, 95:579 (1974).

46. M. L. Dufau, K. J. Catt, and T. Tsuruhara, Biochim. Biophys. Acta, 252:574 (1971).

47. A. R. Means and C. Huckins, in Hormone Binding and Target Cell Activation in the Testis (M. L. Dufau and A. R. Means, eds.), Plenum, New York, 1974, p. 145.

48. K. J. Catt, M. L. Dufau, and J. L. Vaitukaitis, J. Clin. Endocrinol. Metab., 40:537 (1975).

49. C. Y. Lee and R. J. Ryan, J. Clin. Endocrinol. Metab., 40:228 (1975).

50. B. B. Saxena, S. H. Hasan, F. Haour, and M. Schmidt-Gollwitzer, Science, 184:793 (1974).

51. L. E. Reichert and V. K. Bhalla, Endocrinology, 94:483 (1974).

52. S. Schwartz, J. Bell, S. Reichnitz, and D. Rabinowitz, Eur. J. Clin. Invest., 3:475 (1973).

53. T. Tsuruhara, M. L. Dufau, J. Hickman, and K. J. Catt, Endocrinology, 91:296 (1972).

54. F. Haour and B. B. Saxena, Science, 185:444 (1974).

55. K. J. Catt, M. L. Dufau, and T. Tsuruhara, J. Clin. Endocrinol. Metab., 36:73 (1973).

56. L. E. Reichert, G. M. Lawson, F. L. Leidenberger, and C. G. Trowbridge, Endocrinology, 93:938 (1973).

57. W.-K. Liu, K.-P. Yang, B. D. Burleigh, and D. N. Ward, in Hormone Binding and Target Cell Activation in the Testis (M. L. Dufau and A. R. Means, eds.), Plenum, New York, 1974, p. 89.

58. G. Scatchard, Ann. N.Y. Acad. Sci., 51:660 (1949).

59. G. Scatchard, J. S. Coleman, and A. L. Sken, J. Amer. Chem. Soc., 79:12 (1957).

60. S. A. Berson and R. S. Yalow, J. Clin. Invest., 38:1996 (1959).

61. R. P. Ekins, G. B. Newman, and J. L. H. O'Riordan, in Radioisotopes in Medicine: In Vitro Studies (R. L. Hayes, F. A. Goswitz, and B. E. P. Murphy, eds.), U.S. Atomic Energy Commission, Division of Technical Information, Oak Ridge, Tenn., 1968, p. 59.

62. R. S. Yalow and S. A. Berson, in Radioisotopes in Medicine: In Vitro Studies (R. L. Hayes, F. A. Goswitz, and B. E. P. Murphy, eds.), U.S. Atomic Energy Commission, Division of Technical Information, Oack Ridge, Tenn., 1968, p. 7.

63. C. L. Meinert and R. B. McHugh, Math. Biosci., 2:319 (1968).

64. E. E. Baulieu and J.-P. Raynaud, Eur. J. Biochem., 13:293 (1970).

65. J. E. Fletcher, A. A. Spector, and J. D. Ashbrook, Biochemistry, 9:4580 (1970).

66. I. M. Klotz and D. L. Hunston, Biochemistry, 10:3065 (1971).

67. G. Vassent, C.R. Acad. Sci., Paris, 273:113 (1971).

68. G. Vassent, C.R. Acad. Sci., Paris, 273:2161 (1971).

69. H. Feldman, D. Rodbard, and D. Levine, Anal. Biochem., 45:530 (1972).

70. H. A. Feldman, Anal. Biochem., 48:317 (1972).

71. D. Rodbard and R. E. Bertino, in Receptors for Reproductive Hormones (B. W. O'Malley and A. R. Means, eds.), Plenum, New York, 1973, p. 327.

72. H. G. Weber, J. Schildknecht, R. A. Lutz, and P. Kesselring, Eur. J. Biochem., 42:475 (1974).

73. G. D. Knott and D. K. Reece, in Proceedings of the ONLINE '72 International Conference, Vol. 1, Brunel University, England, 1972, p. 497.

74. G. D. Knott and R. I. Shrager, in Computer Graphics: Proceedings of the SIGGRAPH Computers in Medicine Symposium, Vol. 6, No. 4, ACM, SIGGRAPH Notices, 1972, p. 138.

75. R. I. Shrager, Technical Report No. 5, Division of Computer Research and Technology, National Institutes of Health, Bethesda, Md., 1970.

76. M. Berman and M. Weiss, SAAM Manual, U.S.P.H.S. Publ. No. 1703, U.S. Government Printing Office, Washington, D.C., 1967.

77. P. De Meyts, J. Roth, D. M. Neville, Jr., J. R. Gavin, and M. A. Lesniak, Biochem. Biophys. Res. Commun., 55:154 (1973).

78. D. Rodbard and K. J. Catt, J. Steroid Biochem., 3:255 (1972).

79. J.-M. Ketelslegers, G. D. Knott, and K. J. Catt, Biochemistry, 14:3075 (1975).

80. W. J. Moore, Physical Chemistry, Prentice-Hall, Englewood Cliffs, N.J., 1962.

81. D. Rodbard, in Receptors for Reproductive Hormones (B. W. O'Malley and A. R. Means, eds.), Plenum Press, New York, 1973, p. 280.

82. G. Vassent and S. Jard, C.R. Acad. Sci., Paris, 272:880 (1971).

83. D. Rodbard and G. H. Weiss, Anal. Biochem., 52:10 (1973).

84. J.-M. Ketelslegers, G. D. Knott, and K. J. Catt, in Hormone Binding and Target Cell Activation in the Testis (M. L. Dufau and A. R. Means, eds.), Plenum Press, New York, 1974, p. 31.

85. P. Freychet, R. Kahn, J. Roth, and D. M. Neville, Jr., J. Biol. Chem., 247:3953 (1972).

86. H. Glossmann, A. J. Baukal, and K. J. Catt, J. Biol. Chem., 249: 825 (1974).

87. C. Y. Lee and R. J. Ryan, Biochemistry, 12:4609 (1973).

88. S. J. Marx, C. Woodward, G. D. Aurbach, H. Glossman, and H. T. Keutmann, J. Biol. Chem., 248:4797 (1973).

89. S. L. Pohl, H. M. J. Krans, L. Birnbaumer, and M. Rodbell, J. Biol. Chem., 247:2295 (1972).

90. M. L. Dufau and K. J. Catt, Nature, New Biol., 242:246 (1973).

91. M. L. Dufau, E. H. Charreau, and K. J. Catt, J. Biol. Chem., 248:6973 (1973).

92. M. L. Dufau, E. H. Charreau, D. Ryan, and K. J. Catt, FEBS Lett., 39:149 (1974).

93. E. H. Charreau, M. L. Dufau, and K. J. Catt, J. Biol. Chem., 294:4189 (1974).

94. M. L. Dufau, E. H. Charreau, D. Ryan, and K. J. Catt, in Hormone Binding and Target Cell Activation in the Testis (M. L. Dufau and A. R. Means, eds.), Plenum Press, New York, 1974, p. 47.

95. K. J. Catt and M. L. Dufau, in Methods in Enzymology Vol. 37 Part B (B. O'Malley and J. Hardman, eds.), Academic Press, New York, 1974, p. 167.

96. M. L. Dufau, E. Podesta, and K. J. Catt, Proc. Natl. Acad. Sci. U.S.A., 72:1272 (1975).

97. M. L. Dufau and K. J. Catt, FEBS Lett., 52:273 (1975).

98. B. Desbuquois and G. D. Aurbach, J. Clin. Endocrinol. Metab., 33: 732 (1971).

99. P. Cuatrecasas, Proc. Natl. Acad. Sci. U.S.A., 69:318 (1972).

100. M. L. Dufau, C. Mendelson, E. Podesta, D. Ryan, A. Baukal, and K. J. Catt, in Fertility Regulation Through Basic Research, Plenum Press, New York, 1975.

101. T. C. Laurent and J. Killander, J. Chromatog., 14:317 (1964).

102. R. G. Martin and B. N. Ames, J. Biol. Chem., 236:1372 (1961).

103. H. K. Schachman, in Ultracentrifugation in Biochemistry, Academic Press, New York, 1959, p. 239.

104. R. W. Olsen, J. C. Meunier, and J. P. Changeaux, FEBS Lett., 28:96 (1972).

105. P. Cuatrecasas, Proc. Natl. Acad. Sci. U.S.A., 69:1277 (1972).

106. M. L. Dufau, D. Ryan, A. Baukal, and K. J. Catt, J. Biol. Chem., 250:4822 (1975).

107. R. P. C. Shiu and H. G. Friesen, J. Biol. Chem., 249:7902 (1974).

Chapter 10

GONADOTROPIN RECEPTORS

Brij B. Saxena

Cornell University Medical College
New York, New York

I. INTRODUCTION

Specific gonadotropin (FSH, LH, and hCG) receptors have been described in mouse, rat, pig, cow, and human ovary [1-9], and rat [10-12] and human testes [13]. Specific receptors for prolactin (PRL), which may be a component of the gonadotropic complex, have also recently been localized in the ovary [14-19]. The hormones bind specifically to their receptor sites in the target organ and the hormone action begins with the activation of adenylate cyclase, resulting in the formation of cAMP [20-22]. The ubiquitous role of cAMP, however, suggests that hormonal specificity resides in the structure of the receptor and the enzymic profile of the target cells. The molecular details of the hormone-receptor interaction have not been elucidated because of lack of isolation of receptor and the key enzyme, adenylate cyclase. It is not clear whether the receptor and the adenylate cyclase are the components of the same or different molecules. There is evidence for the presence of multiple hormone-receptor sites in the plasma membranes of the fat cells which compete for the same adenylate cyclase [23], and that the hormonal binding to the receptor and the activation of adenylate cyclase can occur independently [24-26]. It is likely that hormonal action may involve more than one protein molecule. Isolation and characterization of the gonadotropin receptor is, therefore, of fundamental importance to the elucidation of the mechanism of gonadotropic hormones at the molecular level, which may provide further understanding and newer means of control of disease due to end-organ failure.

II. PREPARATION OF HORMONES AS LIGANDS

The recognition of the receptor is based on the specific binding of the hormone, which may be determined histologically in sections of target tissue by autoradiography [11, 14, 17], and by the competition between the hormone labeled with a suitable marker and unlabeled hormone for binding to the target-organ receptor in both in vitro and in vivo experiments [27, 28]. Highly purified hormones are labeled with markers like fluoroscein [29] and ferritin [30, 31]; enzymes [32, 33]; and with ^{125}I, ^{131}I, and ^{3}H [12, 34-36] in the localization of receptor sites in the target organ, tissue slices, cells, homogenates, and isolated subcellular components [37]. It is recognized that configurational modifications during labeling procedures should be minimal to achieve specific binding of the labeled hormones to the receptor sites. When using radioisotopes, attempts are made to label the hormone with one atom of the radioisotope per molecule of the hormone. This is achieved by trial and error by changing the concentration of reagents, temperature, and exposure to the radioisotope. The degree of labeling can be determined by electrophoresis [38]. The integrity of native configuration of the labeled hormone is monitored by the degree of retention of biological activity by the labeled hormone.

Highly purified human chorionic gonadotropin (hCG) containing 12,000 IU/mg are received as gifts from Dr. Om P. Bahl (State University of New York, Buffalo, New York) and Dr. Robert E. Canfield, (College of Physicians and Surgeons, Columbia University, New York, New York). Human prolactin (hPRL) containing 26 IU/mg is obtained from the Hormone Distribution Officer (National Institute of Arthritis, Metabolism, and Digestive Diseases, National Institutes of Health, Bethesda, Maryland). Human luteinizing hormone (hLH) and human follicle-stimulating hormone (hFSH) containing 5,292 IU and 5,677 IU of Second International Reference Preparation of human menopausal gonadotropin (2nd IRP-hMG) are isolated from acetone-preserved human pituitary glands supplied by the National Pituitary Agency (Baltimore, Maryland).

A. Labeling of Hormone with Radioactive Iodine

Carrier-free radioactive iodide (^{131}I or ^{125}I) are used for external labeling to prepare labeled hCG, hLH, and hPRL with a low degree of chemical substitution. The advantage of ^{131}I is its high specific activity; however, the percent content of ^{131}I in commercial preparations is usually 35 to 40%, with an isotopic abundance of only 5 to 15% [39], which may often cause difficulties in obtaining suitably labeled tracer for receptor-binding studies. However, 95 to 97% carrier-free ^{125}I with isotopic abundance of 50 to 90% is available and the production of the labeled hormones of lower specific activity and higher biological activity is possible. The lower specific activity of ^{125}I-labeled hormone is compensated for by the counting efficiency, which is twofold greater than ^{131}I and is not dependent on crystal size. The longer half-life of ^{125}I eliminates frequent labeling. The labeling of hLH, hCG, and hPRL with ^{125}I and ^{131}I is performed by the chloramine-T method [40] or lactoperoxidase method [41].

B. Chemistry of Iodination

Phenolate ion is a chemical requirement for accepting iodine in a covalent linkage. Hence, the reaction with tyrosyl residues is the basis of all iodine tagging of proteins. The specific activity and yield of the labeled hormone, therefore, depends upon the number of tyrosyl residues of the protein available for iodination. The uniform introduction of one atom of radioactive iodine per mole of protein in a stable atomic substitution does not drastically affect the biological activity of the labeled hormone. The first step is the conversion of the iodide to a higher oxidation state, usually 0 or -1 by chloramine-T as follows:

$$CH_3C_6H_4SO_2NHCl + 2I^- \longrightarrow CH_3C_6H_4SO_2NH^- + Cl^- + I_2$$

Immediately after this reaction, the excess iodine is reduced to iodide by the addition of sodium metabisulfite in order to avoid overiodination of the protein.

$$I_2 + HSO_3^- + H_2O \rightarrow HSO_4^- + 2H^+ + 2I^-$$

The kinetics of the iodination, using acetyl tyrosine, may be visualized in the following reactions:

(1) O^- / R (phenolate ring) + I_2 $\underset{K-C}{\overset{Kc}{\rightleftharpoons}}$ O / I, H / R (quinoid ring) + I^- Kc

(2) O / I, H (quinoid ring) + base $\overset{Kd}{\rightleftharpoons}$ O^- / I / R (iodophenolate ring) + base$-H^+$

The rate-determining step is the second reaction, involving base-catalyzed displacement of a proton from a quinoid form. The rate of entry of the second iodine atom, to form diiodoacetyl tyrosine, is 30 times slower than the first when the rate measurement is based on the concentration of the respective phenolate ion. Therefore, at neutral pH, where both tyrosine and its monoiodo derivatives are present in their ionized form, the rate of iodination must be equal, since the phenolic pK's are 10.2 and 8.8, respectively. Consequently, with incomplete iodination, the presence of both mono- and diiodotyrosyl residues is to be expected [42-47].

Chloramine-T in excess may oxidize methionine residues to sulfones and contribute to the damage of the protein during iodination. The action of sulfite on the disulfide bonds in a protein, as shown by the equilibrium in the following reaction, is simultaneously a reduction and an oxidation, providing a highly specific means of fission.

$$R{-}S\cdot S{-}R + SO_3^{2-} \rightarrow RS^- + RS\cdot SO_3^-$$

Damage to protein may also occur due to the hydroxyl radical formed by the radiolysis of water, releasing ^{131}I from the labeled hormone and split disulfide bonds [39]. Radiation damage to protein may be reduced by the use of a minimum quantity of isotope, minimum exposure to chloramine-T and sodium metabisulfite, adsorption of the unreacted iodide on ion-exchange resins, and addition of albumin. Peroxidases, on the other hand, have the ability to oxidize iodide [41] and have been used to label proteins with iodine. The oxidation by lactoperoxidase is limited to iodide only, and the enzyme specificity prevents degradation of proteins.

C. Labeling of hCG, hLH, or hPRL with $Na^{125}I$ by Lactoperoxidase Method

Highly purified hCG, hLH, or hPRL are labeled with $Na^{125}I$ (Cambridge Nuclear, Cambridge, Mass.). One millicurie of $Na^{125}I$ in 20 μl of 0.4 M sodium acetate buffer of pH 5.6 is mixed with 10 μg of the hormone in 10 μl of 0.05 M phosphate buffer of pH 7.0 containing 0.15 N NaCl in the reaction vials. An aliquot of 600 ng lactoperoxidase (Sigma RZ = 0.78) in 20 μl of 0.1 M sodium acetate buffer of pH 5.6 and 300 ng of H_2O_2 in 20 μl of H_2O are added to the vial to initiate the reaction. Two additional 150-ng portions of H_2O_2 in 10 μl of water are then added at 2-min intervals. The reaction is stopped at the end of 6 min by dilution with 0.5 ml of phosphate buffer. Suitable aliquots of the crude reaction mixture are analyzed for the determination of specific activity of the hormone as described in Sec. II. E. The crude reaction mixture is purified by gel filtration on a 1 × 30-cm column of Sephadex G-100 equilibrated with 0.9% NaCl containing 0.1% bovine serum albumin (BSA). The labeled hormone is recovered in the unretarded fraction.

D. Labeling of hFSH with ^3H [12]

An aliquot of 500 μg of highly purified hFSH is dissolved in 0.1 ml sodium acetate, pH 5.6. To the solution are added 0.86 μmol of potassium periodate for oxidation, and the reaction is allowed to proceed for 10 min. The reaction is terminated by the addition of 20 μmol of glucose. The solution is dialyzed overnight against 200 ml of 0.15 M K_2HPO_4 of pH 7.5. The dialyzed material is transfered to a tube containing 1.6 μmol [^3H] KBH (6 Ci/mmol; Amersham/Searle Corporation, Arlington Heights, Ill.) in 0.2 ml of 0.01 M KOH. The mixture is stirred at room temperature for 15 min and allowed to stand for 15 min at room temperature. The excess of borohydride is discharged by adjusting the pH between 5.0 and 5.5 with 1.0 N acetic acid. The reaction mixture is dialyzed for 4 hr at room temperature and 15 hr at 4°C. At the end of the dialysis, an aliquot of the dialyzed solution containing tritiated FSH is analyzed for protein by the method of Lowry [48], and the [^3H] FSH is stored in suitable aliquots at -20°C until use.

E. Determination of Specific Activity of Labeled Hormones

Specific activity of the labeled hormones is determined by calculating the percent utilization of the isotope; the distribution of the radioisotope in labeled, damaged, and free radioisotope fractions is determined by chromatoelectrophoresis [38]. Calculations for specific activity are as follows:

$$\%\text{ utilization of } {}^{125}\text{I} = \frac{[\text{total counts (TC)} - \text{counts under } {}^{125}\text{I peak}] \times 100}{\text{TC}}$$

$$\text{SA}[{}^{125}\text{I}]\text{ hormone } \mu\text{Ci}/\mu\text{g} = \frac{\mu\text{Ci } {}^{125}\text{I (used)} \times \%\text{utilization } {}^{125}\text{I}}{\mu\text{g hormone iodinated}}$$

The specific activities of the labeled hormones are also determined by trichloroacetic acid precipitation. Five microliters of the crude reaction mixture are diluted to 5 ml with 0.05 M phosphate buffer, pH 7.5, containing 0.1% BSA. To a 200-μl aliquot of this solution are added 200 μl of a 20% solution trichloroacetic acid to a final concentration of 10% trichloroacetic acid. The precipitated proteins are recovered by centrifugation. The radioactivity with the trichloroacetic acid-precipitable material is considered as protein bound and is used in the calculation of the specific activity of the labeled hormone.

The specific activity of the labeled hCG, hLH, and hPRL is 30 to 35 μCi/μg. The biological activity of the labeled LH and hCG are 3,881 IU (95% confidence limits; 2,317-4,684 IU) and 8,923 IU (95% confidence limits; 5,826-12,250 IU), respectively, as determined by ovarian ascorbic acid depletion test [49]. The specific activity of the [^{3}H] FSH is 0.8 μCi/μg as determined by trichloroacetic acid precipitation and the biological activity of 3,258 IU (95% confidence limits; 2,029-3,858 IU) as determined by Steelman-Pohley assay [50]. The labeled hPRL is examined for its ability to specifically bind to midpregnant mammary tissue homogenates as an index of biological activity [51, 52]. The labeled hormones show 50 to 60% specific binding with excess receptor, and excess of unlabeled hormones competes with the labeled hormone so as to cause greater than 90% reduction in the specific binding of the labeled hormone. It may be pointed out that the calculations of molar concentration of hormones on the basis of the specific activity may be conceived to have some error, since it is assumed that all the molecules were labeled. However, such error is of little practical significance when severalfold changes in the binding capacity of the receptor is encountered.

III. LOCALIZATION OF THE RECEPTOR IN VIVO

Twenty-five-day-old female rats (Holtzman Company, Madison, Wisc.) are superovulated by a subcutaneous injection of 50 IU pregnant mare serum (Equinex, Ayerst Laboratories, Inc., New York, N.Y.) followed by 25 IU hCG (Follutein, E. R. Squibb, New York, N.Y.) in 0.1 ml physiological saline containing 0.1% BSA [3]. On day 6 following hCG injection, each animal is injected via tail vein with 0.6 μg of labeled hCG in 0.5 ml physiological saline containing 1% BSA and sacrificed by cervical dislocation after 90 to 120 min. The ovaries, kidney, liver, and rectus muscle are dissected and weighed. The radioactivity in the whole tissue and in an aliquot of blood is counted in an autogammacounter (Model 5219; Packard Instrument Company, Downers Grove, Ill.). The results are calculated as uptake per 100 mg tissue divided by uptake per 100 μl of blood. As shown in Table 1, ^{131}I-labeled hCG is concentrated in the ovaries and kidneys of normal and superovulated rats, as indicated by ovary-to-

TABLE 1. Distribution of ^{131}I-Labeled hCG in Rats[a]

Immature female	Ratio of molar uptake of ^{131}I-labeled hCG					
	Ovary/Blood	Kidney/Blood	Liver/Blood	Muscle/Blood	Ovary/Muscle	Kidney/Muscle
Normal	4.6 ± 1.1	2.4 ± 0.3	0.3 ± 0.03	0.06 ± 0.005	58.6 ± 9.3	44.3 ± 7.8
Superovulated	34.0 ± 8.4	2.0 ± 0.2	0.2 ± 0.01	0.09 ± 0.01	197.9 ± 74.5	20.7 ± 2.8

[a]Mean ±S.E. of four experiments are given.

blood and kidney-to-blood ratios of greater than one, and liver-to-blood and muscle-to-blood ratios of less than one. The ovarian uptake of labeled hCG in superovulated rats is sevenfold greater than in normal rats. The renal and hepatic uptakes in normal and superovulated rats are similar. Injection of unlabeled hCG as well as injection of labeled hCG and hCG antisera simultaneously inhibited the uptake of labeled hCG by the ovary. These observations suggest the presence of a specific receptor for hCG in the rat ovary and in the cow and human ovary as well [1-9, 53, 54]. Similarly, receptors for other gonadotropins can be localized in vivo.

IV. LOCALIZATION OF RECEPTOR IN VITRO

A. Preparation of Subcellular Fractions of Luteal Cells

The ovaries of 25-day-old superovulated rats and corpora lutea of first trimester of pregnancy from the ovaries of cows obtained fresh from the slaughterhouse are immediately frozen in liquid nitrogen following dissection. The corpora lutea tissue from the ovaries of women operated at The New York Hospital for uterine leiomyoma or benign ovarian cyst are also obtained following surgery and frozen in liquid nitrogen. Ovarian tissue is weighed and pulverized in liquid nitrogen by the aid of a frozen tissue pulverizer (Thermovac, Copaque, N.Y.) and homogenized by 15 to 20 strokes in a glass-teflon homogenizer at 4°C in 10 mM tris-HCl buffer of pH 7.2 containing 0.25 M sucrose, 1 mM DTT, and 1 mM $MgCl_2$. The homogenates are diluted 1:10 (wt/vol) and filtered through four layers of cheesecloth to remove connective tissue and intact cells. The filtrate is fractionated by discontinuous gradient centrifugation to isolate various subcellular components. All sucrose solutions are made in tris-HCl buffer (Diagram 1).

B. Protein Determination

An aliquot of each subcellular fraction is suspended in a known volume of 0.1 M NaOH containing 0.1% SDS, and heated in a boiling water bath for 30 min. An aliquot of the solution is used for protein determination, using BSA as the standard [48]. Fractions containing various subcellular components are stored in suitable aliquots at -20°C.

C. Electron Microscopy

A suitable aliquot of each subcellular fraction is diluted fivefold with tris-HCl buffer and centrifuged at 100,000 g. The pellets are fixed for 24 hr in 6.25% glutaraldehyde in 0.067 M cacodylate buffer of pH 7.3. The

DIAGRAM 1. Preparation of Subcellular Fractions of Luteal Cells

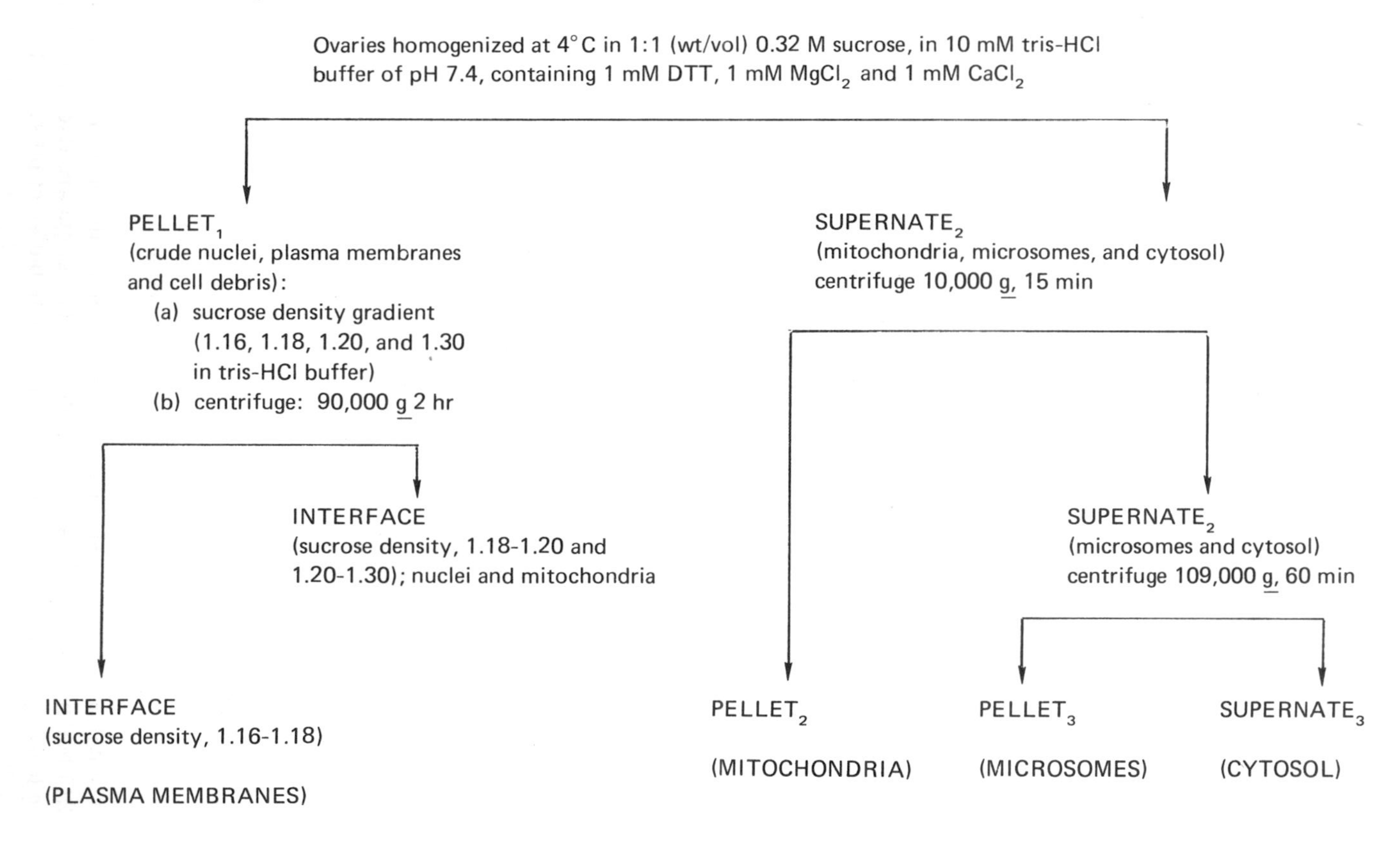

samples are washed for 5 min in chilled 0.25 M cacodylate or phosphate buffer containing 1% OsO_4 of pH 7.3 for 2 hr. Subsequently, all samples are dehydrated by passing through a graded series of alcohol and embedded in either Epon or Araldite. Thin sections (0.06-0.09 μ) are cut and stained in a 4% aqueous uranyl acetate solution and photographed by a Phillips EM-300 electron microscope. As shown in Figure 1, the plasma membrane fraction shows high purity, with 5 to 10% contamination with microsomes, mitochondria, and other cytoplasmic remnants.

D. Specific Binding

One-hundred-microgram protein aliquots of plasma membrane fractions are incubated in 12 × 75-mm disposable glass tubes with ^{125}I-labeled hCG, LH, PRL, and [^{3}H] FSH alone, as well as together with 100 ng of respective unlabeled hormones in a total volume of 320 μl of 10 mM tris-HCl buffer of pH 7.2 containing 0.1% BSA for 20 min at 37°C in a Dubanoff

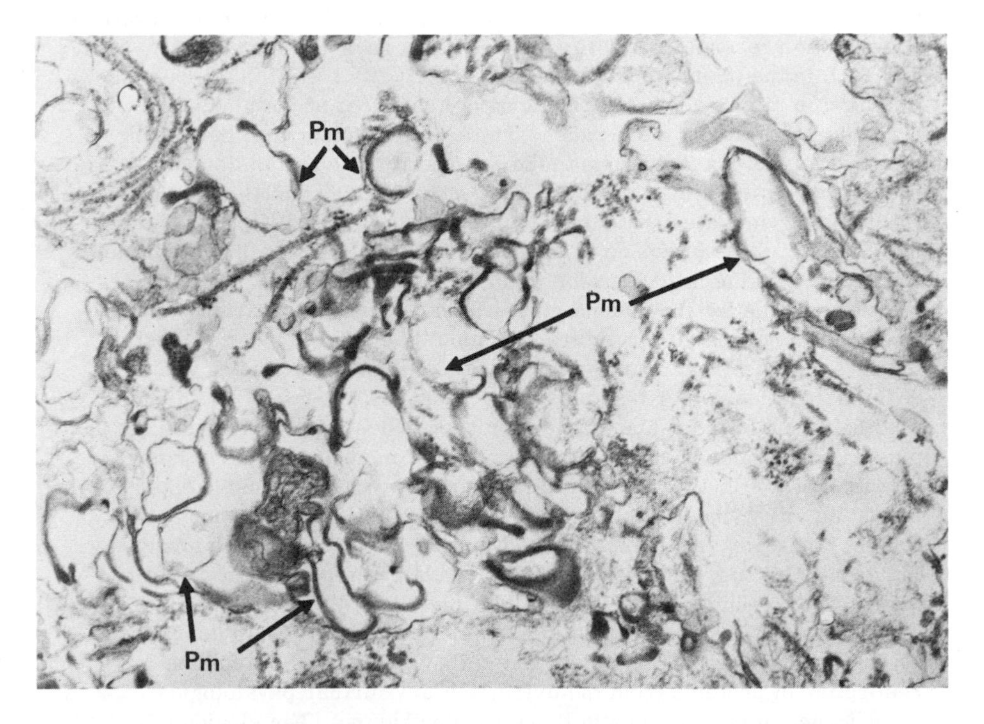

FIG. 1. Electronmicrograph of corpus luteum plasma membranes isolated from the ovaries of superovulated rats. The fraction contains predominantly plasma membranes, with some cytoplasmic remnants.

Shaker. The specificity of binding is tested by the competition with other unlabeled hormones. One-milligram protein aliquots of mitochondrial and microsomal fractions are also examined for specific binding as described. The plasma membrane fraction is centrifuged at 5,000 g for 20 min and the mitochondrial and microsomal fractions are centrifuged at 10,000 g for 15 min and at 109,000 g for 30 min, respectively. The pellets are washed twice with 1 ml of tris-HCl buffer. The supernates are decanted, and the pellets are counted. The specific binding is calculated as the difference between total binding and binding in the presence of excess labeled hormone.

The plasma membranes show the highest specific binding with labeled hCG, LH, FSH, and PRL. The labeled hCG bound to the membranes is displaced by unlabeled LH but the unlabeled FSH and PRL do not compete with labeled hCG and LH. The labeled LH is displaced by unlabeled hCG, suggesting that hCG and LH bind to the same receptor. Similarly, [^{3}H] FSH bound to the plasma membrane is displaced with the unlabeled FSH and not by unlabeled hCG, LH, or PRL; and [^{125}I] PRL bound to plasma membrane by unlabeled PRL but not by hCG, LH, or FSH. These observations suggest the presence of specific and separate receptor for hCG-LH, FSH, and PRL in the plasma membranes. Recently, Han et al. [55] have shown by electron microscopic radioautography that most of the binding is located in the plasma membranes, which is consistent with the finding of other investigators that plasma membranes of the luteal cells are the predominant site of hCG receptor [56]. The mitochondrial and microsomal fractions accumulate considerable amounts of labeled hCG, as well as labeled BSA. The presence of unlabeled hCG inhibits the binding of the labeled hCG to the mitochondrial and microsomal fraction to a much lesser degree than for the plasma membranes, suggesting that the majority of uptake is nonspecific in nature. It should be pointed out that most of the investigators have found varying degree of specific binding with subcellular fractions other than plasma membranes [56-58]. Whether this binding is due to the contamination of other subcellular fractions with plasma membrane or represents the entry of modified forms of hormones into the cell to bind to intracellular receptor awaits conclusive experiment. In vitro, the high specificity of binding is limited to the gonadal tissue. However, hCG administered in vivo is not bound exclusively to the gonadal tissue, but liver and kidney have been found to incorporate significant amounts of ^{125}I-labeled hCG [3, 59, 60]. Morell et al. [61] have shown the liver to be the site of metabolic clearance of desialyzed glycoproteins including hCG, and the kidney is responsible for the resorption of proteins from the urine, but any further physiological significance of the hormone in these tissues has not been elucidated. On the other hand, Lee and Ryan [62] have isolated a protein from ovaries, kidneys, and liver which specifically binds hCG, whereas others [63, 64] could not find such a binding protein from either liver or kidney.

Therefore, the tissue specificity of the in vivo binding of hCG is not clearly defined, nor has it been demonstrated that the in vitro and in vivo gonadal binding sites are identical. This is further supported by the data of Rao and Saxena [3], who find specific binding of [^{125}I]hCG to ovarian subcellular fractions in vivo, whereas binding of [^{125}I]hCG in vitro is nonspecific.

V. CHARACTERIZATION, SOLUBILIZATION, AND PURIFICATION OF THE RECEPTOR

A. Preparation of Plasma Membranes

The plasma membranes have been isolated previously in large quantities from other tissues by modifications of the original method of Neville [65]. Bovine corpora lutea provide a rich and easily available source of plasma membranes for use in the purification of the receptor. Fresh bovine ovaries from first trimester of pregnancy (foetus length from crown to rump up to 22 cm) are obtained fresh from the slaughterhouse and stored in liquid nitrogen until use. The plasma membranes are prepared as shown in Diagram 2. In a typical experiment, up to 100 corpora lutea from fresh bovine ovaries are ground in one liter of chilled 10 mM tris-HCl buffer of pH 7.8, containing 0.25 M sucrose, 1 mM $MgCl_2$, 1 mM dithiothreitol (DTT), and 10,000 IU/l Trasylol (FBA Pharmaceuticals, New York, N.Y.). Homogenization in isotonic sucrose preserves the structural integrity of mitochondria and nuclei, and reduces the contamination of plasma membrane fractions with other subcellular fractions. The tissue is ground and filtered through two layers of cheesecloth to remove connective tissue. The larger particles are reprocessed as above in another portion of 500 ml of the same buffer. During the first step, the tissue is homogenized by 10 to 15 strokes in a Type-C teflon-glass homogenizer with clearance size of 0.12 to 0.17 mm (A. H. Thomas, Philadelphia, Pa.). During the second step, the homogenate is centrifuged at 650 g for 20 min in a refrigerated centrifuge (Sorval, RC2B, Rotor GSA). The 650-g supernate is centrifuged at 13,000 g for 60 min in the same centrifuge. The pellet is rehomogenized and suspended in 50 ml of tris-HCl buffer and injected into the core of a Ti-14 zonal rotor (Beckman Model Spinco L3-50) spinning at 3,000 rpm and containing 500 ml of linear continuous gradient from 30 to 50% (wt/vol) of sucrose and 120 ml of a cushion of 50% (wt/vol) sucrose pumping at the rate of 20 ml/min by the aid of a continuous density gradient pump (Beckman Instruments, Inc.). An overlay of 50 ml of tris-HCl buffer is injected after the sample. At the end of centrifugation at 25,000 rpm for 90 min at 4°C, the centrifuge is decelerated to 3,000 rpm, and the rotor contents are displaced with 55% sucrose at the rate of 20 ml/min. Aliquots of the appropriate fractions,

DIAGRAM 2. Preparation of Plasma Membranes

One hundred fresh bovine corpora lutea, 500 g fresh tissue, (pulverized in liquid nitrogen).

↓

Step 1. HOMOGENIZATION (2.5 liters of 10 mM tris-HCl, pH 7.4; containing 0.25 M sucrose; 1 mM $CaCl_2$, $MgCl_2$, and DTT; and 10,000 IU Trasylol at 4°C.

↓

Filter through two layers of cheesecloth to remove connective tissue and intact cell aggregates (the larger particles may be processed as above).

↓

The homogenate is further ground in a type-C teflon-glass homogenizer with a clearance size of 0.12-0.17 mm by 10 to 15 strokes.

↓

Step 2. FIRST CENTRIFUGATION (650 g for 20 min at 4°C)

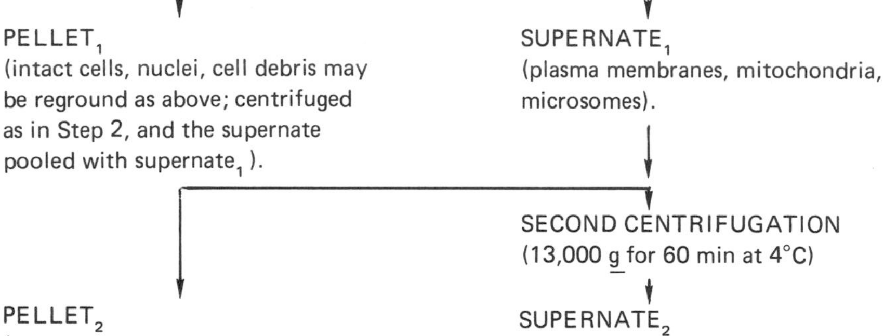

(homogenized and suspended in 500 ml of tris-HCl buffer and injected in 100-ml aliquots into a zonal rotor containing 500 ml of linear sucrose and spinning at 3,000 rpm, gradient of 30-50% (wt/vol); centrifugation for 90 min at 4°C at 25,000 rpm (90,000 g).

↓

ELUTION (20-ml fractions).

↓

Step 3. PLASMA MEMBRANES (aliquoted and lyophilized).

eluted following zonal centrifugation, are immediately frozen in liquid nitrogen for electron microscopy and for the determination of protein, hormonal binding, 5'-nucleotidase [66], and cytochrome-C reductase [67] activities.

As shown in Figure 2, the first peak (fractions 1-10) contains soluble proteins, lipids, and fragments of subcellular components. Two major peaks are then eluted in fractions 11 through 16 and 17 through 29. The second peak, eluted between 40 and 45% sucrose concentration, has high binding capacity for hCG, high 5'-nucleotidase activity, and low

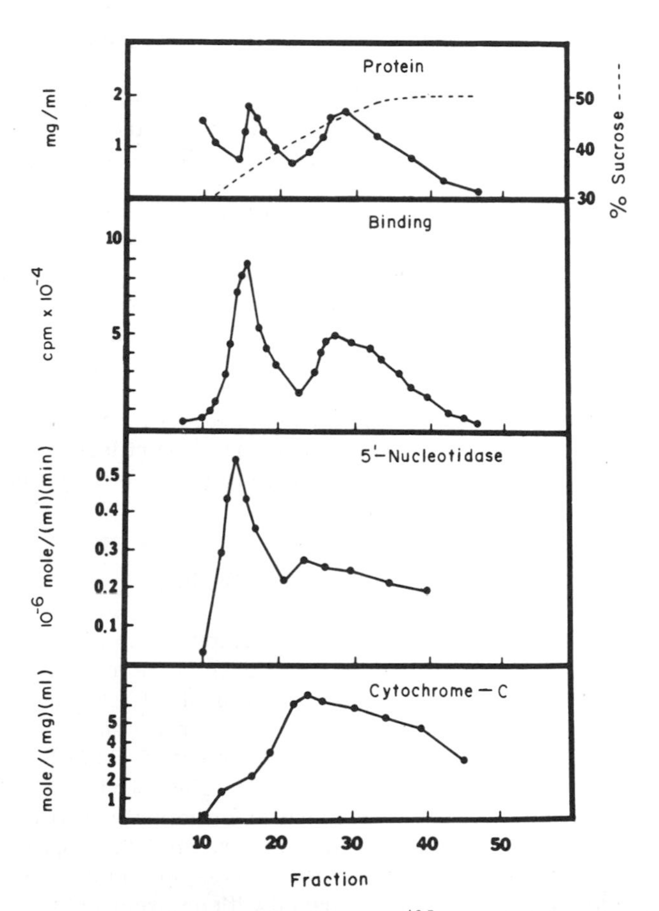

FIG. 2. Distribution of protein binding to [^{125}I]hCG, 5'-nucleotidase, and cytochrome-C reductase in fractions obtained from a preparative continuous sucrose density gradient (30-50% sucrose) centrifugation of plasma membrane-rich pellet obtained from bovine corpora lutea homogenate.

cytochrome-C reductase activity. The third peak also shows binding capacity for [^{125}I]hCG and 5'-nucleotidase activity; however, it has high cytochrome-C reductase activity. The presence of high 5'-nucleotidase activity is the most reliable marker for the plasma membrane, whereas cytochrome-C reductase activity is a good marker for mitochondria [66, 67]. The presence of plasma membrane in the second fraction and mitochondria in the third fraction is also confirmed by electron microscopy. The yield of plasma membranes is 8 to 10% based on the 5'-nucleotidase activity of all fractions obtained by continuous density gradient centrifugation; and the plasma membrane fraction is approximately 80 to 85% pure, with 5 and 10% contamination with microsomes and mitochondria, respectively, as judged by electron microscopy and enzymatic content. The fractions containing plasma membranes are mixed and homogenously suspended. Suitable protein equivalents of the suspension are aliquoted in 10 × 75-mm disposable plastic tubes, lyophilized, and stored in a refrigerator as receptor for the binding assay. The plasma membranes are stable in the lyophilized form for several months. The lyophilized membranes are reconstituted in an appropriate volume of water prior to the performance of the binding assay (Table 3). The presence of sucrose is helpful both in the preparation of homogeneous suspension and reconstitution of the lyophilized membranes. Sucrose at these concentrations does not interfere in the binding. The intact rat testes or homogenate remains 80% active when stored at -80°C for a period of six months. However, partial inactivation of the receptor or [^{125}I]hCG may occur during binding, which may be due to degradation by enzymes present in the crude receptor preparations [68]. This can be avoided by further purification of crude homogenates and the addition of enzyme inhibitors, as described for bovine corpus luteum plasma membranes.

B. Hormone Receptor Binding

The binding of hormones to the receptor is performed in 75 × 100-mm disposable plastic tubes (Falcon Plastics, Beckman Instruments, Inc.). All dilutions are made in 10 mM tris-HCl buffer of pH 7.2, containing 0.1% BSA, 1 mM $MgCl_2$, and 1 mM $CaCl_2$. Plasma membrane fractions equivalent to 20 μg of protein are incubated at 37°C for 45 min with [^{125}I]hCG, with [^{125}I]PRL (1.5×10^{-10} M, specific activity of 30-35 μCi/μg), or with [^{3}H]FSH (1.5×10^{-8} M, specific activity of 0.8 μCi/μg) in a total volume of 300 μl of 10 mM tris-HCl buffer, pH 7.2, containing 0.1% BSA. Each tube also contains 20 IU of Trasylol* in 20 μl of tris-HCl buffer. Similar incubations are performed with an excess of unlabeled hCG, (1×10^{-7} M), PRL (1.5×10^{-9} M), and FSH (1×10^{-5} M), respectively.

---

*Enzyme inhibitor (FBA Pharmaceuticals, New York, N.Y.).

Reagent blanks as well as controls are performed with the homogenates of rat rectus muscle with each experiment. The receptor-bound hormone is separated by filtration through an oxoid membrane of pore size of 0.45 μ (Amersham-Searle). The hormone-receptor complex retained on the filter is washed with 2 to 3 ml of 1% BSA in 0.9% NaCl. Usually, less than 1% of the total radioactivity is absorbed on the filter. The filters are transfered into a glass vial and the radioactivity is counted in a Packard autogammacounter with an efficiency of 51% for ^{125}I. In the case of [^{3}H] FSH, the filters are solubilized in glass vials by the addition of 1 ml of TS-1 (Research Products International Corporation, Elk Grove Village, Ill.) and counted for radioactivity in a Packard scientillation counter with an efficiency of 40% (Packard Instrument Company). Molecular weights of 32,000, 28,000, 40,000, and 20,000 are used in the calculation of molar uptake of hFSH, hLH, hCG, and PRL, respectively.

C. Characteristics of Binding of [^{125}I]hCG to Plasma Membrane Receptor

The high affinity of hormone for the receptor has been demonstrated by Kd values ranging from 10^{-9} to 10^{-11} mole/liter, which is consistent with the ability of the receptors to concentrate hormones from the peripheral circulation with concentration of hormones as low as 10^{-9} to 10^{-11} M. The half-saturation is attained at a concentration of 4×10^{-10} M. The rate constants of association and dissociation of [^{125}I]hCG to and from the receptor have been reported to be 2.17, 5.9, 2.8 × (10^6 M^{-1} sec^{-1}); 2.46 and 1.1 × (10^{-3} sec^{-1}); and 2.1×10^{-5} sec^{-1}, respectively. The dissociation constant calculated from these rates have been 8.0×10^{-10} M, 1.9×10^{-10} M, and 1.13×10^{-9} M. The apparent dissociation constants calculated from the equilibrium data are 1.5×10^{-10} M, 1.9×10^{-10} M, and 3.0×10^{-9} M [1, 8]. The number of hCG-binding sites, as determined by Scatchard analysis [69], are in the order of 6.3×10^{-15} moles/mg ovary and 8.4×10^{-10} moles/mg protein. Bellisario and Bahl [64], using a 2,000-g fraction of rat testis homogenate as the receptor, estimated saturation to be reached at 2×10^{-10} M hCG. A Scatchard plot of the binding data yields an association constant (Ka) of $10^{10} \times M^{-1}$; and a maximum binding of 0.12 μg, or 3 pmol, of hCG per testes is estimated. The second-order rate constants of binding at 25 and 37°C are 8×10^7 M^{-1} and 1.4×10^8 M^{-1} (min^{-1}), respectively. The in vitro dissociation of [^{125}I]hCG follows first-order kinetics, yielding dissociation rate constants of 6×10^{-4} and 2×10^{-3} min^{-1} at 25 and 37°C, respectively. The association constant (Ka) calculated from the association- and dissociation-rate constants is 1.3×10^{11} M^{-1} at 25°C and 7×10^{10} M^{-1} at 37°C. The close values for Ka obtained at both temperatures is due to a two- to threefold increase in association- and dissociation-rate constants at 37°C compared with 25°C. Similarly, for the binding of FSH by rat testes homogenate a Kd of 6.7×10^{-10} M and binding sites in the order of

6.2×10^{-14} moles/mg wet tissue weight are estimated [70]. The number of receptor sites in human ovary has been shown to change with various stages of the menstrual cycle [71-73]. The variations in the labeled hormone, temperature, and system used may affect the estimates of these constants. It has been suggested that endogenous hormone bound to the ovarian receptor may affect the measurement of receptor sites by the use of exogenous labeled hormone.

D. Dissociation of Receptor-Bound Hormone

The mechanism of dissociation of receptor-bound hormone is not understood. It is speculated that dissociation of receptor-bound hormone may be the reversal of steps following during association of hormone to the receptor, leading to a conformational change in the receptor to loosen the binding. At 4°C, the hormone-receptor complex has been shown to be stable, and the bound hormone is not dissociated up to 48 hr; whereas at 37°C the bound hormone is dissociated from the receptor rapidly [10, 74]. The dissociated hormone has been shown to be immunologically and biologically identical to native hormone [62, 75]. It is of interest that $[^{125}I]$hCG eluted from receptors is significantly more active than the original $[^{125}I]$hCG in binding to the receptor, and is biologically active [64]. This could be due to affinity purification or to loss of sialic acid during initial binding [76]. However, disk gel electrophoresis of $[^{125}I]$hCG eluted from the bovine membrane receptor using autoradiography showed identical electrophoretic mobility as the native hCG [77]. It has been postulated that a separate site for the degradation of hormone, for example, insulin, may exist at the receptor site [78].

E. Effect of Various Treatments on Hormone-Receptor Binding

The binding of $[^{125}I]$hCG to plasma membranes is inhibited at acidic and alkaline pH's; however, it can be restored by adjusting the pH close to neutral. Brief exposure of the membranes to temperatures of 50 to 90°C results in irreversible loss of $[^{125}I]$hCG binding. The binding of $[^{125}I]$hCG is not affected by Ca^{2+} concentration in the range of 0.1 to 20 mM; however, high concentrations of $CaCl_2$, $MgCl_2$, and NaCl, as well as the presence of guanidine-HCl and urea, are inhibitory. The binding is not affected or increased by treatment with neuraminidase. Lipase and phospholipases A and C inhibit, whereas phospholipase D did not inhibit [1, 79] the binding of $[^{125}I]$hCG to plasma membranes. The results with other enzymes are conflicting. For example, Gospodarowicz [9] reports that DNase, RNase, trypsin, chymotrypsin, pepsin, and collagenase do not affect the binding. In our studies [1], glucosidase, trypsin, and α-chymotrypsin do not affect the binding, whereas pepsin does. Rao [79] reports that pronase, trypsin, trypsinogen, and chymotrypsinogen inhibit the binding. Among several nucleotides tested, 1 to 2 mM ATP, AMP, adenosine cyclic 3'5'-monophosphate, CTP, and guanosine cyclic 3'5'-monophosphate (cGMP) have significantly inhibited [^{125}I]hCG binding; CTP and cGMP being most effective. These nucleotides also increased the dissociation of bound hormone when added at various times. It should be pointed out that the experiments performed with crude receptor preparations cannot be conclusively extrapolated to the isolated receptor protein. It may be interesting to note that these properties of [^{125}I]hCG binding to the plasma membrane receptor are similar to other hormone-receptor interactions [74].

F. Types of Receptor Sites

In general, during saturation experiments, both high-affinity and low-capacity, as well as low-affinity and high-capacity, receptor sites are recognized. The low-affinity and high-capacity receptor sites may represent nonspecific binding; however, the possibility of saturation of sites which may become available by slow dissociation of both exogenous and endogenous hormone during extended incubation at 37°C may also be considered. The use of very high levels of unlabeled hormone to compete with ^{125}I-labeled hormone bound to low-affinity and high-capacity receptors has cast doubt on their specificity. Cuatrecasas and Hollenberg [80] have shown that nonreceptor materials may exhibit saturability, specificity, affinity, and reversibility similar to specific hormone-receptor interactions; for example, insulin binding to talc, silica, glass, and protein-agarose derivatives. The negative cooperativity in such nonspecific system is due to ligand-ligand interactions, and the Scatchard plots yield nonlinear relationships which can be falsely interpreted as cooperative interactions [78] between receptors or in the presence of additional groups of binding sites. Awareness of such artifactual interactions leads to greater scrutiny, particularly for those interactions observed with high concentrations of hormones where the results are frequently interpreted to indicate second or multiple classes of receptor sites. The possibility exists that only one type of receptor may be present at the target site [81]. It is recognized that in general only 10 to 15% of the total sites may be saturated to invoke hormone action. The increase in the receptor sites has been attributed to the unmasking of existing spare receptors [82-84]; however, there is presumptive evidence that receptor sites may have a high turnover rate and that new de novo receptor sites may be synthesized which may be under endocrine regulation. FSH has been shown to increase the amount of LH or hCG receptors in ovary as indicated by the increased binding of [^{125}I]hCG to granulosa cells, possibly due to de novo synthesis of the receptors [136]. Assuming that receptor may be a glycoprotein, Bahl [85] has shown in vivo, as well as in vitro, the incorporation of

N-acetyl-1-[^{14}C]-D-glucosamine in the plasma membranes of hCG-primed immature rats within 90 min following the administration of N-acetyl-1-[^{14}C]-D-glucosamine associated with the receptor activity, following solubilization of the plasma membranes in 0.25% emulphogene, chromatography on Sepharose 6B, and affinity chromatography in Sepharose 4B covalently bound with hCG. Park and Saxena [86] have provided experimental evidence that increase in the binding sites of LH-RH in the pituitary at estrus is estrogen-dependent.

G. Solubilization of Plasma Membranes

During the past few years, solubilization and purification of receptors for insulin [87, 88], glucagon [81, 89], norepinephrine [90], and a catecholamine [91, 92] have been reported. Recently, efforts to characterize and solubilize gonadotropin (hCG-LH) receptors from gonadal tissue have appeared from various laboratories [1-5, 71, 93-97]. The efforts toward the isolation of the receptor in our laboratory are undertaken in two phases: the solubilization of plasma membranes in urea and guanidine hydrochloride as well as in various ionic detergents like sodium deoxycholate (Schwarz-Mann, Orangeburg, N.Y.) and SDS (Pierce Chemical Company, Rockford, Ill.); and in nonionic detergents like Brij 35 (Schwarz-Mann), Triton X-100 (Rohm and Haas Company, Philadelphia, Pa.), Lubrol-XY (General Biochemicals, Chagrin Falls, Ohio), Igepal-630 (General Aniline and Film Corporation), and Tween-80 (Schwarz-Mann). It may be pointed out that each of these approaches has its disadvantage. For example, solubilization in urea or guanidine-Hcl can result in drastic conformational changes in the receptor protein; and use of detergents can lead to the formation of micelles [98] of various molecular weights, posing the problems of removal of the detergent completely from the purified receptors and accurate determination of protein concentration. Hence, under different conditions of detergent extraction, multiple forms of receptors are isolated [99] which are difficult to interpret.

H. Purification of Guanidine-HCl Solubilized Plasma Membranes

The plasma membrane fraction rich in gonadotropin-binding activity is utilized for solubilization for the isolation of the receptor. Seventy-five milligrams of the plasma membrane protein are solubilized in 1.5 ml of 6 M guanidine-HCl solution. Prior to use, guanidine-HCl (Heico, Inc., Delaware Water Gap, Pa.) is purified by the addition of 1 g of charcoal (Norit A, Eastman Kodak Company, Rochester, N.Y.) to 100 ml of a 6 M guanidine-HCl solution. The charcoal is removed by filtration.

The solubilized membranes are centrifuged at 100,000 g in a Beckman ultracentrifuge L265B of rotor type 50 for 1 hr and at 300,000 g in rotor type SW 50 for 2 hr. The solubilized material is separated into a clear protein fraction at the bottom and a viscous lipid layer at the top of the tube. The absence of nonsedimentable material is considered as the evidence for the solubilization of most of the plasma membranes. The solubilized membranes are also not retained on 0.45 μ pore-sized oxoid filter during filtration. The clear solution is aspirated and concentrated to 5 ml by ultrafiltration (PM-10, Amicon Corporation, Lexington, Mass.). The concentrated material is applied to the column of Sepharose 4B. The column is eluted with 6 M guanidine-HCl, and 2-ml fractions are collected per 15 min. As shown in Figure 3, fraction I (eluted with the void volume) contains predominantly high-molecular-weight material. Fractions II, III, and IV are rich in proteins; whereas fractions I and VI contain residual lipids and pigments. These fractions are dialyzed against water followed

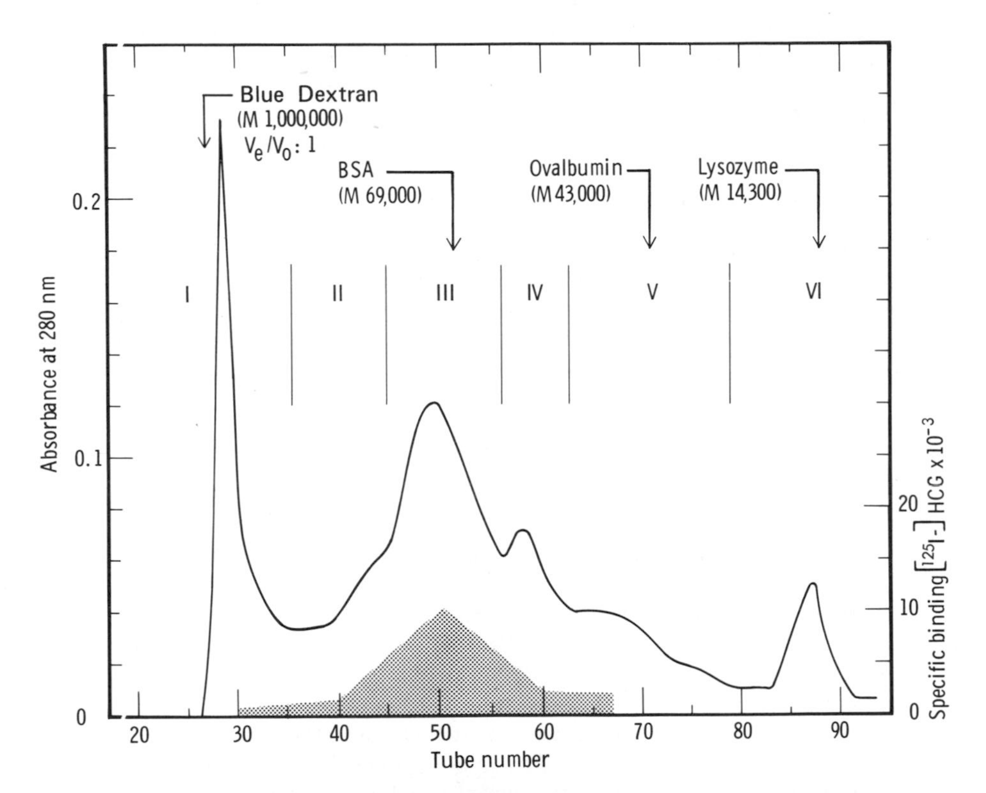

FIG. 3. Purification of 50 mg of plasma membrane proteins of bovine corpora lutea solubilized in 6 M guanidine-HCl on a 2.5 × 100-cm column of Sepharose 4B equilibrated with 6 M guanidine-HCl, pH 5.5, at 4°C. BSA: bovine serum albumin.

by 10 mM tris-HCl buffer, pH 7.2, to eliminate the guanidine-HCl and to recover the receptor-containing protein precipitates in insolubilized form. These precipitates show no specific binding ability with [^{125}I]hCG. However, protein-lipid complexes made with the crude lipid fraction from the sucrose density gradient recovered during the initial purification of the plasma membranes (see Diagram 2) and the lipid fraction obtained from solubilized membranes are able to restore the specific binding of fractions II, III, and IV to [^{125}I]hCG. The maximum binding is present in fraction III. The lipid fractions alone do not show any binding to [^{125}I]hCG [1].

Fraction III from the Sepharose 4B column is equilibrated and concentrated by ultrafiltration using PM-10 membrane (Amicon) in 10 mM tris-HCl buffer of pH 7.2, containing 3 M urea, 1 mM $MgCl_2$; and fractionated by ion-exchange chromatography on DEAE-cellulose as well as isoelectric focusing [1]. The fraction eluted between 0.04 and 0.2 M NaCl is dialyzed to recover the receptor-containing protein in insolubilized form, and shown to contain specific binding with [^{125}I]hCG after reconsitution with the lipids. Efforts toward further purification of the receptor protein by isoelectric focusing in 2 M urea result in the separation of several protein bands. Due to the precipitation of these bands on the column, a discrete isolation of the protein associated with the receptor has not been achieved. The isoelectric point of the fraction containing receptor activity is ~ 4.5.

These observations suggest: (1) That solubilization of plasma membranes in guanidine-HCl resulted in the separation of a major protein component and a minor but vital lipid component of the gonadotropin receptor; (2) The fractionation of the protein component on a column of Sepharose 4B results in the purification of the receptor protein; and (3) It is possible to reconstitute the receptor by mixing protein and lipid component in vitro. Prior to the reconsitution of protein and lipid component, it is necessary to remove all the guanidine-HCl from the lipids, by repeated washing by floatation centrifugation in 10 mM tris-HCl buffer.

I. Purification of Detergent Solubilized Plasma Membranes

Effectiveness of various detergents for solubilization of the membrane receptor is in the following order: Triton > Lubrol > Igepal-630 > SDS > deoxycholate > Brij-35 > Tween-80. Approximately 10 mg plasma protein can be dissolved in 1 ml of 0.5% Triton X-100. After solubilization and centrifugation at 100,000 g for 1 hr or at 300,000 g for 2 hr, the soluble membrane protein is precipitated by making the solution 10% in ammonium acetate and 30% in ethanol. After centrifugation, the precipitates are washed with tris-HCl buffer and the binding ability of the precipitates is tested. The precipitates obtained from plasma membranes solubilized in Triton X-100, Igepal-630, and Brij-35 show significant recovery of the binding activity. These experiments indicate that nonionic detergents like Triton X-100, Brij-35, and Igepal-630 are promising media for the solubilization of plasma membranes and purification of the receptor protein.

Plasma membranes from zonal centrifugation are diluted fivefold with 0.1 M tris-HCl buffer of pH 9.5; containing 1 mM $MgCl_2$, 1 mM DTT, and 10,000 IU/liter Trasylol, and centrifuged at 30,000 g for 20 min to remove sucrose and adsorbed soluble proteins. The pellet is suspended in tris buffer, sonicated, and solubilized in 0.25% Triton X-100 (10 mg protein/ml) and centrifuged at 100,000 g for 1 hr at 4°C. Approximately 80% of the protein and 95% of the starting binding capacity are recovered in the soluble fraction. The soluble material is concentrated by ultrafiltration to a suitable volume and applied to a 10 × 100-cm column of Sepharose 6B equilibrated with 10 mM tris-HCl buffer of pH 9.5 in 0.5% Triton. As shown in Figure 4, a protein fraction, presumably containing

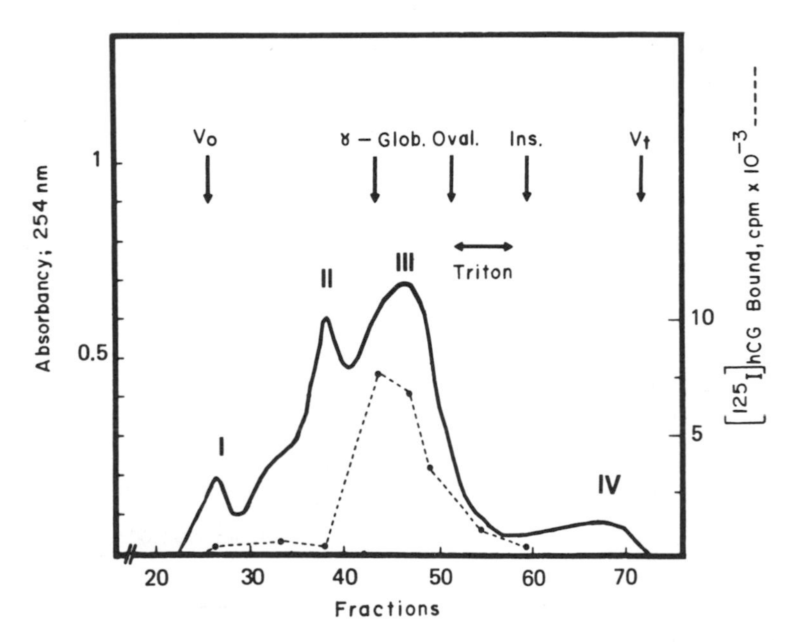

FIG. 4. Purification of bovine corpora lutea plasma membranes, solubilized in Triton X-100 on a 2.5 × 100-column of Sepharose 6B equilibrated in 10 mM tris-HCl buffer of pH 9.5 in 0.5% Triton X-100 at 4°C. Five-milliliter fractions are collected. The eluate is monitored at 254 nm (——) to minimize interference with Triton X-100. Specific binding of [^{125}I]hCG to each fraction (. . . . .) shows maximum receptor binding in fractions 40 to 50, pooled as fraction III.

aggregated material, is recovered in the void volume, which is followed by a second fraction of molecular weight > 150,000. A major third protein fraction contains the highest specific binding capacity (Fig. 4). This fraction is concentrated and further purified by affinity chromatography as described in Sec. V. K. Fraction III is also subjected to SDS disk electrophoresis.

J. Assay of Soluble Receptor by Ammonium Acetate-Ethanol or Polyethylene Glycol

One-hundred-microliter aliquots of plasma membranes solubilized in Triton X-100 and subsequent fractions from Sepharose 6B column and affinity columns are separately incubated for 15 min at 37°C with 50,000 cpm of [^{125}I]hCG in 100 μl of 10 mM tris-HCl buffer, pH 7.2, containing 0.1% BSA; with 100 μl of tris buffer alone; and with excess of unlabeled hCG (100 ng) in a total volume of 300 μl. The tubes are placed in ice. Ten microliters of normal serum and 2 ml of 6.6% ammonium acetate in 66% ethanol, made by mixing 77.5 ml of absolute ethanol and 22.5 ml of 34.5% ammonium acetate, are added to each tube and mixed to precipitate the hormone-receptor complex (Fig. 5). The tubes are centrifuged for

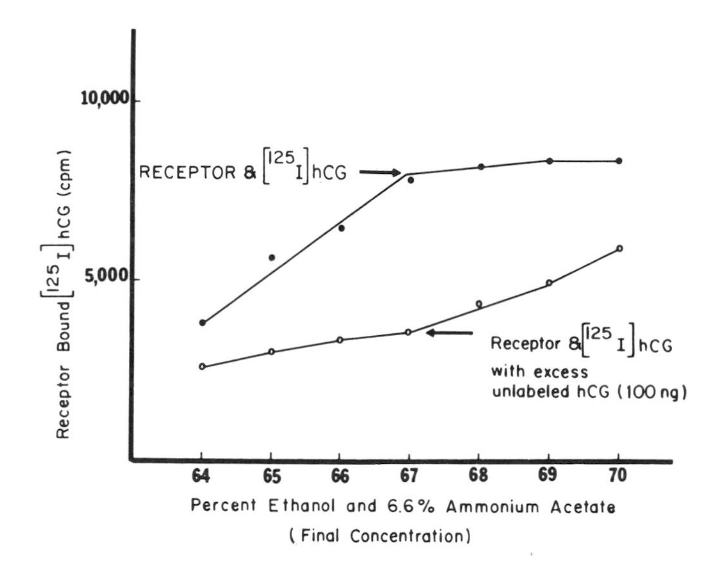

FIG. 5. Precipitation of receptor-bound [^{125}I]hCG, alone, and in the presence of excess unlabeled hCG, at various concentrations of ethanol containing 6.6% ammonium acetate. Maximum precipitation of receptor-bound [^{125}I]hCG occurred at 66% ethanol, with a concomitant minimum precipitation of [^{125}I]hCG released in the presence of excess unlabeled hCG.

10 min at 5,000 rpm. Supernates are aspirated and the pellets containing hormone-receptor complex are counted. The hormone-receptor complex can also be precipitated by polyethylene glycol as described for solubilized insulin receptor [87, 88] and hCG receptor from rat testes [99]. The soluble receptor (50 μl) is incubated for 15 min at 37°C with $[^{125}I]$hCG (50,000 cpm) in 100 μl of incubation buffer, pH 7.2; with 100 μl of buffer alone; and with excess of unlabeled hCG (100 ng) in a total volume of 250 μl. Tubes are put in an ice bath and 0.5 ml of polyethylene glycol (Carbowax 5,000; 25% wt/vol) in 0.1 M sodium phosphate buffer, pH 7.5, containing 5% normal human serum are added. After mixing, tubes are kept in ice for 10 min to precipitate the hormone-receptor complex, centrifuged at 3,000 g for 15 min, decanted, and the precipitate counted. It may be pointed out that the precipitation of the hormone-receptor complex in this system is fairly independent of the concentration of Triton in the incubation medium.

K. Affinity Chromatography

Two grams of the cyanogen bromide-activated Sepharose beads are soaked in 10^{-3} M HCl solution, washed with 0.5 M bicarbonate buffer of pH 8.3, and mixed with 500 μg of hCG in 2 ml of bicarbonate buffer. The coupling reaction is carried out at 4°C for 18 hr. The Sepharose is then washed in a Buchner funnel with 1 M ethanolamine, pH 8.0, for 2 hr, followed by rapid washings with the following buffers: 10 ml of Na bicarbonate, 0.5 M, pH 9.0; 10 ml of Na acetate, 0.1 M, pH 4.0; and 10 ml of sodium phosphate, 0.05 M, pH 7.5. hCG and LH coupled to AH-Sepharose 4B containing $(CH_2)_6NH$ groups, to 17-hemisuccinyl-poly-(L-Lysyl-DL-alanine) agarose, to Concanavalin A-Sepharose, and to cellulose membranes are also used in an effort to increase capacity during affinity chromatography. Affinity chromatography of the receptor-containing fraction from the Sepharose 6B column was performed on a column or in a batchwise procedure with CNBr-Sepharose-LH, AH-Sepharose-LH, and poly-L-lysyl-DL-alanyl-agarose-LH. The binding of the receptor to the resin was carried out at room temperature for 1 hr at 1°C. The elution of the bound receptor was achieved with 10 mM tris-HCl buffer in 0.01% Triton at pH 9.5 and 3.0 Specific binding was tested using the soluble receptor assay. The CNBr-LH Sepharose yielded the best results. The purified receptor has been analyzed for NH_2-terminal residues and amino acid analysis and by SDS gel electrophoresis. The protein content of the receptor protein is calculated from the amino acid analysis. The yield of the receptor protein at various stages can, at present, only be estimated—in view of several factors like the losses, inactivation, protein concentrations, etc. Approximately 10,000-fold purification of corpus luteum homogenate protein was achieved. The yield of the purified receptor protein based on its amino acid analysis is estimated to be 0.01% of total protein in the homogenate.

In contrast, the yields reported from the rat ovaries are 10 to 20 times higher [100].

L. Dansylation of the NH_2-Terminal Amino Acid

Dansylation is achieved by the method of Airhart et al. [101]. After dansylation, the protein is hydrolyzed under vacuum in 5 N HCl for 16 hr at 100°C and dried. Amino acids are dissolved in acetone and applied on polyamide layer (Cheng Chin Trading Company, Taipei, Taiwan). Identification of DNS-amino acid is performed by two-dimensional chromatography in formic acid (90%):H_2O, 3:200, and in benzene:acetic acid, 90:10. A major spot corresponding to proline and two minor spots corresponding to valine and alanine are identified.

M. Amino Acid Composition of the Receptor Protein

The receptor protein eluted from the affinity column is hydrolyzed under vacuum in 5 N HCl (16 hr at 100°C) in pyrex tubes. Amino acid analysis was performed in a Durrum amino acid analyzer (Model D-500). Triton present in the protein hydrolysate is eluted in the position of histidine.

The amino acid composition of the two purified batches is fairly comparable (Table 2). Aspartic acid, glutamic acid, and leucine represent the highest quantities of the amino acid; cysteic acid, methionine, and tyrosine are present in low amounts. The composition does not include 1/2 cystine, tryptophan, and histidine. The latter is masked in the amino acid analysis by the peak of Triton and cannot be quantitated. The amino acid composition also permitted an estimation of protein content of the purified receptor.

N. Polyacrylamide Disk Gel Electrophoresis in Sodium Dodecyl Sulfate (SDS)

Various protein fractions are analyzed by disk gel electrophoresis in 6.5% polyacrylamide in the presence of 1% SDS [102]. Protein bands are visualized by staining in Coomasie blue, and glycoprotein bands are visualized by Schiff's reagent. BSA (mol wt 69,000), egg ovalbumin (mol wt 43,000), and lysozyme (mol wt 14,300) are used as standard proteins (Fig. 6). Disk electrophoretic pattern of intact plasma membrane indicated the presence of at least 20 major protein bands, whereas Schiff's reagent indicated the presence of four glycoprotein bands [1]. From the Sepharose 4B column (Fig. 3), fraction III, which contained most of the receptor protein activity, showed protein bands of molecular weight ranging between 30,000 and 70,000. This estimate was close to the molecular weight of

TABLE 2. Amino Acid Analyses of hCG-LH-Binding Protein Isolated from Bovine Corpus Luteum by Affinity Chromatography[a]

Amino Acid	Batch A-8	Batch A-9
	g/100 g[a]	
Asp	9.6	9.4
Thr	5.7	5.5
Ser	4.4	4.8
Glu	12.5	12.7
Pro	8.2	9.1
Gly	5.1	4.2
Ala	5.2	5.6
Val	5.3	6.0
Met	1.9	1.5
Ileu	5.7	5.2
Leu	9.4	9.8
Tyr	4.6	3.9
Phe	6.5	5.5
Lys	6.6	7.2
His[b]		
Cysteic acid	1.2	1.4
Arg	7.7	7.4

[a]Calculated from total amino acids recovered except histidine.

[b]Eluted with detergent.

fraction III estimated from the Sepharose 6B column (Fig. 4). On the basis of the recovery of the protein, it was estimated that at least 1,000-fold more purification was necessary to obtain a homogeneous protein with receptor activity. The disk pattern obtained after Triton X-100 solubilization and fractionation of the Sepharose 6B column was identical with the pattern obtained for fraction III obtained by guanidine-HCL solubilization of plasma membranes and fractionation on Sepharose 4B column (Fig. 6). One major and two minor bands are observed for the receptor

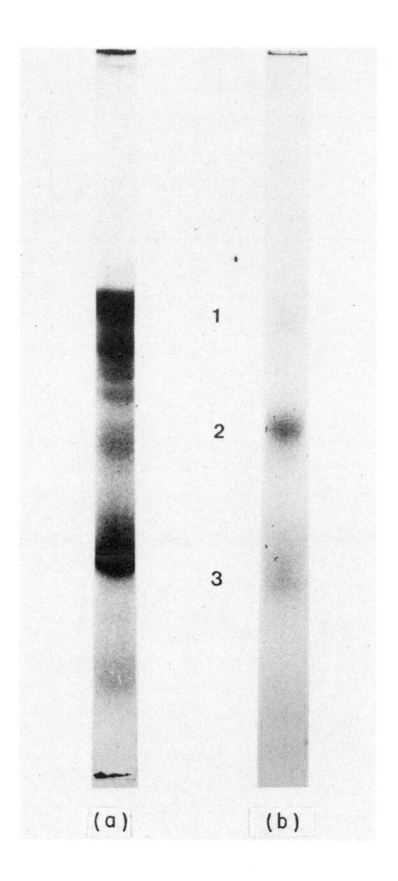

FIG. 6. SDS-polyacrylamide disk-gel electrophoresis of: (a) fraction III showing maximum binding with [^{125}I]hCG, obtained from the purification on Sepharose 6B of bovine corpora lutea plasma membranes solubilized in Triton X-100; (b) fraction showing binding with [^{125}I]hCG eluted from the affinity chromatography of (a) by 10 mM tris-HCl buffer in 0.01% Triton X-100 at pH 9.5.

protein purified by the affinity chromatography. The molecular weight of the major band is close to 80,000, which is in the range of the estimate for the receptor protein purified from plasma membranes solubilized in guanidine-HCl [1]. The two minor bands appear in the molecular weight range of 30,000 to 50,000 (Fig. 6). Gonadotropin receptors purified from rat testes [99] and rat ovaries have been shown to have molecular weights in the range of 200,000 to 300,000.

O. Immunization of Rabbits with Plasma Membrane Proteins

Four albino virgin females are bled to examine if their serum contains any nonspecific substances which may inhibit the binding of [^{125}I]hCG to the plasma membranes. These nonimmune sera are used as controls for the respective animal. The rabbits are immunized with 200 μg plasma membrane protein sonicated in 0.5 ml saline and emulsified in 0.5 ml of complete Freund's adjuvant. The first injection of 0.2 ml is given in toe pads at five sites, followed by weekly intradermal injection of the antigen. The rabbits are bled by ear vein puncture at 3, 6, and 12 weeks of immunization. The serum is stored at -20°C.

The antisera is examined by immunodiffusion in agar gel (Fig. 7). The presence of antibodies to the plasma membrane protein is also examined by the inhibition of binding of [^{125}I]hCG to the receptor. The antisera are also tested for binding with [^{125}I]hCG to check if endogenously bound LH-hCG may have produced antibodies which may cause nonspecific interference with hormone-receptor binding.

Tbe antisera and an aliquot of nonimmune sera are also purified by rivanol precipitation as follows [103]: Rivanol (0.4%) is added to 72 ml

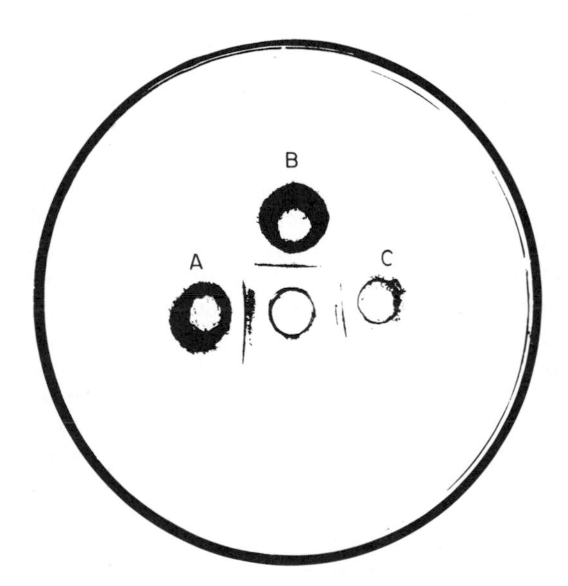

FIG. 7. Immunodiffusion in agar gel. Wells A and B contain antisera from two rabbits immunized with bovine corpora lutea plasma membranes. Well C contains antisera purified by rivanol precipitation. Center well contains solubilized plasma membranes.

of antisera and 7.2 ml of nonimmune sera in a ratio of 35:1 to precipitate albumin and α- and β-globulin, as well as fibrinogen if plasma was used. The precipitate is removed by centrifugation at 3,000 rpm for 20 min. Activated charcoal (Norit 5 × 30) is then added to the supernate in a ratio of 1:40 (vol/wt) to adsorb rivanol. The charcoal is removed by two centrifugation of 20 min at 3,000 rpm. Fine particles of charcoal are allowed to settle in cold. The clear supernates are either concentrated to the original volume of serum or plasma by ultrafiltration through Amicon filter PM-10, or lyophilized and stored at -20°C.

The immune sera in vitro inhibited the binding of the ^{125}I-labeled hCG to the plasma membrane. This effect is even more pronounced with the phospholipase-treated plasma membranes. Immunodiffusion in agar gel between immune sera and purified receptor showed precipitin lines (Fig. 7). It may be concluded that the rabbits immunized with plasma membrane proteins produced antibodies to proteins involved in the binding of the hormone to the receptor. In the absence of pure receptor, these data, however, do not conclusively prove the existence of specific antibodies to the receptor itself.

VI. ROLE OF LIPIDS IN HORMONE-RECEPTOR INTERACTION

Initial experiments strongly indicate the requirement of lipids for the restoration of the [^{125}I]hCG binding following guanidine-HCl solubilization and phospholipase C treatment [1, 79].

Fractionation of the crude lipid fraction. The lipid components of the plasma membranes are extracted with chloroform:methanol:acetic acid: H_2O (25:15:4:2). For the identification of phospholipids, 20 μg of the extract in 20 μl of solvent is separated by thin layer chromatography on 20 × 20-cm silica gel plates using the above solvent system. The plates are chromatographed for 60 min, and various lipid fractions are identified by iodine vapor. For the detection of neutral lipids, two solvent systems are used, other conditions being the same. First, in hexane-benzene system (85:15), the cholesterol esters migrate away from the origin, whereas other lipids remain at the origin. In the second system, in the second dimension, triglycerides, fatty acids, and cholesterol are separated, whereas phospholipids remain at the origin. Using known standards and internal standards with radioactivity, the lipid component of the gonadotropin receptor is shown to contain predominantly phosphatidyl ethanolamine, phosphatidyl serine, lecithin; trace amounts of sphingomyelin, lysolecithin; and small amounts of neutral fats. For preparative purposes, the lipid fraction is extracted with chloroform:methanol (2:1), washed with H_2O three times to remove proteins and guanidine-HCl, and fractionated as described. The phospholipids are extracted from the silica gel by two elutions with chloroform:methanol (2:1), followed by two elutions with chloroform:methanol:H_2O (3:5:2). Other lipids are eluted from silica gel with ether, followed by two elutions of chloroform:methanol (4:1). Aliquots of various lipids are examined for purity by thin layer chromatography. The lipid fraction contains phosphatidyl ethanolamine, phosphatidyl serine, and lecithin as the major components; however, minor quantities of sphingomyelin and lysolecithin are also encountered. Among the neutral lipids, triglycerides, fatty acids, cholesterol, and cholesterol esters are found. The combination with phosphatidyl ethanolamine and phosphatidyl serine restore 50 to 70% of the total binding achieved with crude lipid fraction. Recently, Rao [79] has reported that in an attempt to restore the loss of [^{125}I]hCG binding activity, especially following phospholipase C treatment, phosphatidyl ethanolamine appears to be most effective. It is suggested that phosphatidyl ethanolamine and phosphatidyl serine may represent the vital lipids involved in the receptor structure, and perhaps neutral lipids may assist as an accessory hydrophobic matrix to stabilize the receptor conformation.

VII. ROLE OF CARBOHYDRATES IN THE HORMONE-RECEPTOR INTERACTION

It has been suggested that glycosylation is essential in labeling proteins for intracellular recognition for export [104]. On the other hand, Winterburn and Phelps [105] have proposed that the information for the secretion of proteins is genetically controlled and that the carbohydrates play an extracellular recognitional role in regard to proteins. The concept also implies "antirecognition"; for example, in the protection of a bond from proteolytic attack, making vital the antigenic site; and in preventing self-aggregation of a protein, as in the case of κ-casein. The carbohydrates are ideal in providing recognition sites to the cell, since the multiplicity of structures within a polysaccharide group has greater possibilities than in an equivalent number of amino acids in a polypeptide chain.

Glycoproteins and glycolipids are present in the periphery of the plasma membranes, along with cell-surface glycosyltransferases [106-108]; and are involved in the interaction of cells, with intercellular adhesion, and contact inhibition; as well as macromolecules, including viruses [109-112]. The carbohydrates assist in the secretion of the hormone from the cells [104], and in its transport to the target site [113]. It is conceivable that the carbohydrates on the cell surfaces may be involved in the initial hormone-receptor interaction. The function of sialic acid in the metabolic clearance of glycoproteins from the liver is known [112-114]; however, little information is available on the role of carbohydrate in the hormone-receptor interaction [113].

The concept that the carbohydrate may play a structural rather than a functional role has stemmed from studies on a number of glycoprotein enzymes [115-121] which retain full activity following enzymic removal of sialic acid. Similarly, ribonuclease A and B are equally active although A is devoid of any carbohydrate. On the other hand, hCG and FSH [64] exhibit little activity in vivo, whereas LH and TSH with little sialic acid have full activity. Hence, the role of carbohydrates in biological activity varies with each. In order to delineate the essential requirements of both hCG and the receptor for the maximum binding, the hCG and/or the receptor are treated with proteases, glycosidases, and lipases prior to hormone-receptor interaction. The particular forms of gonadotropin receptors obtained from rat testes [64], rat ovaries [122], and bovine corpus luteum [1, 79] indicate that neuraminidase and glucosidases do not affect the binding of [^{125}I]hCG. The desialyzed, asialo-agalacto, asialo-agalacto-acetyl glucosamine derivatives of hCG show 200, 140, and 85% binding activity, respectively. However, removal of 20% mannosyl residue from hCG produces a dramatic fall in the binding activity [123]. Individual sugars and other glycoprotein hormones such as FSH and TSH, and glycoproteins such as fetuin and α-acid glycoprotein do not bind to the hCG-LH receptors, indicating little involvement of carbohydrates in the hormone-receptor binding. These results clearly demonstrate that at least the first three monosaccharides are not involved in the binding, however, they do suggest an involvement of mannose in the hormone-receptor binding. The involvement of mannose in the interaction of macromolecules to cell surfaces or in intracellular interactions has been observed recently [112-113]. It is interesting that these derivatives are not effective in the stimulation of cAMP, but competitively inhibit the cAMP accumulation by native hCG in the Leydig cell suspension. Some recent evidence showing lack of correlation among hormone binding, cAMP stimulation, and steroidogenesis [124, 125] also suggests that hormone action may involve more than one molecule at the receptor site. It is interesting to note that, in the presence of excess of the derivatives, the cAMP stimulation was completely suppressed, but the steroidogenic response was still maximal. This would indicate that either there is a separate cAMP pool or there is an alternate pathway for steroidogenesis. This inhibitory effect is overcome by increasing the concentration of hCG relative to the derivatives. These results further indicate that removal of the carbohydrates does not alter the affinity of the hormone for the cell, but does interfere with the ability of the hormone to stimulate adenyl cyclase. On the other hand, the glycosidase-treated derivatives of hCG do not inhibit hCG-induced testosterone formation. A dissociation of cAMP production and steroidogenesis has also been demonstrated by Rao and Saxena [3] and Catt and Dufau [126]. On the other hand, complete removal of sialic acid enzymatically from the plasma membranes caused a 62% increase in the binding of hCG, and removal of 70% galactose did not affect the binding. Moyle and Ramachandran [127] have suggested the presence of more than one type of LH-receptor site on Leydig cells, with different affinities for the hormone; the sites which induce steroidogenesis have higher affinity than those which stimulate the accumulation of cAMP. Thus, there appears to be two types of sites on the plasma membranes, one for steroidogenesis and the other for cAMP production. These results would support the idea of more than one type of receptor for hCG. Protease-, glycosidase-, and lipase-treated derivatives of glycoprotein hormones and their receptor sites provide tools for elucidating the molecular mechanisms of hormone action.

VIII. APPLICATION OF GONADOTROPIN RECEPTORS IN BIOLOGY AND MEDICINE

The ability of receptor sites to specifically concentrate hormones has been exploited to develop radioligand-receptor assays of several hormones. In radioimmunoassay, the results may be subject to the quality of the antisera and frequently result in a lack of correlation between radioimmunoassay and bioassay estimates. Evidence has been provided that tissue-receptor assay estimates are in accord with the bioassay [128]. Thus, due to intense stereospecificity, uniformity, and high affinity of the receptor to concentrate hormones in vitro, the radioreceptor assays provide the specificity of the bioassay and sensitivity of the radioimmunoassay. Desialyzed hCG exhibits twofold greater binding activity than the native hCG, indicating that caution must be exercised while assaying hCG preparations varying in sialic acid content. On the other hand, endogenous production of hCG with low or no sialic acid due to genetic or other defects will not sustain a normal physiological function, for example, pregnancy. It would be of interest to measure sialic acid content of hCG from women with habitual or spontaneous abortion which has been linked with chromosomal defects [129].

The radioreceptor assays, similar to radioimmunoassays, are based on the principle of competitive protein binding and require a binding protein or receptor, and highly purified hormone which can be labeled with a suitable marker and is able to specifically bind to the receptor. The binding protein or the receptor for hCG and/or LH can be obtained from the ovaries and testes of various mammalian species. In the present studies, the source of receptor is the plasma membranes obtained from bovine corpora lutea of the first trimester of pregnancy. The corpora lutea from nonpregnant cows can also be used. However, the yield of the receptor from pregnant cows is approximately three- to fivefold greater. The bovine corpora lutea are easily available from the slaughterhouse and the yield of the receptor from the bovine corpora lutea is significantly greater than those of sheep, pig, rat, etc. The preparation of plasma membranes has already been described. Up to 10 to 15 g corpus luteum tissue can be obtained from each pair of pregnant ovaries. From 100 corpora lutea (500 g fresh tissue), 50 to 100 mg plasma protein was recovered. Each assay tube requires 2 to 5 μg plasma membrane protein; hence, this yield is sufficient for several thousand assays.

It may be of interest to point out that the hormone-bound plasma membranes can be treated with tris-HCl buffer of pH 9.0 for 1 min at 40°C, washed, and resuspended in the buffer of pH 7.2 to dissociate 50 to 70% of the labeled hormone. These membranes can be reused in subsequent assays. The hCG containing 12,000 IU/mg is labeled with ^{125}I by the lactoperoxidase method to a specific activity of 30 to 40 μCi/μg. The labeled hCG retains more than 80% of the biological activity. The plasma membrane specifically bind 50 to 60% of the labeled hormones, and more than 90% of the labeled hormone can be displaced by the excess of the respective unlabeled hormone. The plasma membranes have also shown the presence of specific receptors for hPRL. This also provides an opportunity to perform radioreceptor assay of PRL as well.

The protocol for the radioreceptor assay of hCG-LH or PRL is shown in Table 3. The radioreceptor assay is performed in 75 × 100-mm, disposable plastic tubes containing lyophilized receptor which is reconstituted in 100 μl of water. Each standard tube contains the following: (1) 100 μl of standard hCG solution of concentrations of 0.37, 0.75, 1.5, 3.0, 6.0, 12.5, 25, 50, 100, 500, and 1,000 ng/ml; (2) [^{125}I]hCG (15,000-30,000 cpm); (3) 100 μl of plasma from hypophysectomized subjects for LH levels, and pooled plasma obtained from nonpregnant, normally menstruating women during the luteal phase, containing basal levels of LH (~ 3-4 ng/ml) for hCG in pregnancy test; (4) 20 IU of Trasylol in 20 μl of buffer. The unknown plasma are assayed in 100-μl aliquots in duplicates.

The blank, standards, and unknown plasma samples are incubated at 37°C for 20 min. After incubation, the tubes are placed in an ice bath and 1 ml of chilled tris-HCl buffer is added to each tube. The contents of the tubes are mixed on a vortex mixer and centrifuged for 10 min at 5,000 g. The supernates are aspirated, and the radioactivity in the pellet representing [^{125}I]hCG specifically bound to the plasma membranes is counted in an autogammacounter with 51% efficiency for ^{125}I. The entire procedure is performed in 1 hr.

The logit-log transformation of the standard hCG dose-response curve yielded a sensitivity of 0.3 ng or 3 mIU hCG/ml, as shown in Figure 8. Various dilutions of plasma sample from a pregnant woman yield a slope parallel to that of hCG, indicating the validity of the assay. There was no cross-reaction with FSH, TSH, hGH, and hPRL in the assay. There was a 98 to 102% recovery of hCG added to the plasma samples of known

TABLE 3. Protocol for the Radioreceptor Assay of hPRL and LH or hCG Using Plasma Membranes of Bovine Corpora Lutea

Buffer[a]	Unlabeled hormone[b]		Receptors[c]	Control plasma[d]	Labeled hormone[e]		Trasylol[f]
	hPRL or	hCG			^{125}I-hPRL or	^{125}I-hCG	
	50 μl	50 μl	100 μl	50 μl	50 μl	50 μl	20 μl
	50 ng/ml	50 ml	↓	↓	↓	↓	↓
	↓	↓					
	0.2 ml	0.2 ng/ml					
			Plasma samples				
50 μl			100 μl	50 μl	50 μl	50 μl	20 μl

[a] 10 mM tris-HCl buffer of pH 7.0 containing 0.1% BSA, 1 mM $CaCl_2$, 1 mM $MgCl_2$, and 20 IU of Trasylol.

[b] Diluted in buffer.

[c] Highly purified plasma membrane fraction diluted equivalent to 100-150 μg protein.

[d] Plasma from completely hyphophysectomized parents.

[e] Diluted in buffer equivalent to approximately 50,000 cpm for [^{125}I]hPRL and 50,000 cpm for [^{131}I]hCG, respectively.

[f] Trasylol, an enzyme inhibitor (FBA Pharmaceuticals).

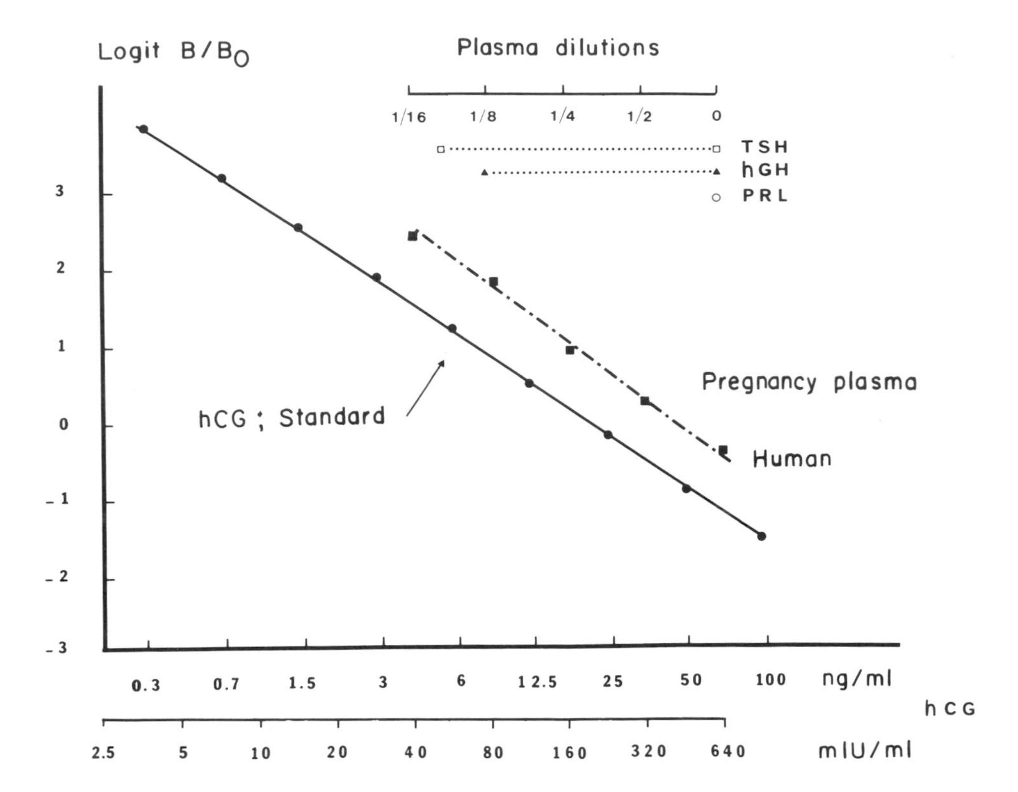

FIG. 8. Computer output of logit-log plot of the dose-response curve for the radioreceptor assay of hCG: TSH (plasma from hypothyroid patients), hGH (plasma from acromegalic subjects), PRL (plasma from women with postpartum lactation). Units of Second International Reference Preparation equivalent to corresponding nanograms of hCG are indicated on the axis.

hormonal concentration. The intra- and interassay variation in the estimates of hormonal levels of plasma pools was 6 and 11%, respectively. The radioreceptor assay, however, does not discriminate between LH and hCG and also shows nonspecific interference with serum proteins. To eliminate the specific interference by basal levels of LH and nonspecific effects of serum proteins, serum from nonpregnant, normally menstruating women during luteal phase was added to each test tube of the standard so that all tubes received a fixed amount of total serum volume of 100 μl. This manipulation resulted in the selective measurement of hCG in the presence of basal levels of LH and was invariably confirmed by the clinical manifestations, for example, pregnancy and nonpregnancy in approximately 1,000 cases, and in two cases of choriocarcinoma. This was

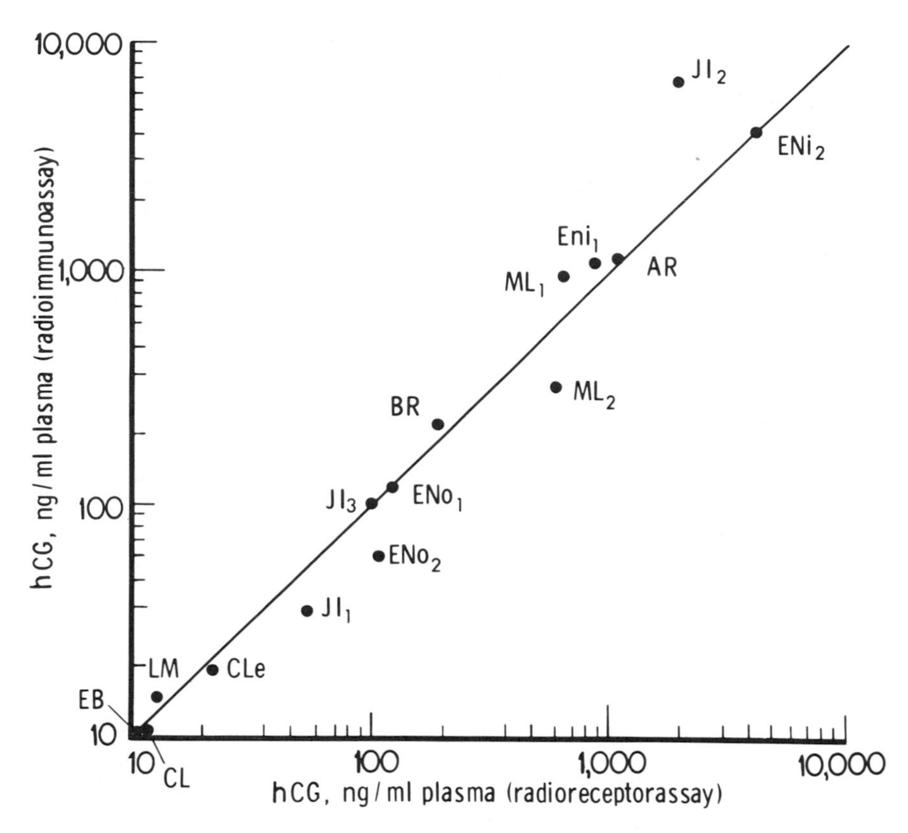

FIG. 9. Correlation of radioreceptor assay of hCG and radioimmunoassay of hCG, using antisera prepared against hCG-β, in 15 cases of early spontaneous abortion (1 ng of hCG is equivalent to 12 mIU).

further examined by comparing the hCG levels in 15 subjects with threatened abortion determined by radioimmunoassay using antisera specific to hCG-β [130] and those determined by the present radioreceptor assay. As shown in Figure 9, patients with low levels of hCG (< 10 ng/ml), both systems estimated lower levels; however, radioreceptor assay measured levels higher than those by radioimmunoassay. There was a reasonable correlation in the range of 10 to 1,000 ng/ml hCG. However, for patients having hCG levels above 2,000 ng/ml, the radioimmunoassay measured 2 to 5 times higher values, indicating the presence of immunologically active, biologically inactive hCG in serum. Thus, the radioreceptor assay provides an earlier, more rapid, and better discriminator for threatened abortion than radioimmunoassay.

During routine pregnancy tests, all unknown samples are compared with a pool of plasma from nonpregnant and pregnant women for accuracy and quality control. It may be pointed out that during early pregnancy, the hCG-LH levels are two- to threefold higher than basal LH levels in nonpregnant women during the luteal phase, and that LH does not rise during early pregnancy. Hence, basal levels of LH do not interfere with the pregnancy tests in 1,000 women with pregnancy of varying durations from one week after conception to one week after the missed period [131]. The clinical application of the assay is summarized in Table 4. The findings of abnormally low hCG levels provides the diagnosis of an ectopic pregnancy (Fig. 10). The site of uterine or extrauterine pregnancy is confirmed by the clinical and pathological findings. The radioreceptor assay is positive in all the 13 ectopics. This compares with only 50% positive in the conventional hemagglutination determinations [132]. However, the ectopic may not produce any clinical abnormalities and may not be recognizable even when the tubes are directly discerned at laparctomy. Similarly, threatened abortions are diagnosed as early as 8 to 10 days after conception [133] by levels of hCG lower than those during normal pregnancy of the same duration (Fig. 11). The radioreceptor assay thus provides a much-needed diagnostic test in the management of early spontaneous abortion.

TABLE 4. Some Uses of the Radioreceptor Assay of hCG-LH

Clinical application	No. of cases
1. Routine pregnancy test Positive: 114 Negative: 293	407
2. In suspected early ectopic pregnancies	33
3. Monitoring of early threatened abortions	32
4. Detection of early pregnancy for miniabortions	55
5. Evaluation of therapy for infertility cases	180
6. Patients with intrauterine devices	44
7. Premenopausal women	16
8. Hydatidiform mole, choriocarcinoma, and other tumors	27
9. Other human experimental studies on early pregnancy phenomena	46
Total	840

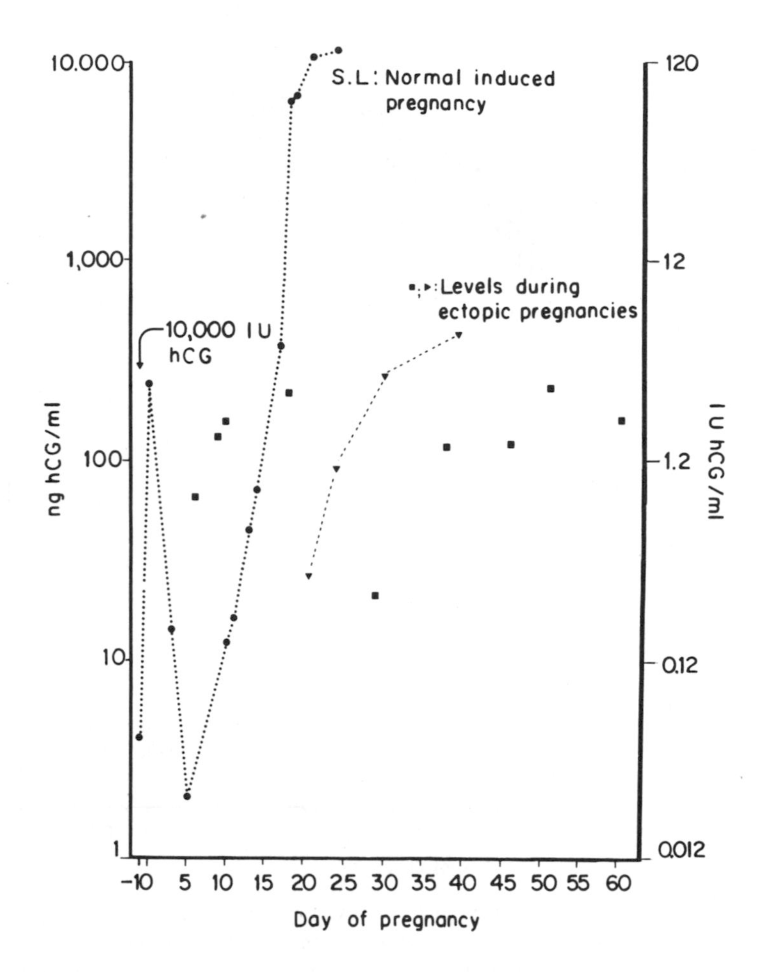

FIG. 10. Comparison of the plasma levels of hCG during hCG-induced normal intrauterine pregnancy (S. L.) with hCG levels in 10 patients with ectopic pregnancy. The plasma levels of hCG during early induced or natural pregnancy were similar [21, 22]. The day of pregnancy was based on the last menstrual period reported by the patients. In one of the 10 patients (. . . . .), the radioreceptor assay was performed on days 21, 24, 30, and 42 of pregnancy. The initial radioreceptor assay on day 21 was associated with a negative Pregnosticon test.

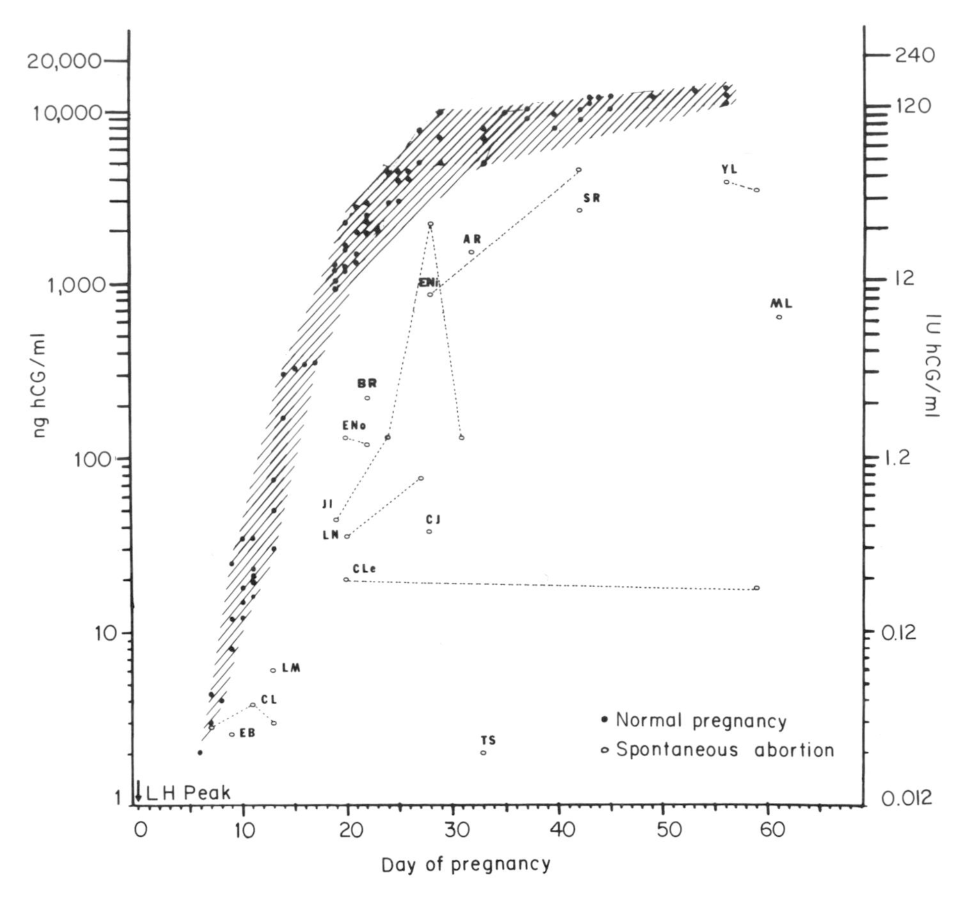

FIG. 11. Comparison of plasma hCG levels in normal pregnancy with plasma hCG levels in 15 patients who spontaneously aborted. The day of pregnancy is based on the last menstrual period and length of cycle as reported by the patients. International units of Second International Reference Preparation of hCG equivalent to nanograms of hCG are indicated on the ordinates.

The miniabortion, formerly described as menstrual extraction, has been performed frequently in the absence of pregnancy. The conventional and available pregnancy tests do not become positive until day 12 to 14 after the missed menses at the 95% level. The receptor assay on the other hand is actually 100% accurate at the time of the missed period, showing over 100 ng/ml plasma of hCG (1,200 mIU/ml) in the positive specimens. The radioreceptor assay now permits accurate diagnosis of pregnancy and eliminates the early abortion procedure in the absence of pregnancy.

The infertility group includes 20 patients with a positive diagnosis of a pregnancy as early as day 6 to 8 following fertilization. It is also true that when a low level of hCG occurs in the inevitable abortion, the positive radioreceptor assay may occur at or about the time of the missed menses. Patients have been followed on clomid, hMG, and hCG. The elevation of LH for one day at midcycle has pinpointed the time of ovulation. Repeated testing of these infertile patients has resulted in the detection of very early abortions, some of which are clinically confused with the menses. A group of patients attempting to conceive by artificial insemination has been followed daily with this assay. Seventeen patients have been studied; five LH peaks have been detected, and one patient has demonstrated a high assay value indicating pregnancy. Other applications of the radioreceptor assay have included the detection of both spontaneous and induced ovulation, and of pregnancy following natural fertilization or artificial insemination. Application of this test has promise in the diagnosis of choriocarcinoma, various malignant tumors, embrymal testicular tumors, malignant teratomas, fourth ventricular tumors, metastic tumor of the spine, ectopic production of hCG [134]; and also in the establishment of early pregnancy in women receiving radiation, and in sociological problems of rape.

As shown in Figure 12, the LH from various species and serum from pregnant monkey, rabbit, rat, and dog yielded slopes parallel to hCG standard. This lack of species specificity is advantageous in using the radioreceptor assay in laboratory animal experimentation to obtain reliable levels rapidly and to avoid the cumbersomeness of setting up homologous radioimmunoassay for each species or searching heterologous system by trial and error. The radioreceptor assay has been utilized successfully in two biological studies. First, in a study designed to analyze the effects of LH-RH in the release and/or biosynthesis of LH by the rat pituitary slices in in vitro cultures and, second, in the study on the in vivo LH release in rabbits following chemical lesions with 5,6-dihydroxytryptamine. Radioligand-receptor assays using homogenates of rat testes have permitted a meaningful comparison of activities of hLH and FSH with FSH and LH from other species during purification procedures [135].

Currently in the author's laboratory gonadotropin receptors are being used in the purification of TSH from LH. The receptor for LH-hCG, PRL and GH are being employed to adsorb endogeneous hormones from serum pools. These hormone-free sera are used in the standards for homologous radioimmunoassays and radioreceptorassay to improve the efficacy of these systems. Similar uses of receptors can be extended to other hormones [136].

The future studies on gonadotropin receptors may be visualized in the following areas: (1) elucidation of the subunit nature of the binding component and adenylate cyclase; (2) physico-chemical characterization

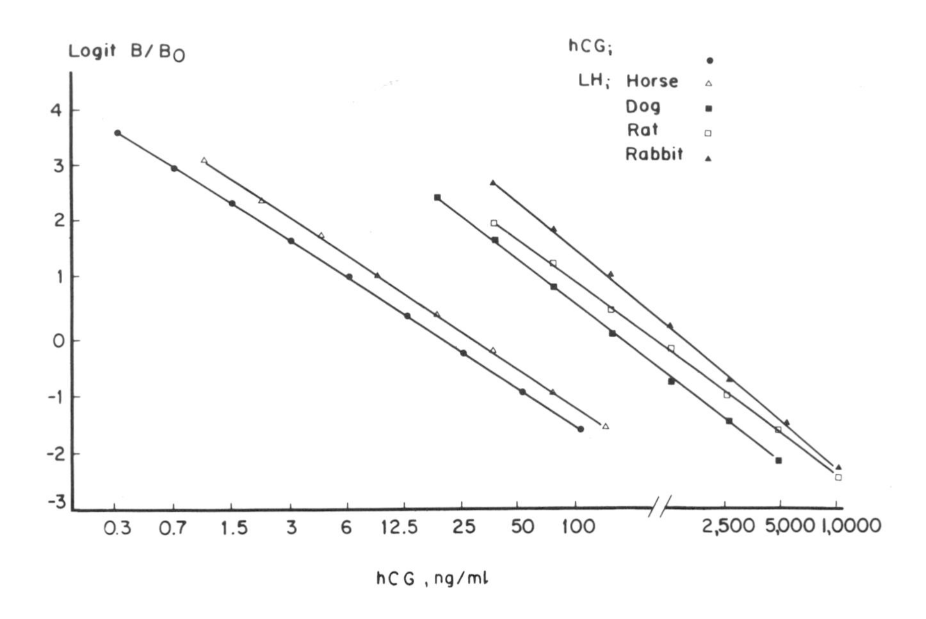

FIG. 12. Competitive inhibition of the binding of [^{125}I]hCG by hCG and LH from various species in the radioreceptor assay. The biological potencies of human, horse, dog, and rabbit LH were 8.9, 5.5, 0.023, and 0.9 units, respectively, in terms of NIH-LH-Sl by ovarian ascorbic acid depletion assay [49].

and structure-function relationship of various components (protein, lipid, and carbohydrate) of the isolated "receptor"; and (3) the use of isolated receptor in in vitro models to elucidate, at the molecular level, various steps of hormone-receptor interaction. These studies should provide information on the mechanism of hormone action, which is vital in the understanding, diagnosis, and treatment of diseases due to end-organ failure.

ACKNOWLEDGMENTS

The participation of Drs. Ch. V. F. Haour, P. Rathnam, R. Landesman, T. Rosal, and Mrs. T. Saito in the experimental portions of the work, as well as in the preparation of the manuscript, is gratefully acknowledged. These studies have been supported by National Institute of Health grants HD 06543, CA 13908, and contract 72-2763; Ford Foundation grant GT-670-0455; and Rockefeller Foundation grant GA-HS 7506. The author is a Career Scientist Awardee, Health Research Council, City of New York, Contract I-161.

REFERENCES

1. F. Haour and B. B. Saxena, J. Biol. Chem., 249:2195 (1974).
2. H. Rajaniemi and T. Vanka-Perrtulas, Endocrinology, 90:1 (1972).
3. Ch. V. Rao and B. B. Saxena, Biochim. Biophys. Acta, 313:372 (1973).
4. C. H. Lee and R. J. Ryan, in Receptors for Reproductive Hormones (B. W. O'Malley and A. R. Means, eds.), Plenum, New York 1973, p. 419.
5. S. Kammerman, R. E. Canfield, J. Kolena, and C. P. Channing, Endocrinology, 91:65 (1972).
6. F. F. Cole, J. C. Weed, G. T. Schneider, J. B. Holland, W. L. Geary, and B. F. Rice, Amer. J. Obstet. Gynecol., 117:87 (1973).
7. C. P. Channing and S. Kammerman, Endocrinology, 92:531 (1973).
8. D. Gospodarowicz, J. Biol. Chem., 248:5042 (1973).
9. D. Gospodarowicz, J. Biol. Chem., 248:5050 (1973).
10. K. J. Catt, T. Tsuruhara, and M. L. Dufau, Biochim. Biophys. Acta, 279:194 (1972).
11. D. M. de Kretser, K. J. Catt, and C. A. Paulsen, Endocrinology, 80:332 (1971).
12. A. R. Means and J. Vaitukaitis, Endocrinology, 90:39 (1972).
13. T. Saito and B. B. Saxena, unpublished data.
14. S. Carlson, S. Kollander, and E. R. A. Muller, Acta Obstet. Gynecol. Scand., 51:175 (1972).
15. R. W. Turkington, G. C. Majumder, N. Kadohama, and J. H. MacIndoe, Recent Progr. Hormone Res., 29:417 (1973).
16. R. P. G. Shiu and H. G. Friesen, Biochem. J., 140:301 (1974).
17. A. R. Midgley, Jr., Advan. Exptl. Med. Biol., 36:365 (1973).
18. T. Saito and B. B. Saxena, Acta Endocrinol., 80:126 (1975).
19. C. P. Channing and S. Kammerman, Biol. Reprod., 10:179 (1974).
20. E. W. Sutherland, G. A. Robinson, and R. W. Butcher, Circulation, 37:279 (1968).
21. J. M. Marsh, J. Biol. Chem., 245:1596 (1970).
22. K. Savard, J. M. Marsh, and B. F. Rice, Recent Progr. Hormone Res., 21:285 (1965).

23. L. Birnbaumer and M. Rodbell, J. Biol. Chem., 244:3477 (1969).

24. G. S. Levey, J. Biol. Chem., 246:7405 (1971).

25. S. L. Pohl, L. Birnbaumer, and M. Rodbell, J. Biol. Chem., 246: 1849 (1971).

26. L. Jarett, M. Reuter, D. W. McKeel, and R. M. Smith, Endocrinology, 89:1186 (1971).

27. J. M. Hammond, L. Jarett, I. K. Mariz, and W. H. Daughaday, Biochem. Biophys. Res. Commun., 49:1122 (1972).

28. J. R. Gavin, III, J. Roth, J. A. Archer, and D. N. Buell, J. Biol. Chem., 248:2202 (1973).

29. S. E. Monroe and A. R. Midgley, Jr., Proc. Soc. Exptl. Biol. Med., 130:151 (1969).

30. A. E. Castro, A. C. Seiguer, and R. E. Mancini, Proc. Soc. Exptl. Biol. Med., 133:582 (1970).

31. A. E. Castro, A. Alonso, and R. E. Mancini, J. Endocrinol., 52: 129 (1972).

32. B. K. Van Weeman and A. H. W. M. Schuurs, FEBS Lett., 15: No. 3 (1971).

33. B. K. Van Weeman and A. H. W. M. Schuurs, FEBS Lett., 24: No. 1 (1972).

34. J. Bockaert, S. Jard, and F. Morel, Amer. J. Physiol., 219:1514 (1970).

35. M. S. Soloff, T. L. Swartz, and M. Saffran, Endocrinology, 91:213 (1972).

36. R. J. Winand and L. D. Kohn, Proc. Natl. Acad. Sci. U.S.A., 69: 1711 (1972).

37. P. Cuatrecasas, Ann. Rev. Biochem., 43:169 (1974).

38. B. B. Saxena, H. Demura, H. M. Gandy, and R. E. Peterson, J. Clin. Endocrinol. Metab., 28:519 (1968).

39. R. S. Yalow and S. A. Berson, Trans. N.Y. Acad. Sci., 28:1033 (1966).

40. W. M. Hunter and F. C. Greenwood, Nature, 194:495 (1962).

41. Y. Miyachi, J. L. Vaitukaitis, E. Nieschlag, and M. B. Lipsett, J. Clin. Endocrinol. Metab., 34:23 (1972).

42. W. L. Hughes, in Labelled Proteins in Tracer Studies, European Atomic Energy Commission, Brussels, 1966, p. 3.

43. P. Freychet, J. Roth, and D. M. Neville, Jr., Biochem. Biophys. Res. Commun., 43:400 (1971).

44. R. J. Lefkowitz, J. Roth, and W. Pricer, Proc. Natl. Acad. Sci. U.S.A., 65:745 (1970).

45. Y. Miyachi, R. S. Mecklenburg, and M. B. Lipsett, Clin. Res., 21:45 (1973).

46. I. D. Goldfine, J. Roth, and L. Birnbaumer, J. Biol. Chem., 247: 1211 (1972).

47. M. A. Lesniak, J. Roth, and P. Gorden, Nature, New Biol., 241: 20 (1973).

48. O. H. Lowry, N. J. Rosebrough, A. L. Farr, and R. J. Randall, J. Biol. Chem., 193:265 (1951).

49. A. F. Parlow, in Human Pituitary Gonadotropins (A. Albert, ed.), Charles C. Thomas, Springfield, Ill., 1961, p. 300.

50. S. L. Steelman and F. M. Pohley, Endocrinology, 53:604 (1953).

51. R. W. Turkington, J. Clin. Invest., 50:94 (1971).

52. R. P. C. Shiu, P. A. Kelly, and H. G. Friesen, Science, 180:968 (1973).

53. C. Y. Lee and R. J. Ryan, Proc. Endocrine Society Meeting, San Francisco, Abstract No. 146, 1971.

54. B. J. Danzo, A. R. Midgley, Jr., and L. J. Kleinsmith, Proc. Soc. Exptl. Biol. Med., 139:88 (1972).

55. S. Han, H. Rajaniemi, M. I. Cho, A. N. Hirshfield, and A. R. Midgley, Jr., Endocrinology, 95:576 (1974).

56. H. Rajaniemi, A. N. Hirshfield, and A. R. Midgley, Jr., Endocrinology, 96:579 (1974).

57. M. L. Dufau, K. J. Catt, and T. Tsuruhara, Biochim. Biophys. Acta, 252:574 (1971).

58. K. J. Catt, M. L. Dufau, and T. Tsuruhara, J. Clin. Endocrinol. Metab., 32:860 (1971).

59. P. Coulson, T. C. Liu, P. Morris, and J. L. Gorski, in Gonadotropins (B. B. Saxena, C. G. Beling, and H. M. Gandy, eds.), Wiley, New York, 1972, p. 227.

60. D. M. de Krestser, K. J. Catt, H. G. Burger, and G. C. Smith, J. Endocrinol., 43:105 (1969).

61. A. G. Morell, G. Gregoriadis, I. H. Scheinberg, J. Hickman, and G. Ashwell, J. Biol. Chem., 246:1461 (1971).

62. C. Y. Lee and R. J. Ryan, in Receptors for Reproductive Hormones (B. W. O'Malley and A. R. Means, eds.), Plenum, New York, 1973, p. 419.

63. V. K. Bhalla and L. E. Reichert, Jr., J. Biol. Chem., 249:43 (1974).

64. R. Bellisario and O. P. Bahl, J. Biol. Chem., 250:3837 (1975).

65. D. M. Neville, Jr., Biochim. Biophys. Acta, 154:540 (1968).

66. C. S. Song and O. Bodansky, J. Biol. Chem., 242:694 (1967).

67. S. J. Cooperstein and A. Lazarow, J. Biol. Chem., 189:665 (1951).

68. B. J. Danzo, K. M. J. Menon, A. R. Sheth, and A. R. Midgley, The Physiologist, 14:129, Abstract (1972).

69. G. Scatchard, Ann. N.Y. Acad. Sci., 51:660 (1949).

70. V. K. Bhalla and L. E. Reichert, Jr., J. Biol. Chem., 249:43 (1974).

71. C. H. Lee, C. B. Coulam, N. S. Jiang, and R. J. Ryan, J. Clin. Endocrinol. Metab., 36:148 (1973).

72. S. Wardlaw, N. H. Lauersen, and B. B. Saxena, Acta Endocrinol., 79:568 (1975).

73. F. E. Cole, J. C. Weed, G. T. Schneider, J. B. Holland, W. L. Geary, and B. F. Rice, Amer. J. Obstet. Gynecol., 117:87 (1973).

74. J. Roth, Metabolism, 22:8 (1973).

75. M. L. Dufau, K. J. Catt, and T. Tsuruhara, Proc. Natl. Acad. Sci. U.S.A., 69:2414 (1972).

76. O. P. Bahl, L. Marz, and W. R. Moyle, in Hormone Binding and Target Cell Activation in the Testes, (M. L. Dufau and A. R. Means, eds.), Plenum, New York, 1974, p. 125.

77. B. B. Saxena, unpublished data.

78. P. Freychet, R. Kahn, J. Roth, and D. M. Neville, Jr., J. Biol. Chem., 247:3953 (1972).

79. Ch. V. Rao, J. Biol. Chem., 249:2864 (1974).

80. P. Cuatrecasas and M. D. Hollenberg, Biochem. Biophys. Res. Commun., 63:No. 1 (1975).

81. M. Blecher, N. A. Giorgio, and C. B. Johnson, in The Role of Membranes in Metabolic Regulation (M. A. Mehlman and R. W. Hanson, eds.), Academic Press, New York, 1972, p. 367.

82. P. Cuatrecasas, J. Biol. Chem., 246:6532 (1971).

83. G. Sayers and S. Seeling, in Proc. Endocrine Society Meeting, Chicago, Abstract No. 156, 1973.

84. M. L. Dufau, E. H. Charreau, and K. J. Catt, J. Biol. Chem., 248:6973 (1973).

85. O. P. Bahl, M. R. Pandian, W. R. Moyle, and Y. Kobayashi, in Advances in Fertility Regulation through Basic Research, Plenum, New York, in press 1975.

86. K. Park and B. B. Saxena, Federation Proc., 34:240, Abstract (1975).

87. P. Cuatrecasas, J. Biol. Chem., 247:1980 (1972).

88. J. R. Gavin, III, D. L. Mann, and J. Roth, Biochem. Biophys. Res. Commun., 49:870 (1972).

89. F. Krug, B. Desbuquois, and P. Cuatrecasas, Nature, New Biol., 234:268 (1971).

90. G. S. Levey, J. Biol. Chem., 246:7405 (1971).

91. R. J. Lefkowitz, E. Haber, and D. O'Hara, Proc. Natl. Acad. Sci. U.S.A., 69:2828 (1972).

92. D. Keller and A. Goldfien, Clin. Res., 21:202 (1973).

93. S. Kammerman and R. E. Canfield, Endocrinology, 90:384 (1972).

94. C. P. Channing and S. Kammerman, Endocrinology, 92:531 (1973).

95. B. J. Danzo, Biochim. Biophys. Acta, 304:560 (1973).

96. H. Rajaniemi, Endocrinology, 90:1 (1972).

97. K. J. Catt, M. L. Dufau, and T. Tsuruhara, J. Clin. Endocrinol. Metab., 34:860 (1972).

98. C. W. Dwiggins, Jr., R. J. Bolen, and H. N. Dunning, J. Physiol. Chem., 64:1175 (1960).

99. E. H. Charveau, M. L. Dufau, and K. J. Catt, J. Biol. Chem., 249:4189 (1974).

100. C. Y. Lee and R. J. Ryan, in Hormonal Proteins and Peptides, (C. H. Li, ed.), Academic Press, New York, 1974, p. 975.

101. J. Airhart, S. Sibiga, H. Sanders, and E. A. Khairallah, Anal. Biochem., 53:132 (1973).

102. G. Fairbanks, T. L. Stock, and D. F. H. Walbach, Biochemistry, 10:2606 (1971).

103. J. Horejsi and R. Smetana, Acta Medica Scand., 155:Fasc. 1 (1956).

104. E. H. Eylar, J. Theoret. Biol., 10:89 (1965).

105. P. J. Winterburn and C. P. Phelps, Nature, 236:147 (1972).

106. R. J. Winzler, in International Review of Cytology (G. H. Bourne, J. F. Danielli, and K. W. Jeon, eds.), Vol. 29, Academic Press, New York, 1970, p. 86.

107. S. Roth, E. J. McGuire, and S. Roseman, J. Cell. Biol., 51:536 (1971).

108. S. Roth and D. White, Proc. Natl. Acad. Sci. U.S.A., 69:485 (1972).

109. G. L. Nicholson and M. Lacorbiere, Proc. Natl. Acad. Sci. U.S.A., 70:162 (1973).

110. R. S. Turner and M. M. Burger, Nature, 244:509 (1973).

111. T. Tsuruhara, M. L. Dufau, J. Hickman, and K. J. Catt, Endocrinology, 91:296 (1972).

112. A. G. Morell, R. A. Irvine, I. Sternlieb, I. H. Scheinberg, and G. Ashwell, J. Biol. Chem., 243:155 (1968).

113. W. E. Pricer, Jr. and G. Ashwell, J. Biol. Chem., 246:4825 (1971).

114. E. V. Van Hall, J. L. Vaitukaitis, G. T. Ross, J. W. Hickman, and G. Ashwell, Endocrinology, 88:456 (1971).

115. A. Szewchuk and G. E. Connell, Biochim. Biophys. Acta, 83:218 (1964).

116. D. C. Cashman, J. B. Laryea, and B. Weissman, Arch. Biochem. Biophys., 135:387 (1969).

117. S. Saraswathi and B. K. Bachhawat, Biochem. J., 107:185 (1968).

118. H. Fritz, I. Eckert, and E. Werle, Hoppe-Seyler's Z. Physiol. Chem., 348:1120 (1967).

119. O. Svensmark and P. Kristensen, Biochim. Biophys. Acta, 67:441 (1963).

120. F. Margolis and P. Feigelson, Biochim. Biophys. Acta, 89:357 (1964).

121. H. Tuppy, V. Weisbauer, and E. Wintersberger, Mh. Chem., 94:321 (1963).

122. C. Y. Lee and R. J. Ryan, Biochemistry, 12:4609 (1973).

123. O. P. Bahl and L. Marz, in Gonadotropins and Gonadal Function, (N. R. Mougdal, ed.), Academic Press, New York, 1974, p. 460.

124. M. L. Dufau, K. J. Catt, and T. Tsuruhara, Endocrinology, 90: 1032 (1972).

125. M. L. Dufau, K. Watanabe, and K. J. Catt, Endocrinology, 92:6 (1973).

126. K. J. Catt and M. L. Dufau, in Receptors for Reproductive Hormones (B. W. O'Malley and A. R. Means, eds.), Plenum, New York, 1973, p. 379.

127. W. R. Moyle and J. Ramachandran, Endocrinology, 93:127 (1973).

128. L. E. Reichert, Jr., F. Leidenberger, and C. G. Trowbridge, Recent Progr. Hormone Res., 29:497 (1973).

129. J. G. Boue, A. Boue, and P. Lazar, Amer. J. Obstet. Gynecol., 116:806 (1973).

130. J. Vaitukaitis, G. D. Braunstein, and G. T. Ross, Amer. J. Obstet. Gynecol., 113:751 (1972).

131. R. Landesman and B. B. Saxena, Proc. American Fertility Society Meeting, Los Angeles, April 1975 (Abstract).

132. L. Wide and C. A. Gemzell, Acta Endocrinol., 35:261 (1960).

133. T. P. Rosal, B. B. Saxena, and R. Landesman, Fertil. Steril., 26:1105 (1975).

134. G. D. Braunstein, J. L. Vaitukaitis, P. P. Carbone, and G. T. Ross, Ann. Intern. Med., 78:39 (1973).

135. F. Leidenberg and L. E. Reichert, Jr., Endocrinology, 92:646 (1973).

136. B. B. Saxena, personal communication.

137. A. J. Zeleski and A. R. Midgley, Program of the 55th Meeting of the Endocrine Society, Abstract 41, Chicago, June 1973.

Chapter 11

INSULIN AND GROWTH HORMONE RECEPTORS IN HUMAN CULTURED LYMPHOCYTES AND PERIPHERAL BLOOD MONOCYTES

Pierre De Meyts

Diabetes Branch
National Institutes of Arthritis, Metabolism and Digestive Diseases
National Institutes of Health
Bethesda, Maryland

I. INTRODUCTION

The development of methods for labeling polypeptide hormones at high specific activity while preserving their biological properties [1, 2] has stimulated direct studies of the binding of these hormones to their specific membrane receptors in various in vitro preparations of target tissues [3, 4]. However, the physico-chemical characterization of hormonal receptors could not be applied readily to the study of their physiological regulation and of clinical disorders in humans, largely because of the difficulties in obtaining sufficient quantities of human target tissues under appropriate physiological conditions. A major breakthrough in this respect was accomplished with the discovery by Gavin, Roth, and co-workers [5] of insulin receptors on human circulating cells (especially, as demonstrated later, blood monocytes [6, 7]) as well as in human lymphocytes in established cultures [5, 8]. The latter provided in vitro models of insulin-sensitive human cells which are more easily studied. The receptors for insulin on cultured lymphocytes and peripheral monocytes were found to be indistinguishable by a variety of criteria from receptors studied in more classical targets such as liver and adipose tissue [6, 7, 8]. Some human cultured lymphocyte lines also proved to be an abundant source of receptors for growth hormone [9, 10] and, more recently, for calcitonin [11].

I will review in this paper the basic methods for studying the binding of insulin and growth hormone to human lymphocytes in culture and to blood peripheral monocytes, illustrated with some of the representative results which have emerged from the study of these particular models. The methods used to study receptors exploit the competition of the native hormone with a tracer of highly purified, radioactively labeled hormone for the specific receptor molecule. Although receptors have been used to measure hormones in so-called "radioreceptor assays" [12-16], in this review, I will concentrate on the methods used to study the properties of the receptors themselves. I will, thus, discuss successively the characteristics of the cell types used as a source of receptors, the techniques for obtaining biologically active hormones labeled at high specific activity, the optimal conditions for incubating cells with hormones, and the methods used to measure and quantify binding and competition.

II. THE CELL PREPARATIONS

A. Cultured Lymphoid Cell Lines

1. General Characteristics and "Normalcy" Criteria

Lymphoid cell lines derived from human peripheral blood have been established in apparently permanent suspension culture from a large number of normal individuals [17, 18] and patients with varied benign [19] and malignant [20-22] lymphoproliferative disorders. These cells grow vigorously as free-floating pleiomorphic forms with a fundamental lymphoid character on light and electron microscopic analysis [23], reaching, in certain conditions, densities up to 3×10^6 cells per ml with near 100% viability. More than a thousand such cell lines have been established [24, 25]; and many of them are currently maintained in many laboratories around the world (some have been in continuous culture for more than ten years), and have been made available from cell line banks and commercial sources.

The important question arises whether these cell lines are to be considered as representative of "normal" or "transformed" cells, and, if transformed, whether they are potentially or actually "malignant." Transformation has been defined as "an induced inheritable change in the properties of a cell, accompanied by the loss of regulatory controls of cell growth" [26]. In the case of viral transformation, the criteria for transformation of cells generally include the following [27]: (a) loss of contact inhibition, (b) altered morphology, (c) increased growth rate, (d) increased capacity to persist in serial subcultures, (e) chromosomal abnormalities, (f) emergence of new antigens, (g) capacity to form neoplasms, and increased resistance to reinfection by the transforming virus. Let us examine these criteria in the case of cultured lymphocytes.

a. Contact Inhibition

The criterion of contact inhibition is irrelevant in the case of human lymphoid cell lines where cells grow in suspension.

b. Morphology

The cell population resembles the immature, "blast-like" transformed cells seen after the stimulation in vitro of peripheral lymphocytes by phytomitogens and antigens [28]. Moore and co-workers [29] have studied the morphology of 400 cell lines from nearly 5,000 original, temporary and long-term cultures derived from about 1,000 individuals. In their opinion, "there are no microscopic or ultramicroscopic characteristics that permit separation of cultured human lymphoblastoid cells derived from normal individuals and from patients with leukemia, Burkitt's lymphoma and infectious mononucleosis." Further, there seemed to be no difference between lymphoblasts stimulated with phytomitogens and those of established cell lines. A typical distribution of cells was 85% lymphoblasts, 10% medium and small lymphocytes, and 5% multinucleated cells. Some stem cells and phagocytic cells are also present, but they maintain lymphoid morphology and undoubtedly differentiated from lymphoblastoid cells. No granulocytic or erythroblastic cells were seen in long-term cultures. A majority of the cell lines have retained normal karyotypes. Moore emphasizes that in a cell line derived from a patient with a malignant disease, the cell population does not necessarily consist of abnormal cells. To explain the absence of development of cell types other than lymphoblastoid from the original blood cells inoculum, Moore theorized that only cells in the lymphocyte series were capable of dedifferentiation into primitive forms of precursor cells that could become self-sustaining. An alternative suggestion was that the permanent cell lines developed from one or more stem cells present in the initial inoculum and that such stem cells differentiated into lymphoblastoid forms only because of the environmental conditions of culture. Further evidence indicates that the cultures are in fact clones, i.e., derived from a single cell, and represent "relatively immature but immunoglobulin committed cells" [30].

c. Increased Growth Rate

Cultured human lymphoid cells are capable of attaining doubling times of about 20 hours, the minimal time observed for human cells [31].

d. Increased Capacity to Persist in Serial Subculture

Normal human cells in primary culture cannot subsist beyond approximately 40 to 50 population doublings [32]. On the contrary, established lymphocyte cultured lines have continued to multiply rapidly for months and years, and "thus appear to have an infinite life" [29].

e. Chromosomal Abnormalities

Studies by Glade and Hirshhorn [23] have demonstrated the general stability of the diploid human chromosomal pattern in permanently established lymphoid cell lines. Maintenance of the identity of the genome of cultured lymphocytes with the donor's genome has been proved by the persistence, in some cases, of unusual morphological features present in the donor (e.g., Chediak-Higashi syndrome), the synthesis of a variety of enzymes with wide genetic polymorphism, and the identity of HL-A surface antigens of host cells and lymphoid cells in long-term culture. Cytologic studies have detected abnormalities in some cell lines derived from patients with leukemia and Burkitt's lymphoma, but not from normal individuals [29]. The spread of chromosome sets around the mode of 46 and the few instances of polyploidy are similar to those seen in temporary cultures of lymphoblasts stimulated with PHA but well within the limits of normalcy set by the Tissue Culture Association Subcommittee [29, 33].

f. Emergence of New Antigens

Membrane antigens reflecting the presence of the Epstein-Barr virus are present in most human cultured B-type lymphoid cell lines (see Sec. II. A. 2) [34].

g. Capacity to Form Neoplasms

There is little indication that cultured cell lines have true oncogenic capacities. Reinfusion to patients of cultured autologous lymphocytes showed no evidence of subsequent development of leukemia [35]. However, Adams and co-workers [36] reported induction of tumors upon heterotransplantation to ALS-immunosuppressed neonatal Syrian hamsters, of human lymphoid cells isolated in continuous culture from patients with Hodgkin's disease, Letterer-Siwe disease, infectious mononucleosis, and from normal subjects. The tumors, of the immunoblastoma type, "were progressively growing, fatal, serially transplantable, locally invasive and metastatic, and thus possessed the behavioral, gross and histologic characteristics of malignancy." Adams suggested that it may be a general characteristic of such cell cultures to possess malignant potential demonstrable by heterotransplantation methods.

In summary, by a variety of criteria, it is evident that human cultured lymphoid cell lines are not strictly normal cells but represent transformed lymphocytes. Most of the characteristics of malignant cells are, however, usually absent (no chromosomal abnormalities, no demonstrated tumorigenic potential outside the findings of Adams et al.). Exceptions will be discussed in Sec. II. A. 2. Most of the lines conserve the functional properties of normal peripheral lymphocytes that have been stimulated by mitogens or antigens; these cells synthetize various products, some of which are secreted in the medium, which are of biological significance in the normal defense mechanisms of the immunocompetent host [23]: immunoglobulins (see also [30]), mediators of cellular immunity, C'3, interferon, enzymes, and histocompatibility antigens. The cultured lymphocyte might have undergone the same "de-repression" as an immunocompetent lymphocyte after contact with the initiating antigen or mitogen, and many of the complex biochemical changes observed under controlled conditions in vitro may parallel activities which are typical of the immune response in vivo [30]. The lymphoid tissue culture system may provide thus an "in vitro" model for studying the activated lymphocyte [30].

2. Nature of the Transforming Agent in Cultured Lymphoid Cell Lines

Evidence is accumulating that all long-term lymphocyte cell lines represent viral transformants [37]. Most established human lymphoblastoid cell lines of the bone marrow- or "bursa"-derived, B-type, irrespective of origin, have been found to carry the genome of the Epstein-Barr herpes virus (EBV), the causal agent of infectious mononucleosis [38-42]. This virus was initially isolated from Burkitt's lymphoma [43], and an increasing body of evidence seems to implicate this virus as an etiologic agent in this and maybe other human malignancies [38-44]. Even "purified" human T-lymphocytes (thymus-derived) suspensions gave rise, after EBV infection, to EBV-genome-carrying lines with B-cell characteristics [44]. Even in this case, thus, the cell line probably developed from a few contaminating B-cells, which is understandable since only B- (but not T-) lymphocytes carry receptors for the EBV [45]. Even when isolated from patients with benign or malignant lymphoproliferative disorders, the lymphocyte lines seem to have grown, not from the tumoral cells, but from EBV-carrying normal B-lymphocytes. Typical examples are the cell lines IM-9 (which we use routinely for insulin binding) and IM-10 which were derived from patients with multiple myeloma; in culture, these cell lines synthetize immunoglobulin molecules distinct from the patients' myeloma proteins [25]. By contrast, most Burkitt's lymphoma-derived cell lines appear to represent the malignant lymphoma cells in culture [46]. The rare B-cell lines growing in culture but not carrying the EBV, such as 8226 and 266B1, are also likely to represent the malignant myeloma cell in culture. Cell line 8226 produces a λ Bence-Jones protein, as did the myeloma tumor of the patient from whom the original cells were obtained [47]. Moreover, an abnormal karyotype and additional features of malignancy are present [47]. Three cell lines (at least two being clearly B-cells) apparently lacking the EB viral genome have recently been reported [44]; they were derived from patients with malignant lymphoma and again presumably represent the malignant cell itself. Cultured T-cell lines, EBV negative, have been derived only from patients with acute lymphoblastic T-cell leukemia or the leukemic phase of lymphosarcoma and probably represent the malignant T-cells [25, 48].

It has been suggested [25] that the "malignant" cell transformation is distinct from the "EB-viral" transformation necessary for normal B-lymphocytes to become established as permanent cell lines.

This somewhat lengthy development will prove useful for the discussion in Sec. II. C. 2 of the relationship between the presence of insulin receptors and lymphocyte transformation.

3. Cultured Lymphocytes as Biohazards

The ubiquitous presence of the EB virus, as well as the still largely unknown oncogenic potential of the cell lines themselves, obliges investigators to consider lymphoid cell lines as a potential hazard. Thus, it is essential that persons studying such cells take appropriate precautions [37, 49].

4. Preparation and Care of Established Human Lymphoid Cell Lines

The establishment of lymphoid cell lines from aliquots of peripheral blood represents a specialized work and the methods will not be detailed here (see Refs. 50 and 51). Unless interested by the transformation process itself, the investigator willing to study hormonal receptors in lymphoid cell lines is better off screening the cell lines already established in various laboratories and commercial sources. The regular care of an already established cell line is easy, provided that a minimum of precautions (mostly concerned with sterility) are respected. All manipulations of the cells, except when used for receptor binding, are made in a laminar flow hood.

The cells grow in suspension at 37°C in 250-ml, sterile Falcon disposable plastic tissue culture flasks containing 50 ml of a prescribed growth medium. Cells grown in Eagle's minimal essential medium (MEM) or RPMI 1640 (see any textbook of tissue culture for composition) were found to have identical hormone-binding patterns. The medium contains 100 units/ml penicillin and 100 μg/ml streptomycin, and is supplemented with 10 to 20% fetal calf serum (Grand Island Biological) according to the desired cell density at stationary phase of growth. The line IM-9 reaches a density of 10^6 cells per ml 48 to 72 hr after feeding with a medium containing 10% fetal calf serum. Glutamine is omitted from the medium for storage (4°C) purposes; 10 ml of a 3% solution of glutamine is added to the medium just prior to use. The cells are "fed" twice weekly by dividing the cultures 1:3 and adding warmed fresh medium (37°C). They are used for binding studies at 48 to 72 hr after feeding, when in a late exponential or early stationary phase of growth [30]. A progressive change in the color of the medium [which contains a pH indicator (phenol red)], from red to yellow, is normal and reflects active cell metabolism.

It is prudent to store some cell samples in liquid nitrogen, since accidents like contamination of the culture may occur. Freezing cells may be accomplished following a procedure adapted from Glade and Broder [50] and Moore [51]. Approximately 2 to 5×10^6 cells in the log phase of growth are centrifuged at 600 g for 5 to 10 min and resuspended (5×10^6 cells per ml) in fresh complete sterile medium containing 20% fetal calf serum at 0°C with 10% glycerol or 12.5% dimethylsulfoxide (DMSO), sealed in glass ampules or plastic tubes, slow frozen (1°C/min) to -40°C, and then placed in a liquid nitrogen storage tank. To reactivate the culture, the sample is defrosted and incubated in a 37°C water bath. The ampule must be broken or the tube open sterily. The cells are washed three times with cold medium plus 10% fetal calf serum and seeded in regular feeding medium.

5. Screening of Existing Cell Lines for Insulin and Growth Hormone Binding

A large number of established human lymphoid cell lines have been examined by Gavin [25] for insulin and human growth hormone receptors using direct binding of ^{125}I-labeled hormones with the incubation procedures described in Sec. IV. Lines of both B- and T-lymphocytes were studied. B-cell lines were characterized by membrane staining with a fluorescent goat antiserum to human γ-globulin (Meloy Laboratories). T-cell lines were all positive for HTLA [52], and their T-cell properties have been described (see Ref. 25 for bibliography). More details about cell markers for B- and T-cell as well as an up-to-date review of the general properties of both types of cells can be found in an excellent recent book [53].

Representative data are given in Tables 1 and 2. The data can be summarized as follows [25]: Insulin binding was present in most of the

TABLE 1. Insulin and hGH Binding to Cultured Lymphoid Cells[a]

Cell line		Source	Insulin bound[b] (ng/10^7 cells)	hGH bound[c] (ng/10^7 cells)
Class I.	8866	Acute myelogenous leukemia	0.47	0.39
	1120	Normal	0.50	not tested
	IM-9	Multiple myeloma	0.52	0.38
Class II.	IM-10	Multiple myeloma	0.28	0.21
	8205	Chronic myelogenous leukemia	0.12	0.18
	Levins	Infectious mononucleosis	0.30	0.17
Class III.	Daudi	Burkitt's lymphoma	0.04	0.002
	Molt-4[d]	Lymphoblastic leukemia	0.04	0.000
Class IV.	Namalva	Burkitt's lymphoma	0.01	0.000
	8226	Myeloma	0.01	not tested

[a]Data from Gavin [25]; used with permission.

[b]1.5×10^7 cells were incubated at 15°C for 75 min with an insulin concentration of 2 ng/ml.

[c]2×10^7 cells were incubated at 30°C for 90 min with an hGH concentration of 5 ng/ml.

[d]Molt-4 is the only T-cell line tested which has had any insulin binding. All other cell lines in this table are B-cells.

TABLE 2. Cell Volumes and Surface Areas of B- and T-Cell Lines[a]

Cell line	Type	Average volume (μ^3)[b]	Average surface area (μ^2)[c]	Relative insulin binding (%)[d]	Relative hGH binding (%)[d]
IM-9	B	1,478	628	100	100
4265	B	1,444	618	41	71
SB	B	3,138	1,037	69	not tested
Naliaka	B	1,513	637	0	0
8402	T	798	416	0	0

[a] Data from Gavin [25]; used with permission.

[b] Determined from Coulter counter volume plot.

[c] Calculated from (b) using formula for volume and surface of a sphere.

[d] Cell line IM-9 taken as 100% standard.

cell lines studied (Table 1). Interestingly, the pattern of insulin binding did not represent a statistical distribution (see Note number 1), but the cell lines clustered around distinct binding patterns, which have been assigned arbitrary categories by comparison with one of the best-binding cell lines (IM-9), used as a reference (the maximum tracer binding by this cell line was considered 100%). Cells classified in Class I bound 75 to 100% as much insulin as IM-9; Class II: 25 to 74%; Class III: 7 to 24%. Class IV was reserved for cell lines without displaceable insulin binding. The same pattern was observed for hGH binding, which could be classified along the same criteria. In general, less hGH was bound per cell, but the relative patterns of binding were similar to those of insulin, and cells, with a few exceptions, fell in the same category for both types of hormones. An interesting feature noted by Gavin [25] is that the Burkitt's lymphoma-derived cell lines as a group (also EBV carriers but probably malignant) bound little or no insulin or hGH.

The T-cell lines studied (which also have malignant characteristics) had no displaceable binding of insulin, except Molt-4, which fell in the low range of Class III. The T-cells had slightly smaller surface area (Table 2), but even when correction for this difference was made by increasing the number of cells, no specific binding could be detected. Table 2 shows that differences in surface area did not account for observed differences in binding. A "catalogue" of all cell lines tested by Gavin as of May 1974 can be found in Ref. 25. As they pointed out, a Class I or Class II cell line should be selected for studying hormone-receptor interactions, but cell lines of Class IV, lacking receptors, can serve as controls in searching for a hormone-mediated membrane alteration or metabolic effect. The potential importance of the lymphocyte as a model cell for biochemical studies has recently been stressed by Parker et al. [54].

B. Peripheral Blood Mononuclear Cells

1. Blood Mononuclear Cells Preparation

Peripheral blood (300-500 ml) from normal volunteers is drawn into acid-citrate-dextrose solution and centrifuged at 1500 g for 3 min at 20°C. The buffy coat is removed, diluted 1:1 with isotonic saline, and fractionated on Ficoll-Hypaque gradient according to the following method, after Boyüm [56]:

Ficoll (Pharmacia Fine Chemicals): Dissolve 9 g in 100 ml distilled water (= 9%). Since powder goes into solution slowly, usually prepare the day before.

Hypaque-M, 90%, sterile aqueous solution (Winthrop): Warm opaque solution in water bath until clear and prepare a 33.9% (vol/vol) solution in distilled water.

These two stock solutions can be prepared in advance and kept refrigerated, but should be warmed to room temperature before use.

Gradient: 10 parts of 33.9% Hypaque, 24 parts of 9.0% Ficoll. Mix together well; add 3 ml of the gradient to each of 10 or more 16 × 150-mm plastic tubes. The gradient has to be prepared daily.

Layer 10.0 ml of the buffy-coat saline suspension carefully, so as not to break the gradient. This careful layering can be accomplished by using a Pasteur pipet to layer on 1 to 2 ml of buffy coat-saline solution. Then with a syringe (50 ml) and a 18-gauge needle, complete the filling of the tubes with the buffy coat- saline suspension. Centrifuge the tubes at 400 *g* for 36 min at 20°C.

After centrifugation, four layers can be distinctly seen (top to bottom): (1) plasma saline, (2) mononuclear cells, (3) Ficoll-Hypaque, and (4) granulocytes and red blood cells.

The thick mononuclear cell layer, along with the upper one-third of Ficoll-Hypaque layer, is removed with a Pasteur pipet and distributed into 50-ml, conical centrifuge tubes. Dilute the isolated mononuclear cells with cold normal saline (~30 ml). Centrifuge at 250 *g* for 16 min at 4°C; carefully decant the supernatant and resuspend in normal saline; recentrifuge at 250 *g* for 8 min. Discard the supernatant and resuspend the cell pellet in the assay buffer solution to a cell concentration of 50×10^6 mononuclear cells per ml as determined by hemocytometer or Coulter counting. Viability, as assessed by trypan blue dye exclusion, should be greater than 95%. These procedures are detailed in Sec. IV. A.

The average composition of this cell population [6] is $22.2 \pm 2.6\%$ (range 10 to 36) monocytes; $8.6 \pm 1.7\%$ (range 2 to 16) B-lymphocytes; and $68 \pm 2.6\%$ (range 58 to 79) nonphagocytic, nonimmunoglobulin-bearing cells (T-lymphocytes). The preparation also contains a few contaminating granulocytes (~1%) and a variable number of platelets. Red blood cells contamination rarely exceeds one red blood cell per ten white blood cells. The average specific insulin binding to this mixed population is $5.1 \pm 0.7\%$ when 0.1 ng of [^{125}I]insulin is added to 2×10^7 cells.

2. Identification of the Insulin-binding Cells

Gavin and co-workers demonstrated [5] that peripheral blood mononuclear leucocytes from normal blood donors, purified on Ficoll-Hypaque gradients according to Boyüm [56], bear receptors for insulin. This cell preparation thus constitutes an ideal tool for the study of insulin receptors in physiological and pathological states in humans. In contrast with cultured lymphocytes, however, this cell population remains somewhat heterogeneous; an important methodological question is to properly identify the cell type responsible for insulin binding. Since lymphocytes comprise about 80% of the cells in Boyüm's preparation, it was initially assumed that the binding observed was to receptors on lymphocytes [5, 8]. However, using a similar preparation after purification by passage on tightly packed nylon wool, Krug et al. did not detect any significant insulin binding to the remaining cells [57]. Since that purification method removes B-lymphocytes [58-60], Olefsky and Reaven [61] suggested that the insulin binding measured in preparations of mononuclear leucocytes might have been to receptors on B-lymphocytes.

The question was recently clarified by Bianco, Schwartz, and co-workers [6, 7], who showed that monocytes are the predominant insulin-binding cell in peripheral mononuclear leucocytes preparations. These preparations contain a significant proportion of monocytes (~ 20%) when prepared by Ficoll-Hypaque gradients [62]. Monocytes are depleted by nylon wool purification [63, 57].

By various depletion and enrichment experiments, Bianco, Schwartz, and co-workers showed that specific insulin binding correlated with the number of phagocytic cells (monocytes) [6, 7], identified by morphological criteria [64] in cytocentrifuge smears with Wright's or Giemsa's stain, and by the fundamental criterion of latex particle ingestion [65] (Fig. 1 and Table 3). The binding did not correlate with the number of immunoglobulin-bearing cells (B-lymphocytes) or other cell types (Table 3). Further proof was brought by direct visualization of the cells binding [^{125}I]insulin, by autoradiography. These cells were predominantly monocytes both by morphological and functional criteria. Monocyte counts (see Sec. II. B. 3) are now routinely performed in this group in all peripheral mononuclear blood cells preparations used for insulin binding. It must be pointed out that ~10% of the binding detected by autoradiography was to cells which were large nonphagocytic mononuclear cells; these could be nonadherent monocytes, but also blast-transformed or activated B-lymphocytes [7].

3. Monocyte Counts

Binding of insulin to mononuclear cells should be related to the actual number of monocytes in the preparation. We count them by three methods involving different properties of monocytes: (1) morphological criteria [64], (2) latex beads ingestion [65], and (3) esterase staining [66].

The counts obtained by these three methods correlate very well; the counts obtained by esterase staining are usually 110 to 115% of the counts obtained by latex ingestion, possibly due to the presence of a minority of nonphagocytic monocytes.

TABLE 3. Binding of [^{125}I] Insulin to Mononuclear Leucocytes Separated on Glass Wool and Sephadex G-10 Adherence Columns[a]

Cell populations[b]	Specific insulin binding (%)[c]	Monocytes (%)	B-Lymphocytes (%)	Nonimmunoglobulin-bearing, nonphagocytic cells (%)
Glass Wool Columns				
Unseparated cells	5.9 ± 2	27.5 ± 8.5	4 ± 2	68.5 ± 10.5
Nonadherent cells	1.2 ± 0.3	1.3 ± 0.3	4 ± 3	94.7 ± 3.3
Sephadex G-10 Columns				
Unseparated cells	4.7 ± 0.5	23.2 ± 4.9	8.3 ± 3	65.5 ± 3.5
Nonadherent cells	0.79 ± 0.03	2.0 ± 1.5	6.3 ± 1.3	90.7 ± 2.3
Adherent cells	7.9 ± 1.4	39.0 ± 3	8.7 ± 2	50.5 ± 1.5

[a]Data from Schwartz, et al. [7]; used with permission.

[b]Unseparated cells are the mononuclear leucocytes derived from Ficoll-Hypaque gradient separation of human peripheral blood buffy coats. The nonadherent cells were those mononuclear leucocytes which were not retained by the glass wool or Sephadex G-10 matrix after washing the columns with 50 ml of medium. The adherent cells from Sephadex G-10 were those cells which were retained after washing but removed by subsequent mechanical agitation.

[c]$4\text{-}8 \times 10^7$ cells were incubated with 0.1-0.5 ng of [^{125}I] insulin for 3 hr at 22°C, pH 8, in the presence or absence of 50 μg/ml of unlabeled insulin. The percentage of [^{125}I] insulin specifically bound was calculated by subtracting the percentage of [^{125}I] insulin bound in the presence of unlabeled insulin (nonspecific) from the percentage bound in the absence of unlabeled insulin (total) (see Sec. IV. C).

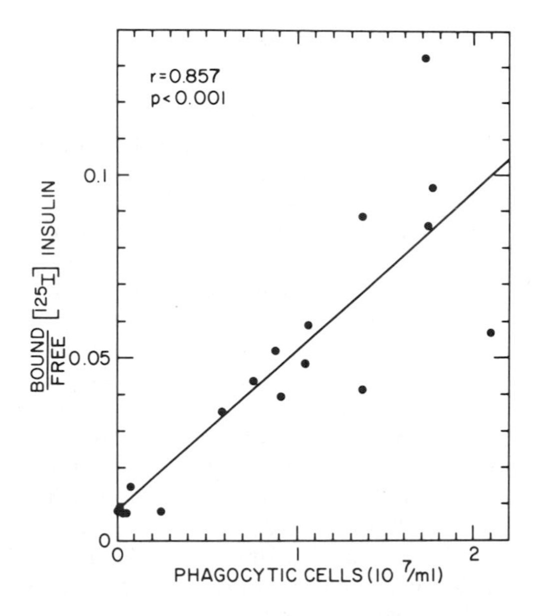

FIG. 1. Correlation between the quantitative level of insulin binding and the number of monocytes in Boyüm's preparations of blood mononuclear cells. The data of all experiments in which both insulin binding and the number of monocytes were determined were pooled; these include unseparated cell populations as well as the separated populations from both glass wool and Sephadex G-10 experiments. Correlation coefficients (r) were calculated under the assumption that the paired data points were a random sample from a bivariate normal distribution. The best fitting line was drawn through the data points using a linear regression analysis of Y (insulin binding) in X (cell number). From Schwartz et al. [7], used with permission.

Besides morphological identification, the latex bead method is the easiest and is performed as follows (modified from Ref. 65):

A 200-μl aliquot from a 5×10^7 cells per ml sample is added to 3 ml 2% bovine serum albumin in phosphate buffer-saline, spun at 600 g for 10 min; the pellet is taken up in 0.5 ml of 100% fetal calf serum with 10 μl of washed latex beads suspension (Uniform latex particles, 1.1 μ diam, Dow diagnostics). The cells are incubated with the beads for 30 to 40 min in a shaking bath at 37°C, then spun at 600 g for 10 min at 4°C. The pellet is taken up in 1 ml phosphate buffered saline and layered over 3 ml 100% fetal calf serum and spun at 4°C for 10 min at 600 g; the cells containing phagocytized beads pellet, whereas the unphagocytized beads float at the top of the calf serum layer. The cells from the pellet are examined in a phase contrast microscope; a cell containing four beads or more is considered to be a monocyte.

C. Biological Relevance of Hormonal Receptors on Lymphocytes and Monocytes

1. Identification of the Binding Molecule as a Receptor

Our ignorance—still deep—concerning the molecular nature of the events that mediate the action of hormones on target cells has tended in the recent past to be compensated by the development of a proliferative semantics, which eventually led to a "molecular linguistics" [67] of hormone action far more perfected than the "molecular biology." Even identification of the first step in this verbose chain of events (in our case, binding of the polypeptide hormone to a specific surface receptor) raises some problems. Which criteria will one require to be fulfilled to accept that a "binding protein" is a "receptor"? The "purist" approach [68] is to consider as a receptor only a binding molecule which not only specifically "recognizes" the hormone, but also the binding to which can be demonstrated to actually initiate biological events characteristic of the hormone studied. Let us examine the current status of hormonal "receptors" on leucocytes.

Although lymphocytes and monocytes are not known to be major targets for insulin and growth hormone, data in the literature suggest that these hormones have biological effects on leucocytes (Table 4). It has now been clearly demonstrated that certain hormones and mediators of inflammation act in vivo to regulate the character and intensity of inflammatory and immune responses; this regulation is mediated by a general inhibitory action of cAMP on immunologic and inflammatory functions of leucocytes [55], including lymphocyte activation and mitogenesis [54]. In particular, several authors have suggested that insulin is involved in inflammatory and hypersensitivity reactions (Table 4). However, we have not, so far, directly measured biological effects of either insulin or growth hormone on cultured human lymphocytes or peripheral blood monocytes. Thus, the identification of the molecules which specifically bind insulin and growth hormone in the membrane of these cells as "receptors" might have been reasonably questioned. The following several lines of evidence suggest a more pragmatic approach.

a. Identification by Physico-chemical Criteria

The "receptor" for insulin in lymphocytes and monocytes shows a striking resemblance to the "receptor" described in other tissues, including purified liver plasma membranes [4], fat cells [4], placental cells [88], and turkey red blood cells [89], with respect to factors as varied as pH and

TABLE 4. Effects of Insulin and Growth Hormone on Leucocytes, Lymphoid Organs, and Immune Responses

Tissue or organ	Source	Treatment	Response studied	Effect observed (Ref.)
Leucocytes	Normal humans	Insulin in vitro	Glucose utilization	None [69, 74] Increase [70, 71, 73] None (short incubation) [72] Slight increase (prolonged incubation) [72]
			Lactate production	None [69, 70, 74] None (short incubation) [72] Slight increase (prolonged incubation) [72]
			Glycogen synthesis	None [72]
	Diabetic humans	None	Glucose utilization	Decrease [70, 73]
			Lactate production	None [70]
		Insulin in vitro	Glucose utilization	None [73, 74] Increase [69, 70, 71]
			Lactate production	None [70] Increase [69]
			Glycogen synthesis	Increase [71]

(continued)

TABLE 4 (continued)

Tissue or organ	Source	Treatment	Response studies	Effect observed (Ref.)
		Insulin in vivo	Glucose utilization	Increase [72]
			Lactate production	Increase [72]
			Glycogen synthesis	Increase [72]
Polymorphonuclear leucocytes	Normal humans	Insulin in vitro	G-6-P independent fraction of glycogen synthetase	Increase [75]
			Phagocytic rate	Increase [76]
Peripheral blood lymphocytes and cultured lymphocytes, RPMI 1788	Humans	Insulin in vitro	Membrane ATPase activity	Increase [77a]
			Glucose uptake	Increase [77a]
Leukemic lymphoblasts in culture	Humans	Growth hormone in vitro	Synthesis of RNA, DNA, and proteins	Increase [77b]
Leukemia cells in culture	Humans	Insulin in vitro	Growth in defined or unsupplemented media	Increase [21]
Thymocytes	Rat	Insulin in vitro	AIB influx	Increase [78]

			Glucose uptake	Increase [79]
Lymphoid tiss	Rat	Insulin	Growth	Increase [80]
Thymus	Dwarf mice	None	Proliferation	Defective [81]
		Growth hormone	Proliferation	Increase [81]
Spleen	Rat	Insulin or growth hormone with or without antigenic stimulus	DNA synthesis	Increase [82]
Lymph nodes	Rat	Insulin or growth hormone plus antigenic stimulus	DNA synthesis	Increase [82]
Systemic	Humans	Insulin	Anaphylaxis Anaphylactoid reaction Tuberculin reaction	Increase [83]
	Guinea pig	Alloxan	Tuberculin reaction	Decrease [84]
			Skin sensitivity to PPD	Decrease [85]
		Alloxan plus insulin	Skin sensitivity to PPD	Restored [85]

(continued)

TABLE 4 (continued)

Tissue or organ	Source	Treatment	Response studied	Effect observed (Ref.)
	Rat	Alloxan	Anaphylaxis Skin sensitivity to PPD	Decrease [85]
		Alloxan plus insulin	Anaphylaxis Skin sensitivity to PPD	Restored [85]
	Albino rat	Insulin	Tuberculin reaction	Increase [86]
	Rat	Insulin	Anaphylactoid reaction to Dextran	Increase [87]
	Diabetic rat	None	Anaphylactoid reaction to Dextran	Decrease [87]
		Insulin	Anaphylactoid reaction to Dextran	Restored [87]

temperature dependencies of binding; equilibrium and kinetic binding characteristics and negative cooperativity, as well as a variety of other more general functional criteria (reversibility, saturability) (for review see [4]). Thus, the bulk of evidence indicates that the same binding molecule is being described in the various tissues studied.

b. Identification by Criteria of Structural Specificity

Detailed studies have shown that the insulin "receptor" in cultured lymphocytes [8] and peripheral monocytes (Bar et at., manuscript in preparation) exhibits a high degree of structural specificity toward a wide range of insulin analogs, which vary 300-fold in their relative affinity for the receptor. This specificity was superimposable on the specificity measured by the relative potencies of these analogs in stimulating glucose oxidation in fat cells, a typical cellular effect of insulin (Table 5). This structural specificity is also identical for insulin "receptors" in all the other tissues studied. This is in contrast with the distinct patterns of specificity displayed by anti-insulin antibodies as measured in radioimmunoassays [90], and in contrast with the peculiar specificity of the insulin-degrading system [91], which excludes these two types of molecules as candidates for being the binding structure on lymphocytes and monocytes.

A high degree of structural specificity was also demonstrated for the human growth hormone receptor in lymphocytes [10]. Human growth hormone preparations bind to this receptor in direct proportion to their biological potencies. The affinity of human placental lactogen is low and nonprimate growth hormones do not bind, well in agreement with the growth hormone bioactivity of these preparations in humans [10].

It does not seem unreasonable to identify the hormone-binding molecule as the receptor when these two types of criteria, physico-chemical similarity with a more "legitimate" receptor and structural specificity, are fulfilled, even if biological effects have not been demonstrated in the same cell where the binding is studied.

The finding of the "receptor" may then constitute a stimulus for investigating possible new biological properties of the hormone studied. The role of insulin and growth hormone in the immune response is a nice example of a new, exciting field of research triggered by the discovery of receptors for these hormones on lymphocytes. Some work already done on the potential role of insulin in lymphocyte mitogenesis and transformation deserves some comments.

TABLE 5. Potency of Insulin Analogs Relative to Porcine Insulin[a]

Analog	In vitro bioassay (%)	Receptor binding: liver membranes (%)	Receptor binding: lymphocytes (%)
Porcine insulin	100	100	100
Human insulin	100	100	100
Fish insulin	50	53	48
Porcine proinsulin	11	4.5	4.5
Split proinsulin	14	5.7	5.8
Desdipeptide proinsulin	14	22	20
Desnonapeptide proinsulin	24	27	27
Bovine desoctapeptide	1.5	1.9	1.2
Bovine des Ala-des Asp.	1.4	1.7	1.0
Guinea-pig insulin	1.0	1.1	1.9
NSILA-s	0.3	0.4	0.4

[a]Data from Freychet, Gavin, Kahn, Gorden, and Roth (see also Refs. 201 and 202).

2. Hormone Receptors and Lymphocyte Transformation

Two groups have studied this relationship [57, 95]. In the first study, the appearance of receptors for insulin on lymphocytes transformed by the plant lectin Concanavalin A (Con A) [57] was reported. This was correlated with a possible role of insulin in the processes of cell growth and division [57, 92]. Without taking issue with some interesting speculation arising from such studies, I am not convinced, for several important methodological reasons, that the reported data demonstrated the presence of insulin receptors in Con A-transformed lymphocytes. First, the high dose of Concanavalin A used may have been in the toxic concentration range [93] (40 μg/ml for 5 to 7×10^5 purified lymphocytes). Powell and Leon [94] reported Con A to be maximally effective on hyman lymphocytes at 10 μg/ml for 1.3×10^6 lymphocytes. Determination of cell viability would be an important control, since an increased proportion of dead cells and cellular debris might induce a higher level of insulin binding (mostly nonspecific) in the "transformed" cell population. The level of "specific" binding reported [57] appears to be at the limit of detection; for example, in Figure 1 of Ref. 57, a peak of 2×10^{-16} moles/ml bound from a tracer 0.35 to 0.70 nM, corresponds to a specific binding of approximately 0.03 to 0.06% of the [^{125}I]insulin. The amount of insulin binding reported in transformed lymphocytes and in the cultured lymphoid cell line RPMI 6237 [57] is in the range of binding found in some of the cell lines screened by Gavin [25] which were considered not to bind insulin significantly (category IV, Table 1). The competition for this [^{125} I]insulin binding by unlabeled insulin (Fig. 4, Ref. 57) followed a bizarrely shaped curve, with about as much displacement of [^{125} I]insulin in the range defined by the authors as nonspecific (>5 μg/ml) than in the "specific" range of concentrations. This curve differs strikingly from competition curves reported for other insulin receptors.

Finally, the criteria which should be the most imperative to be fulfilled for identifying a binding protein as a receptor were not studied; no data were reported concerning either the physico-chemical properties of the "receptor" in transformed lymphocytes, or its structural specificity toward insulin analogs or other closely related substances.

Contrasting results were found in the second study [95]. In this study, peripheral mixed mononuclear cells were placed into culture either with addition of Concanavalin A or without ("control cells"). Transformation was monitored by cell number, mean cell volumes (Coulter), and with a single DNA quantitation by flow microfluorimetry (FMF) analysis using the mithramycin method. The results were: (1) Insulin specific binding fell from 4% in freshly obtained control cells to 0.5% after 24 hr of culture. Thereafter no specific binding was observed on control cells. This is explainable by the progressive disappearance of the insulin-binding monocytes subpopulation. (2) Con A (40 μg/ml) led to a 50% or greater fall in cell number within 12 hr; 20 μg/ml had no such acute toxic effect. (3) Mean cell volume increased by 24 hr in a subpopulation of Con A-treated cells. This became pronounced over five days with appearance of a bimodal volume distribution. (4) FMF analysis revealed that initially and 24 hr after Con A, all cells were in G_1 phase; at 48 hr, a population of cells in S phase were seen and at 72 hr, cells were observed in both S and G_2 phases. (5) No specific binding of insulin or growth hormone was observed in Con A-treated cells at any time.

It must be noted that Con A is a T-cell transformant; the observations of Rabinowitz et al. [95] were thus concordant with the general pattern observed in lymphoid cell lines [25], where lines of the T-type usually also fail to bind insulin. It is interesting to correlate this with the fact that, according to Moore and Minowada [24], all normal lymphoid cell lines have B-cell characteristics, but most T-cell lines have malignant characteristics (see Sec. II. A. 5). It is clear from the presence of insulin and growth hormone receptors on cultured lymphoid cells of the B-type, that there might be a relationship between certain types of lymphocytes transformation and the appearance of hormonal receptors; it would be of interest to study the kinetics of the appearance of hormonal receptors on peripheral lymphocytes in the presence of B-cell transformants. In that case, the same rigorous criteria for receptor identification applied to insulin receptors in other tissues should be assessed.

III. LABELING THE HORMONES

A. Carrier-free versus Stoichiometric Monoiodination

The methods used in the labeling of polypeptide hormones for receptor studies have been recently reviewed [3, 4, 96] and, thus, no detailed comparative analysis will be attempted here. We will only briefly discuss the methods of radioiodination. The chloramine-T method of Hunter and Greenwood [97], which yields immunologically reactive iodinated peptides of very high specific radioactivity, has been the most widely used in receptor studies. However, the original method had to be modified to avoid the deleterious effects of the excess of chloramine-T or metabisulfite, or of overiodination of the hormone [3], on the biological activity and reactivity of the hormone with cell receptors. Roth and his co-workers have formulated two generally applicable methods that reproducibly yield monoiodo ^{125}I-labeled hormones of high specific radioactivity suitable for hormone-receptor studies. Both methods have been successfully applied to insulin [91, 96, 98] and other peptides [3, 4]. One approach is to label a small minority (10% or less) of the molecules with one I atom, followed by chemical separation of the uniodinated molecules from the monoiodohormone. The final product ("carrier-free" monoiodohormone) has one I atom per molecule but was never exposed to the vigorous conditions of the standard chloramine-T method. The second approach ("stoichiometric iodination") introduces directly an average of 0.2 to 0.8 atoms of I per hormone molecule under special conditions, followed by a traditional purification step. A variable percentage of uniodinated hormone subsists in the preparation, but diiodination is largely avoided. In both methods, deleterious side reactions are minimized by the following modifications [3]: (1) The concentrations of reactants are maximized by reducing the volume as much as possible, e.g., hormone solution is near the limit of its solubility; a small volume of concentrated $H_2PO_4^-$ is substituted for a larger volume of neutral phosphate buffer to neutralize the $Na^{125}I$-NaOH and buffer the iodination solution. (2) The amount of radioactive iodide at the start is two to five times greater than that intended to be incorporated, so that the $[I^-]$ never falls significantly during the iodination. (3) Chloramine-T is added in limiting amounts in multiple small additions; after each addition the degree of iodine incorporation is measured on an aliquot of the iodination solution by TCA precipitation of the labeled hormone or Dowex adsorption of the free I^- and the size of the next addition is based on the finding. (4) Metabisulfite is limited to an amount twice that of the chloramine-T added. When chloramine-T is used in limiting amounts, metabisulfite may be omitted entirely, and albumin picks up residual reactive I, forming $[^{125}I]$albumin, which can be removed readily (the batch of albumin used must be checked carefully for hormone degrading activity).

Similar principles have also been applied to modify the lactoperoxidase iodination method [99, 100] for preparing monoiodinated insulin [101-103]. The biological and chemical properties of monoiodoinsulin have now been extensively characterized [97, 101-104]. It must be pointed out that iodination of all tyrosines in a molecule may not be uniform [4, 101, 104] and thus, iodination with an average of one I per molecule does not usually result in a homogeneous monoiodohormone. Although the preparation of carrier-free monoiodohormone is obviously preferable for chemical characterization and initial bioactivity studies, it is much more time-consuming and not necessarily superior to the stoichiometric iodination method for most everyday use in the study of hormone-receptor interactions. Actually, carrier-free monoiodoinsulin prepared by Sodoyez et al. [102, 103] and stoichiometrically labeled insulin prepared by us were compared in an interlaboratory experiment and found indistinguishable in terms of receptor reactivity in both cultured lymphocytes and rat liver membranes. Protocols for the stoichiometric iodination of insulin and growth hormone [9, 10] are described in the following Section, III. B.

B. Protocols for Iodination (Stoichiometric Monoiodination)

To determine in each particular case the desirable level of iodination, the following calculations have to be performed. The theoretical specific radioactivity in μCi/μg for one radioactive atom per hormone molecule is given by the following formula [96]:

$$\frac{130 \times 10^6}{\text{mol wt of hormone monomer} \times \text{t } 1/2 \text{ of isotope (days)}}$$

Carrier-free [^{125}I]monoiodo-growth hormone would thus have a specific activity of 110 μCi/μg, and insulin 360 μCi/μg. As a first approximation, Roth advocates [96] 0.2 I atoms incorporated per tyrosyl moiety, up to a maximum of 1.0 I per molecule. In this way, the relative number of diiodotyrosyl groups produced will be minimized. So, for growth hormone (8 tyrosyls), $8 \times 0.2 = 1.6$, so the maximum level desirable is 1.0 per molecule. For insulin (4 tyrosyls), $4 \times 0.2 = 0.8$, would be the level of I per molecule desirable. Practically, we aim to label both hormones at about 0.5 I per molecule, or half the theoretical specific activity of the carrier-free monoiodohormone.

Knowing the specific activity of the Na^{125}I, it is easy to calculate the desirable incorporation in a given iodination:

Example. The lot of Na^{125}I furnished has 655 μCi/μl. We want to label insulin at a final specific activity of 180 μCi/μg (we tolerate 150-225); then if we react together 5 μl of Na^{125}I (=3275 μCi) and 10 μg of insulin, we will try to incorporate, by stepwise additions of chloramine-T: $180 \times 10/3275 = 54\%$ of the Na^{125}I at most.

The following protocols are useful for insulin and hGH.

1. Human Growth Hormone (after Roth [96])

a. Materials

[A] Monosodium phosphate 0.2 M in water (pH unadjusted).

[B] Sodium phosphate buffer, 0.3 M, pH 7.4.

[C] Na^{125}I at concentration $\geq$ 300 μCi/μl in 0.1 NaOH.

[D] Human growth hormone (immunochemical grade, National Pituitary Agency, National Institutes of Arthritis, Metabolism and Digestive Diseases, N.I.H., Bethesda, MD 20014), dissolved at 5 mg/ml in water.

[E] Chloramine-T (Eastman Kodak, Rochester, N.Y.), 30 mg powder in a 10-ml test tube (keep away from light).

[F] Human serum albumin (HSA) dissolved at 1 mg/ml in a 1:10 solution of sodium phosphate buffer (solution [B]). One milliliter of this is pipetted into each of ten tubes (12 × 75 mm) placed in a rack to form two parallel rows of five tubes each.

[G] Trichloroacetic acid, 10 g/100 ml in water (=10%).

[H] Sephadex G-100 column (1.5 × 90 cm) equilibrated with solution [F].

b. Procedure

The letters in brackets refer to the listed materials.

Step 1. Immediately before starting step 2, dilute chloramine-T by 10 ml of phosphate buffer [B], mix well, then further dilute 1 to 100 (10 μl to 1 ml phosphate buffer). The concentration is now 30 μg/ml.

Step 2. Add, in sequence at room temperature, to a small test tube with a conical bottom:

~5 μl monosodium phosphate [A] (volume equal to that of $Na^{125}I$ used)

~5 μl $Na^{125}I$ (volume containing 2.2 mCi)

4 μl human growth hormone (20 μg)

5 μl of chloramine-T (mix well)

The final concentration of reagents is now:

5×10^{-5} M $Na^{125}I$

5×10^{-5} M human growth hormone

2.5×10^{-5} M chloramine-T

Step 3. After at least 2 to 5 min, determine the percent of ^{125}I incorporated in growth hormone. Transfer from the iodination solution a minute aliquot with the tip of a Pasteur pipet to one of the tubes containing 1 ml of 1 mg/ml HSA [F] and mix well (dilution #1). From dilution #1, transfer a small aliquot with the tip of a new Pasteur pipet into a second tube containing 1 ml HSA [F] (dilution #2). To dilution #2, add 1.0 ml of 10% TCA [G] (final concentration of TCA = 5%) and mix well. The albumin forms a visible precipitate with which the [^{125}I] human growth hormone coprecipitates. Sediment the precipitate in a table-top centrifuge for 30 sec. Separate and count both the supernatant (free ^{125}I) and the precipitate [^{125}I]-human growth hormone). The double-step dilution is aimed at avoiding saturation of the counting capacity of the gammacounter by too high radioactivity, which yields biased estimates of incorporation. If more than 40% of the radioactivity is incorporated, go to the next step. If 10 to 40%

is incorporated, add another 5-μl aliquot of chloramine-T and repeat step 3. If less than 10% is incorporated, add 10 μl of chloramine-T and repeat step 3.

Step 4. Add 1 ml of 1 mg/ml HSA [F] to the iodination mixture and transfer to the Sephadex column. Elute with 1 mg/ml HSA [F]. The intact [^{125}I]human growth hormone, which appears as a discrete peak at about halfway between the void and the salt, is free of iodide and other undesirable iodinated components. Transfer a minute aliquot of the final product to 1 ml HSA [F] and add 1 ml 10% TCA [G]. The TCA precipitability should be > 90%. Freeze the [^{125}I]human growth hormone in small aliquots.

2. Insulin (after Kahn and Goldfine, unpublished data)

a. Materials

[A] Veronal buffer 0.05 M, pH 8.6.

[B] Sodium phosphate buffer 0.3 M, pH 7.4.

[C] Na^{125}I at concentration ≥ 300 μCi/μl in 0.1 N NaOH.

[D] Porcine insulin (zinc-insulin crystals, purchased from Elanco), 1 mg/ml in 0.01 M HCl.

[E] Chloramine-T (Eastman Kodak, Rochester, N. Y.), 40 mg powder in a 10-ml test tube (keep away from light).

[F] Bovine serum albumin (BSA) dissolved at 1 mg/ml in a 1:10 dilution of solution [B]. One milliliter of this is pipetted into each of ten tubes (12 × 75 mm) placed in a rack to from two parallel rows of five tubes each.

[G] Trichloracetic acid, 10 g/100 ml in water (= 10%).

[H] Sodium metabisulfite (Aldrich), 20 mg powder in a 10-ml test tube (keep away from light).

[I] 100 ml 2.5% BSA (25 mg/ml) in phosphate buffer (solution [B]) adjusted to pH 7.4.

[J] Six milliliters of 12% BSA (120 mg/ml) in phosphate buffer (solution [B]).

[K] Cellulose powder (Whatman, standard grade); small column 1.5 in. high in a Pasteur pipet.

[L] Six test tubes (12 × 75 mm) labeled 1 through 6.

[M] Four test tubes (12 × 75 mm) labeled A through D, containing 2 ml Krebs-Ringer phosphate buffer without albumin, pH 7.4.

b. Procedure

The letters in brackets refer to the listed materials.

Step 1. Wash the cellulose column with several milliliters of veronal buffer [A] and keep moist with the addition of veronal.

Step 2. Immediately before starting step 3, dilute chloramine-T [E] with 10 ml of phosphate buffer [B], mix well, then further dilute 1 to 100 (10 λ to 1 ml phosphate buffer). The concentration is now 40 μg/ml.

Step 3. Add, in sequence at room temperature, to a small test tube with a conical bottom:

35 μl of phosphate buffer [B]

5 to 10 μl of $Na^{125}I$ (volume containing ~3 mCi)

10 μl of insulin (10 μg)

10 μl of chloramine-T solution (mix well)

The final concentration of reactants is now:

2.8×10^{-5} M insulin

2.5×10^{-5} M $Na^{125}I$

2.5×10^{-5} M chloramine-T

Step 4. After at least 2 to 5 min, determine the percent of ^{125}I incorportated in insulin by TCA precipitation (procedure strictly identical as for growth hormone: see Sec. III. A. 1). Usually, 40 to 60% of the iodine is reacted with insulin and, therefore, TCA precipitable. If necessary, add another 5 to 10-λ aliquot of chloramine-T to achieve the desired incorporation, and repeat the TCA precipitation.

Step 5. Dilute the metabisulfite [H] with 10 ml of phosphate buffer [B], mix well, then dilute further 1 to 10 (100 μl to 1 ml phosphate buffer). The concentration is now 200 μg/ml. Add 5 μl to the labeling vial.

Step 6. Add 100 μl of the 2.5% BSA [I] and transfer the mixture to the cellulose column. Wash the column with six 1.5-ml aliquots of veronal [A] to elute any free I or damaged insulin and collect in the tubes labeled 1 through 6 [L].

Step 7. Elute the [^{125}I] insulin with four 1.5-ml aliquots of the 12% BSA [J] and collect in the tubes labeled A through D [M].

Step 8. Check tube B, which contains the bulk of purified [^{125}I] insulin, for TCA precipitability, by removing a small aliquot of tube B and adding it to one of the tubes containing 1 mg/ml BSA [F]. Distribute 50-μl aliquots from tube B in 12 × 75-mm test tubes and freeze (typical yield of this labeling is 70 tubes containing 50 μl with 100,000 to 150,000 dpm/μl of [^{125}I]insulin). Use within one month. Tube C is usually kept frozen unfractionated; TCA precipitability is as good as tube B, ≥ 95%, but C is more diluted.

Each time a new batch of insulin is used for labeling, the labeled product should be checked once by gel filtration for possible high molecular weight insulin-like contaminants (proinsulin, etc.). If these are present, it is better to use a gel filtration for purifying the label (Sephadex G-75).

IV. STUDY OF THE REACTION OF HORMONES WITH RECEPTORS

A. Reaction of Hormones with Cell Receptors under Optimal Conditions

Competitive assays using radioactive ligands have been used extensively over the last 15 years to measure and characterize hormones and other substances of biological interest. The general principles have been extensively discussed [105, 106].

In this section, I will describe the methods used to study the binding of insulin to human cultured lymphocytes and blood peripheral monocytes, and of human growth hormone to cultured lymphocytes. In typical experiments, cells (lymphocytes or monocytes) are incubated simultaneously in assay buffer with labeled and unlabeled hormone for a time sufficient to reach a steady-state of binding. The cell-bound insulin is then separated from the free insulin in the supernatant by centrifugation.

1. Materials

a. Assay Buffer

HEPES (N-2-hydroxyethylpiperaxine-N'-2-ethanesulfonic acid [107]): 100 mM; NaCl 120 mM; KCl 5 mM; $MgSO_4$ 1.2 mM; Na acetate 15 mM; glucose 10 mM; EDTA 1 mM, and bovine serum albumin 1 mg/ml.

This concentration of HEPES was found necessary to maintain the pH during the prolonged incubations with peripheral mononuclear cells. A lower concentration of HEPES (25 mM) slightly increases hGH binding to IM-9-cultured lymphocytes (Lesniak, personal communication).

The pH is adjusted to 7.0 for growth hormone assay with cultured lymphocytes, 8.0 for insulin assay with monocytes, and 7.6 for insulin assay with cultured lymphocytes. The composition of this buffer is optimized for avoiding cell agglutination and insulin nonspecific adsorption on glassware.

b. Cell Stock Solutions

(1) Cultured lymphocytes: At stationary phase of growth (72 hr after feeding) the flasks contain $\sim 1 \times 10^6$ cells per ml [30], which can be maintained for 2 to 3 days without further feeding. For binding studies, cells are centrifuged at 600 g for 10 min at room temperature; the medium is decanted and discarded; the cells are washed once in assay buffer and recentrifuged, afterwards the cell pellet is resuspended in assay buffer (volume calculated according to the total number of incubation tubes and to the final cell concentration per tube desired: Sec. IV. A. 4). (2) Blood mononuclear cells: The preparation has been described in Sec. II. B. 1.

c. Hormone Stock Solutions

These are prepared in assay buffer in Falcon plastic ware. (1) The labeled hormone is diluted to a concentration such that addition of 50 μl per incubation tube will yield a final concentration of 0.1 ng/ml (insulin) or 0.5 ng/ml (hGH). (2) Unlabeled procine insulin is first dissolved in HCl or acetic acid 0.01 N at 1 mg/ml; serial dilutions are then performed in assay buffer at each of the following concentrations: 0.01, 0.1, 1.0, 10.0, and 100 μg/ml. (3) Unlabeled hGH is dissolved in water at 5 mg/ml; serial dilutions are performed in assay buffer at 0.1, 1.0, 10.0, and 100 μg/ml.

d. Incubation

Incubation is performed in a series of 12 × 75-mm Falcon plastic tubes to which has been added:

For growth hormone assay

assay buffer, 0-50 μl, (to give final volume of 500 μl)

[^{125}I]hGH, 50 μl, 0.50 ng/ml final concentration

unlabeled hGH, 0-50 μl, 0-10 μg/ml final concentration

cells (cultured lymphocytes IM-9), 400 μl, 20×10^6/ml final concentration (always add cells to the tube last)

For insulin assay

assay buffer, 0-50 μl, (to give final volume of 500 μl)

[^{125}I]insulin, 50 μl, 0.2 ng/ml final concentration

unlabeled insulin, 0-50 μl, 0-10 μg/ml final concentration

cells

cultured lymphocytes IM-9, 400 μl, 2.5×10^6/ml final concentration

peripheral mononuclear cells, 400 μl, 40×10^6/ml final concentration (always add cells to the tube last)

If the purpose of the experiment is assay of hormone in samples, 50 μl of the sample replaces the unlabeled insulin. Important: Carefully resuspend the cells before removal of each aliquot.

2. Incubation Conditions

For growth hormone-cultured lymphocytes, tubes are gently shaken continuously in a constant temperature bath at 30°C for 90 min.

For insulin-cultured lymphocytes, incubation is at 15°C (to minimize degradation: see Sec. IV. B. 1) for 90 min; shake the tubes gently every 10 min.

For insulin-mononuclear blood cells, incubation is at 22°C for 3 hr. Shake gently every 15 min to be sure no durable sedimentation of cells occurs. Do not vortex these cells.

3. Separation of Bound and Free Hormone

The method used was originally described by Rodbell et al. [108]. After incubating for 90 min, duplicate 200-μl aliquots are transferred to individual microfuge tubes (total capacity, 400 μl) containing 200 μl of chilled (4°C) assay buffer. Before removing each aliquot, care is taken to be certain that the cells are homogeneously suspended. The microfuge tubes are spun for 1 min in a Beckman Model 152 microfuge. The bulk of the supernatant is aspirated by vacuum and discarded. The tubes are then inverted, and the last traces of visible buffer are removed from around the cell pellet by capillary action with a fine-tipped pipet without disturbing the cell pellet. The tip of the tube, containing the cell pellet, is excised, and the radioactivity in the cell pellet is counted. This represents the bound radioactivity. Free radioactivity is found by subtracting bound from total. Total radioactivity is usually measured by pooling the incubation medium still remaining in the incubated tubes and counting the radioactivity of a 200-μl aliquot; again care is taken so that the cells are homogeneously resuspended before removing the aliquot.

Bound radioactivity is usually related to free (B/F ratio) or to total (B/T, or $B/T \times 100 = \%$ bound) radioactivity. If labeled and unlabeled hormone behave identically, the specific activity of hormone is identical in both phases (cell-bound and supernatant) and the fraction of radioactive hormone in each phase reflects the fraction, in that phase, of the total hormone incubated.

4. Cell Counts

To allow interassay comparison, it is important to refer the data to the actual cell concentration in each experiment; this is also important for the determination of parameters like receptor concentration per cells, etc. The easiest method is the electronic counter (Coulter Electronics, Hialeah, Fla.), which performs a single count in about 20 sec with a counting error of 1 to 2% [109]. However, the only feasible method in most laboratories is the hemocytometer count. Under the best conditions, this method is subject to no less than 10% variation [109].

5. Cell Viability

Cell viability is usually measured by "vital" staining; we use the method based on trypan blue dye exclusion [21, 110]. "Correction" of data for cell viability may introduce artifacts since the extent of binding impairment (if any) after "cell death" has not been precisely measured. Rather than correcting, we strongly suggest not to use a cell preparation where cell viability is less than 90%. Cell viability is measured by counting on a hemocytometer an aliquot of a well-mixed suspension of 20 μl of the original cell stock solution in 480 μl of normal saline plus 100 μl trypan blue (0.4% = 0.4 g in 100 ml). The suspension is allowed to stand for more than 5 min but less than 15 min. Total cells and unstained cells are counted.

In this section, I have described the optimal conditions for hormone binding to cells and establishing a dose-response curve suitable for radioreceptor assay (Fig. 2). The conditions may naturally be varied according to the needs of a particular experiment. The various parameters determining optimal binding are detailed in Sec. IV. B.

B. Physico-chemical Parameters Determining Optimal Steady-state Binding

1. Temperature

When the time course of specific binding to cells was determined at several temperatures, insulin exhibited a different temperature dependency than growth hormone. When temperature is increased, the initial rate of insulin binding to human cultured lymphocytes [8] and peripheral monocytes

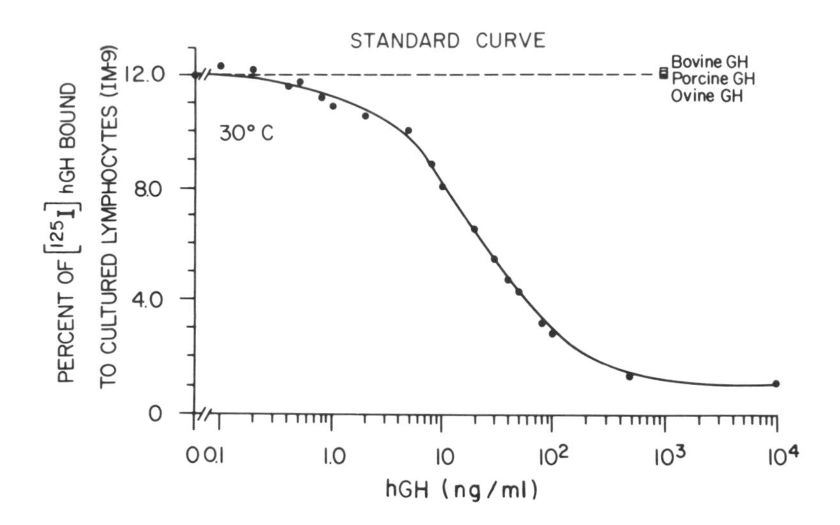

FIG. 2. [^{125}I]hGH bound to human cultured lymphocytes (IM-9) as a function of hGH concentration. Lymphocytes (2×10^7 cells/ml) were incubated for 90 min at 30°C with [^{125}I]hGH (1×10^{-11} M) with increasing concentrations of unlabeled human GH or nonprimate GH. Duplicate 200-μl aliquots were processed as described in Sec. IV. A. 3. The radioactivity bound to the cells (total binding), expressed as a percentage of the total radioactivity in the system, is plotted as a function of the concentration of unlabeled GH. hGH used was preparation 1394. From Lesniak et al. [10], used with permission.

(Bar et al., manuscript in preparation) increases, the steady-state level of binding is reached faster, but the actual level of steady-state binding decreases (Fig. 3). This is due to a greater increase in the rate of dissociation (Fig. 4) than in the rate of association at the higher temperature, and probably reflects a temperature-dependent conformational change of a destabilizing type linked to cooperative interactions among receptors (see Sec. V). Identical characteristics have been described for the insulin receptor in liver plasma membranes [112].

Other factors which affect temperature dependence of steady-state binding are degradation of the free hormone and "degradation" of receptors during incubation (see Sec. IV. B. 4 and 5), which increase with temperature.

The steady-state level of human growth hormone binding to human cultured lymphocytes increased with temperature up to 30°C; in contrast with insulin, growth hormone receptors do not feature cooperative interactions (see Sec. V). Even at 30°C, growth hormone degradation by cultured lymphocytes is trivial [10].

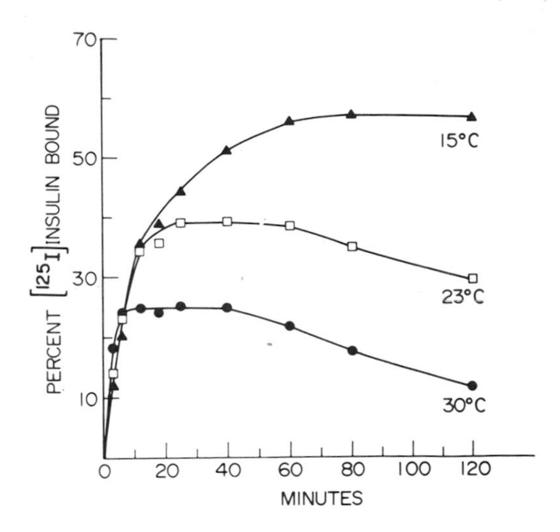

FIG. 3. [^{125}I] Insulin binding to cultured human lymphocytes (4265) as a function of time and temperature. Cells (1.2×10^7 per ml) were incubated with 5×10^{-11} M labeled insulin in assay buffer, pH 7.6 (see Sec. IV). Data have been corrected for nonspecific binding (see Sec. IV. C). From Gavin et al. [8], used with permission.

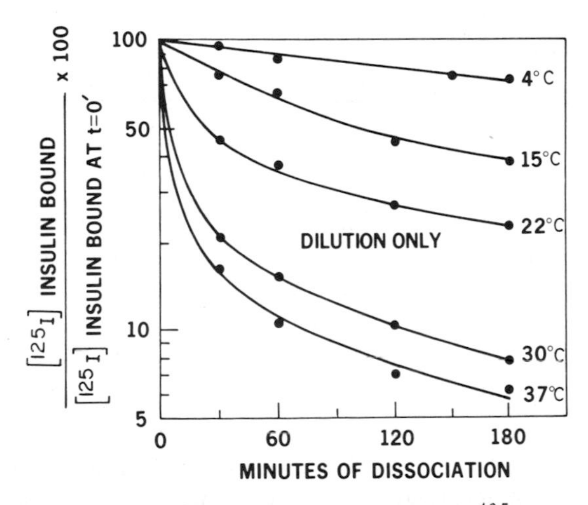

FIG. 4. Effect of temperature on dissociation of [^{125}I] insulin from receptors. [^{125}I] Insulin (10^{-11} M) was incubated with human peripheral mononuclear cells (7.2×10^7 cells) at 23°C for 3 hr at pH 8.0 in assay buffer. The cells were then sedimented, resuspended up to the initial volume, and 200 μl-aliquots distributed in tubes containing 10 ml of assay buffer (100-fold dilution) maintained at temperatures ranging from 4 to 37°C. Duplicate tubes were centrifuged at the times indicated to monitor dissociation. The radioactivity on the cells, expressed as a percentage of the radioactivity present at t = 0 min, is plotted as a function of the time elapsed after the dilution of the system. From De Meyts, Bianco, and Roth [177].

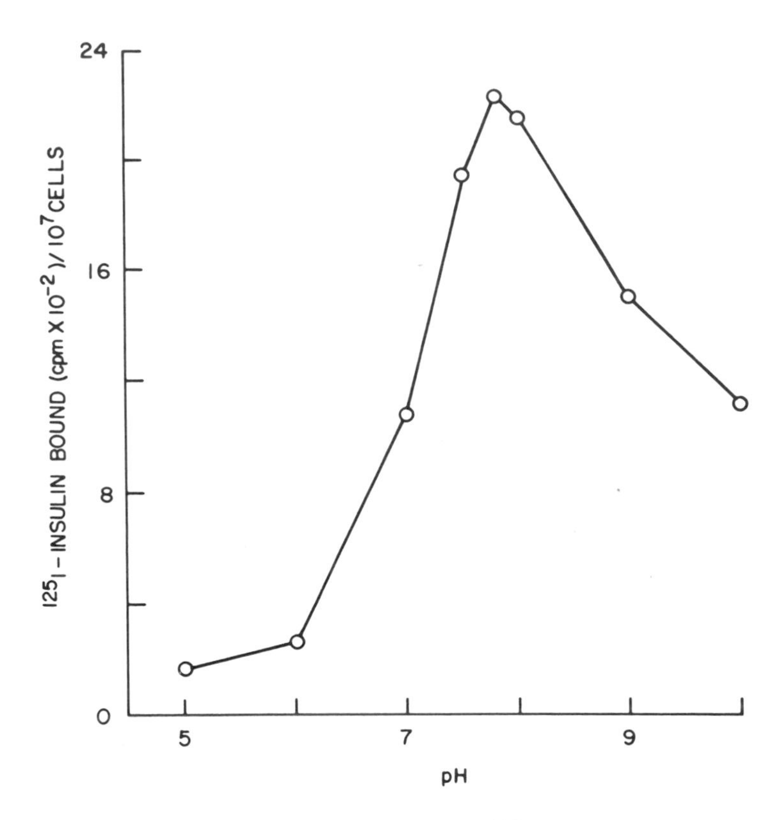

FIG. 5. Effect of pH on specific binding of [^{125}I] insulin to human cultured lymphocytes (4265). Cells (10^7 per ml) were incubated with 10^{-11} M labeled insulin in assay buffer adjusted to appropriate pH. From Gavin et al. [8], used with permission.

2. pH

Insulin binding at steady-state to human cultured lymphocytes is optimal at pH 7.8 (Fig. 5), an identical optimum was found in liver plasma membranes [112]. In contrast, growth hormone binding varied little (Fig. 6) over a pH range of 6.4 to 8.8 [10].

3. Ionic Environment

Studies on the effects of ionic environment have not been performed in detail for lymphocytes and monocytes. Ca^{2+} and Mg^{2+}, over a 10-fold range concentration (1 to 10 mmol), did not affect binding of labeled

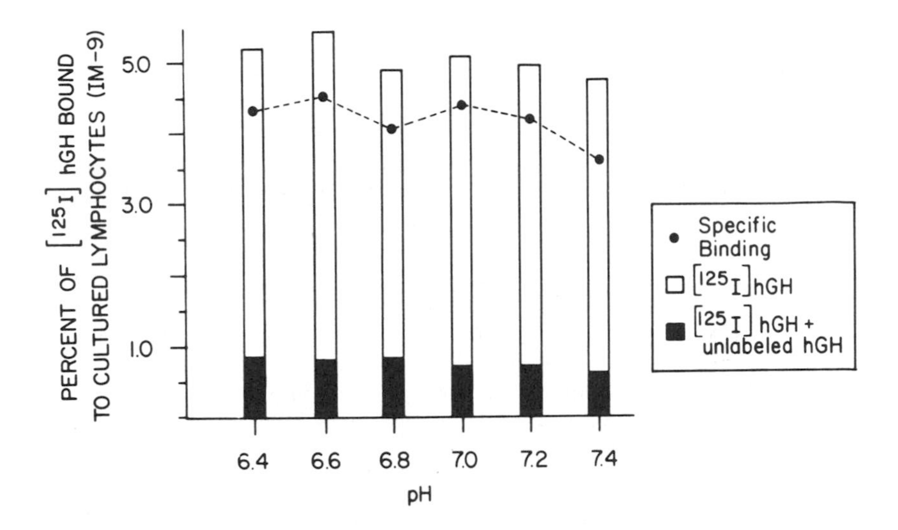

FIG. 6. Effect of pH on total, specific and nonspecific binding (see Sec. IV. C) of [^{125}I]hGH (preparation 1394) to human cultured lymphocytes (IM-9). Cells (2×10^7 per ml) were incubated with 10^{-11} M labeled hGH for 90 min at 30°C in assay buffer adjusted to appropriate pH. Nonspecific binding was determined in the presence of 50 μg/ml unlabeled hGH. From Lesniak, M. A. (unpublished data).

insulin or growth hormone to cultured lymphocytes, or of insulin to monocytes. More marked variations in ionic strength have been shown to affect the binding of insulin to receptors in liver and fat cell membranes [112, 113]: High-ionic-strength media (2 M [Na]) markedly enhance insulin binding. Kahn showed that two processes were responsible for this phenomenon: an inhibition of insulin degradation and an increase in the affinity of the receptors for insulin [112].

4. Hormone Degradation

a. Insulin

Degradation of insulin during incubation is present in all receptor systems studied. The nature of the degradative process has been attributed to proteolysis [114, 115] or reductive cleavage [116-118]. Freychet et al. demonstrated that hormone binding and hormone degradation are independent processes [91]. Several methods are routinely being used to assess the extent of hormone damage during incubation with lymphocytes or monocytes.

(1) Precipitability in trichloroacetic acid (TCA). Aliquots of the cell-free supernatant that has been incubated with the cells is added to 1 ml of sodium phosphate buffer 0.03 M, pH 7.4, containing 1 mg/ml of bovine serum albumin. An equal volume of 10% TCA (final concentration 5%) is added to precipitate the [^{125}I]insulin. The tube is centrifuged for 1 min in a table-top centrifuge. Radioactivity in the precipitate and in the supernatant is counted.

(2) Adsorption to talc. One aliquot of the cell-free supernatant is added to 2.5 ml of Veronal buffer 0.05 M, pH 8.6, containing human serum albumin at 2.5 mg/ml and rabbit fraction II at 0.1 mg/ml. One 50-mg talc tablet (Ormont Drug, Englewood, N. J.) is added to the medium containing each supernate aliquot, vigorously mixed in a vortex mixer, then centrifuged for 5 min at 2,500 rpm. Under these conditions, undamaged hormone adsorbs to the talc [119].

(3) Binding to anti-insulin antibodies. The conditions have to be adapted to the antiserum used. Typically [91], aliquots (10-20 μl) of the cell-free supernate are transferred to 0.5 ml of chilled Veronal buffer 0.05 M, pH 8.6, that contains 2.5 mg/ml human serum albumin, 0.1 mg/ml rabbit fraction II, 5 mM EDTA, 0.25% guinea pig serum, and guinea pig anti-insulin serum in a final dilution of 1:1,000 in 0.05 M Veronal buffer, pH 8.6. After 4 hr at 4°C, 0.5 ml of cold medium is added, and the mixture is centrifuged for 20 min at 2,800 rpm in a 4°C refrigerated centrifuge. The radioactivity of both the supernate and the precipitate is counted.

(4) Rebinding to fresh cells. An aliquot of the cell-free supernatant is added to fresh cultured lymphocytes and incubated in the assay as usual.

The binding of the incubated labeled hormone is compared with an equal tracer size of fresh labeled hormone unexposed to cells, and submitted to the same procedures to assess degradation.

The two latter methods (binding to antibodies and especially rebinding to receptors) are the most sensitive to detect subtle forms of degradation.

Cultured lymphocytes do not usually appreciably degrade insulin: at 30°C, $< 15\%$ degradation is seen at 90 min of incubation using rebinding to fresh lymphocytes. At 15°C, the temperature at which routine studies are performed, no degradation of the labeled insulin is seen (Fig. 7). Similarly, little degradation is found during incubation of insulin with peripheral monocytes.

In case significant degradation is found, increasing the concentration of serum albumin in assay medium from 0.1% to 1% is sometimes enough to suppress it (providing degradation is not due to a contaminant of the serum albumin, which should always be checked) (unpublished observations).

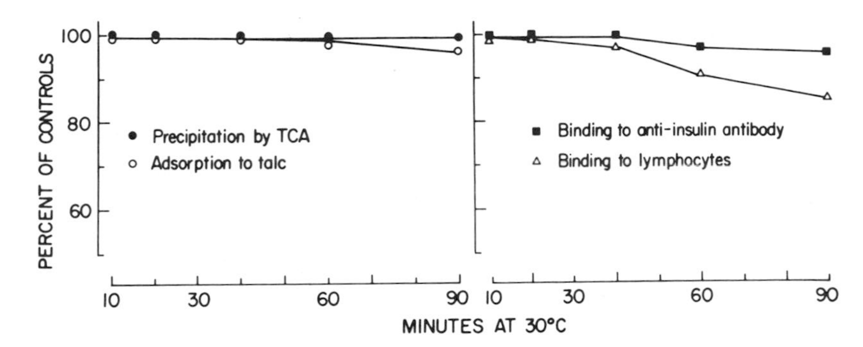

FIG. 7. Degradation of [^{125}I]insulin exposed to human cultured lymphocytes (4265). Cells (2×10^7 per ml) were incubated with 10^{-10} M [^{125}I]-insulin. Degradation was assessed as described in Sec. IV. B. 4. From Gavin et al. [8], used with permission.

b. Growth Hormone

TCA precipitability, talc adsorption, and rebinding to fresh cells are routinely used to estimate degradation of growth hormone. The methods are identical to those already described except that 175 mg of talc is used to adsorb [^{125}I]hGH.

Degradation of human growth hormone by human cultured lymphocytes in 90 min at 30°C is trivial, and only 10% is degraded in 4 hr [10].

5. Receptor Degradation

During prolonged incubation of insulin at 30°C with receptors in rat liver membranes, a falloff in steady-state was observed which was due, in part, to degradation of free insulin, but also in part to a progressive loss in the binding capacity of the membranes [112]. This latter phenomenon has been called "receptor degradation," which may be a poor choice of terms since little is known about the actual fate of the receptors. "Receptor degradation" is usually assessed as follows: Cells are preincubated in the absence of hormone for various periods of time in the regular assay buffer. Following this preincubation, [^{125}I]hormone is then added for the standard 90-min incubation. Decrease in binding capacity of the cells during the preincubation can thus be followed as a function of preincubation time. This phenomenon is absent in cultured lymphocytes at 15°C, but a loss of insulin binding occurs when lymphocytes are incubated at 23°C, and is especially impressive at 30°C at cell concentrations of 2×10^7/ml. In the case of cultured lymphocytes, it was demonstrated that the insulin receptors are released into the incubation medium [8] by an unknown mechanism (possibly by "shedding," which is a common way for lymphocytes to release surface proteins [120]). This property was used to isolate insulin receptors from lymphocytes in soluble form [121] without the use of detergents [122]. Insulin degrading activity is also released in soluble form in certain conditions (De Meyts, unpublished data).

In standard incubation conditions, no loss of binding capacity for growth hormone from cultured lymphocytes occurs, but growth hormone receptors are also released from the cells in soluble form at high cell concentrations (McGuffin et al., manuscript in preparation).

6. Enzymatic Treatment

Treatment of cultured lymphocytes with RNase, DNase, or neuraminidase did not affect insulin binding [8]. Tryptic digestion destroys the insulin binding in cultured lymphocytes, as in liver membranes [90] and fat cells [113]. Growth hormone receptors in cultured lymphocytes are even more sensitive to trypsin [10].

C. Quantitative Analysis of Hormone-receptor Binding at Steady-state

Mathematical treatment of the binding of small molecules by macromolecules, including hormone binding to antibodies and to receptors, has already been the subject of a number of papers [123-128a, 112]. We feel, however, that it is appropriate in this "methods" series to review some of the methods routinely used for analyzing hormone-receptor interactions, since a critical number of assumptions on which equilibrium analysis is based have too rarely been submitted to critical examination and validation.

1. Specific Versus Nonspecific Binding

As seen in Figure 2, the radioactivity bound to the cells does not approach zero even when the [^{125}I]hormone is incubated with very high concentrations of unlabeled hormone, but stays at a low constant value. This phenomenon indicates that, in that range, hormone binding increases linearly with hormone concentration (Fig. 8). This binding is usually referred to as "nonspecific" binding and reflects the presence, besides the receptors, of a compartment which is not saturable in the practical range of hormone concentrations. It can be physically represented as solubility in another phase [111] or as a population of binding sites with "zero affinity, infinite capacity" (see Note number 2).

The fraction (bound/total or B/T) of radioactive hormone bound to cells in the presence of a given concentration of unlabeled hormone represents "total binding"; and the total concentration of hormone bound equals the B/T of radioactive hormone times total hormone concentration incubated (labeled and unlabeled).

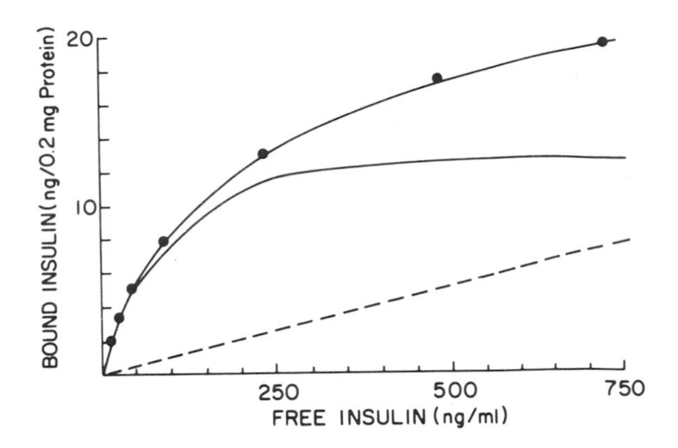

FIG. 8. Definition of specific and nonspecific binding. Saturation curve for insulin binding to mouse liver plasma membranes at 20°C. Unlabeled insulin is mixed with 0.1 ng/ml [^{125}I] insulin to give indicated insulin concentrations. ●: bound equals percent tracer bound times total insulin. This curve (total binding) is resolved into linear (nonspecific binding; - - - -) and hyperbolic (specific binding; ———) components. From Neville et al. [111], used with permission.

The fraction (B/T) of hormone nonspecifically bound is determined by measuring the radioactivity bound to cells in the presence of a supersaturating concentration of unlabeled hormone (ideally several orders of magnitudes higher than the dissociation constant of specific binding); this fraction ("nonspecific binding") is the same for all hormone concentrations since nonspecific binding is assumed to be a linear function of hormone concentration.

Thus, it is easy to correct the binding data to yield specific binding by subtracting the fraction of radioactivity nonspecifically bound from the total binding, and the concentration of hormone specifically bound is equal to the B/T of radioactive hormone (corrected) times the total hormone concentration incubated (labeled plus unlabeled).

In the previous sections of this review, as well as in the following, "binding" always refers to "specific binding" unless otherwise specified.

2. Assumptions for Equilibrium Analysis

a. Labeling

Labeled and unlabeled hormone behave identically. Thus, the measurement of the labeled species truly reflects the behavior of the whole hormonal pool.

b. Maximum Bindability

If not all of the tracer is bindable, as is sometimes the case for impure polypeptide preparations, the nonbindable part will not contribute to the equilibrium between bound and free; correction should be made for this in the estimation of free hormone after assessment of the "maximum bindability" of the tracer by an excess of receptors [128b]. This factor is negligible in the case of monoiodoinsulin, which is more than 95% bindable to an excess of lymphocytes ([103]; De Meyts, unpublished data).

c. B/F

Bound and free hormone can be perfectly separated without perturbing the equilibrium.

d. Equilibrium

Full equilibrium is reached. The achievement of a "plateau" of binding is usually taken as evidence of equilibrium in the binding reaction but it may reflect, in fact, that a steady-state (i.e., a zero net rate of change) is reached among multiple processes taking place. (Other than association of hormone to receptors and dissociation of hormone from receptors, one might consider degradation of hormone, degradation and release of receptors, endocytosis of hormone and/or of hormone-receptor complexes, receptor capping, etc.) The implications of this difference between "equilibrium" and "steady-state" have been pointed out by Roth [129]. The thermodynamics of nonequilibrium reactions requires a different treatment [130].

e. Reversibility

The reaction is fully reversible, The thermodynamics of irreversible processes also requires appropriate treatment [131]; in fact, this assumption is somehow linked to the precedent: stationary states play a role in the thermodynamics of irreversible processes similar to that played by states of equilibrium in classical thermodynamics [130]. The dissociation of some hormone-receptor systems is so slow that full reversibility can hardly be verified and sometimes can be reasonably questioned.

f. Homogeneity

The hormone is present in a homogeneous form (for example, the insulin monomer).

g. Valence

The hormone is univalent; that is, one hormone molecule can react with only one binding site. Hormone molecules bind independently: no ligand-ligand interactions are present.

h. Binding Sites

The binding sites are independent; that is, no site-site (cooperative) interactions are present between binding sites. As will be developed in Sec. V, this is not the case for insulin receptors.

3. The Binding Isotherm

If these critical assumptions are fulfilled, then the reaction of hormone with receptor may be described as a simple, bimolecular, reversible second-order ⇌ first-order reaction, and the condition of equilibrium deduced from the law of mass action [132] is

$$[H] + [R] \underset{k_d}{\overset{k_a}{\rightleftharpoons}} [HR] \tag{1}$$

$$K_a = \frac{k_a}{k_d} = \frac{[HR]}{[H]\,[R]} \tag{2}$$

$$K_d = \frac{k_d}{k_a} = \frac{1}{K_a} \tag{3}$$

where [H] is the concentration of free hormone; [R] is the concentration of free receptor-binding sites; [HR] is the concentration of hormone-receptor complex; k_a is the association-rate constant; k_d is the dissociation-rate constant; K_a is the equilibrium constant [association constant, affinity constant, (units are liters/mole or M^{-1})]; and K_d is the dissociation constant (units are moles/liter or M).

Also

$$[R_o] = [HR] + [R] \tag{4}$$

and

$$[H_o] = [HR] + [H] \tag{5}$$

where $[R_o]$ (or "B_{max}") is the total concentration of receptor-binding sites and $[H_o]$ is the total concentration of hormone.

The fractional saturation of the receptors with the hormone is defined as the fraction of occupied binding sites or $[HR]/[R_o]$, often represented in the literature as $\overline{Y}$. $\overline{Y}$ varies from 0 to 1.0.

The fraction of free receptors $[R]/[R_0]$ thus equals $(1 - \overline{Y})$. Since

$$\overline{Y} = \frac{[HR]}{[R_o]}, \quad [HR] = \overline{Y}\,[R_o] \tag{6}$$

Since

$$(1 - \overline{Y}) = \frac{[R]}{[R_o]}, \quad [R] = (1 - \overline{Y})\,[R_o] \tag{7}$$

The law of mass action tells us that the rate of association equals

$$k_a\,[H]\,[R]$$

or, substituting [R] from Eq. (7)

$$k_a\,[H]\,(1 - \overline{Y})\,[R_o] \tag{8}$$

Similarly, the rate of dissociation equals

$$k_d\,[HR]$$

or, substituting [HR] from Eq. (6),

$$k_d\,\overline{Y}\,[R_o] \tag{9}$$

At equilibrium, association and dissociation occur at equal rates so that

$$k_d\,\overline{Y}\,[R_o] = k_a\,[H]\,(1 - \overline{Y})\,[R_o]$$

Dividing through by $[R_o]$

$$\begin{aligned} k_d\,\overline{Y} &= k_a\,[H]\,(1 - \overline{Y}) \\ &= k_a\,[H] - k_a\,[H]\,\overline{Y} \end{aligned}$$

Dividing through by $\overline{Y}$

$$k_d = \frac{k_a\,[H]}{\overline{Y}} - k_a\,[H]$$

therefore

$$\frac{k_a\,[H]}{\overline{Y}} = k_d + k_a\,[H]$$

or

$$\bar{Y} = \frac{k_a\ [H]}{k_d + k_a\ [H]}$$

Dividing the numerator and the denominator by k_a

$$\bar{Y} = \frac{[H]}{k_d/k_a + [H]}$$

and from Eq. (3)

$$\bar{Y} = \frac{[H]}{K_d + [H]} \tag{10}$$

Dividing the numerator and the denominator by K_d

$$\bar{Y} = \frac{[H]/K_d}{1 + \{[H]/K_d\}}$$

Since

$$\frac{1}{K_d} = K_a,\ \bar{Y} = \frac{K_a\ [H]}{1 + K_a\ [H]} \tag{11}$$

Equations (10) and (11) are the "binding" equivalents of the classical Michaelis-Menten equation (in the case where no product is formed) [133].

A plot of $\bar{Y}$ as a function of [H], or "saturation" plot, is sometimes called a "binding isotherm" or "Langmuir isotherm," because the equation is formally equivalent to Langmuir's adsorption isotherm [134]. It is also called a "Michaelis-Menten plot," although these authors actually used a logarithmic abcissa (in this case, log [H]) [133]. Equation (10) or (11) is a rectangular hyperbola passing through the origin, with a horizontal asymptote $\bar{Y} = 1.0$. Since Eqs. (10) and (11) can be readily transformed into

$$[HR] = \bar{Y}\ [R_o] = \frac{[H]\ [R_o]}{K_d + [H]}$$

$$= \frac{K_a\ [H]\ [R_o]}{1 + K_a\ [H]} \tag{12}$$

in practice, one usually plots [HR] versus [H] instead of $\bar{Y}$ versus [H] ("bound" versus "free", or B versus F in the radioreceptor jargon); the curve has the same form and the horizontal asymptote is $[HR] = [R_o]$. An interesting feature of this plot is that for $[HR] = [R_o]/2$, $[H] = K_d$.

Sometimes, K_d has been derived from the concentration of total hormone, producing half-"displacement" of $[^{125}I]$hormone in the standard curve, or from the "half $[R_o]$" ordinate in a plot of [HR] versus $[H_o]$. These two approximations are erroneous [127], and one has to consider the concentration of free hormone.

Various methods for linearizing the plot describing the binding function have been proposed and all can be easily derived from equations (10) and (11). I will discuss those most utilized in this field.

To interpret these linear equations, one needs only to remember that when a straight line is represented by the equation

$$y = a \times + b$$

the slope of the line is a, the ordinate intercept b, and the abcissa intercept - b/a.

4. The Lineweaver-Burk Plot [135]

Reversing Eq. (12) yields

$$\frac{1}{[HR]} = \frac{K_d + [H]}{[H]\ [R_o]} = \frac{K_d}{[H]\ [R_o]} + \frac{[H]}{[H]\ [R_o]} \tag{13}$$

or

$$\frac{1}{[HR]} = \frac{K_d}{[R_o]} \cdot \frac{1}{[H]} + \frac{1}{[R_o]} \tag{14}$$

A plot of 1/[HR] (= 1/B) against 1/[H] (= 1/F) is, thus, a straight line with a slope of $K_d/[R_o]$, an ordinate intercept of $1/[R_o]$, and an abcissa intercept of $- 1/K_d$ or $- K_a$. It is also called a "double reciprocal" plot.

5. The Scatchard Plot [123, 136]

Since

$$\bar{Y} = \frac{[HR]}{[R_o]} = \frac{[H]}{K_d + [H]} \tag{10}$$

therefore

$$[HR]\ (K_d + [H]) = [H]\ [R_o]$$

Developing,

$$[HR]\ K_d + [HR]\ [H] = [H]\ [R_o]$$

or

$$[HR]\ K_d = -\ [HR]\ [H] + [H]\ [R_o]$$

Dividing through by [H]

$$\frac{[HR]}{[H]} K_d = -\ [HR] + [R_o] \tag{15}$$

and dividing through by K_d

$$\frac{[HR]}{[H]} = -\frac{1}{K_d}[HR] + \frac{[R_o]}{K_d} \tag{16}$$

Thus, a plot of [HR]/[H] (= B/F) against [HR] (= B) is a straight line with a slope of $-1/K_d$ ($= -K_a$), an ordinate intercept $= [R_o]/K_d$, and an abcissa intercept $= [R_o]$ (Fig. 9).

From the value of $[R_o]$, the number of receptor sites per cell can be derived, knowing the number of cells in the incubation tube, according to the following formula:

$$\text{number of sites per cell} = \frac{[R_o]\ (M) \times 6.022 \times 10^{23}}{\text{number of cells per liter}}$$

since there are 6.022×10^{23} molecules in one mole (Avogadro's number).

The binding of human growth hormone to receptors in human cultured lymphocytes yields such a linear Scatchard plot [10] (Fig. 9): The affinity constant derived from this analysis is $1.3 \times 10^9\ M^{-1}$, and there are 4,000 binding sites per cell (cell line IM-9).

Scatchard plots for the interaction of polypeptide hormones with receptors are linear in some cases (growth hormone-lymphocytes [10], calcitonin-lymphocytes [11], gonadotropins-testis [141]); but in many cases, nonlinear plots with upward concavity have been reported, including ACTH, oxytocin, glucagon, catecholamines, TSH (see Ref. 4 for review). In the case of insulin binding to both cultured lymphocytes and peripheral

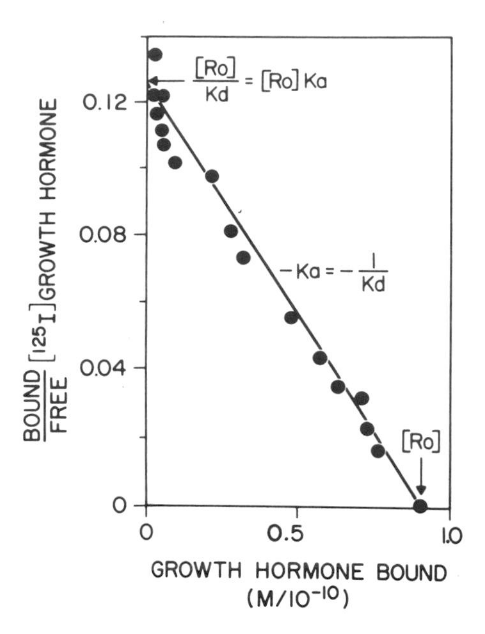

FIG. 9. Linear Scatchard plot for human growth hormone binding to cultured human lymphocytes (IM-9). $[^{125}I]$hGH (2×10^{-11} M) was incubated at 30°C with 20×10^6 cells for 90 min with a range of concentrations of unlabeled human GH. The bound/free ratio of $[^{125}I]$hGH is plotted as a function of the hGH bound. Nonspecific binding has been subtracted. Adapted from Lesniak et al. [10].

monocytes, the Scatchard plot is curvilinear with a concavity upward (Fig. 10).

This has been generally attributed to heterogeneity of the receptor sites with two or more classes of "orders" which differ in binding affinity [136-138, 125]. The affinity constants and binding capacities can be derived by "curve peeling" [124, 112]; as has been stressed, it is necessary to subtract the slope of the "low-affinity site" from the curve before validly estimating the slope and abcissa intercept of the "high-affinity site" [125]. More accurate fitting procedures based on iterative computation methods [139, 140] are available.

In the case of insulin receptors, different affinity constants and different orders of sites have been reported (see Ref. 4 for review), due to methodological differences in analysis or to different assumptions. For example, Cuatrecasas [142] considers only the initial slope of the binding curve, the remainder being considered as nonspecific. This approach is questionable, since the assumption that the curvilinearity derives from

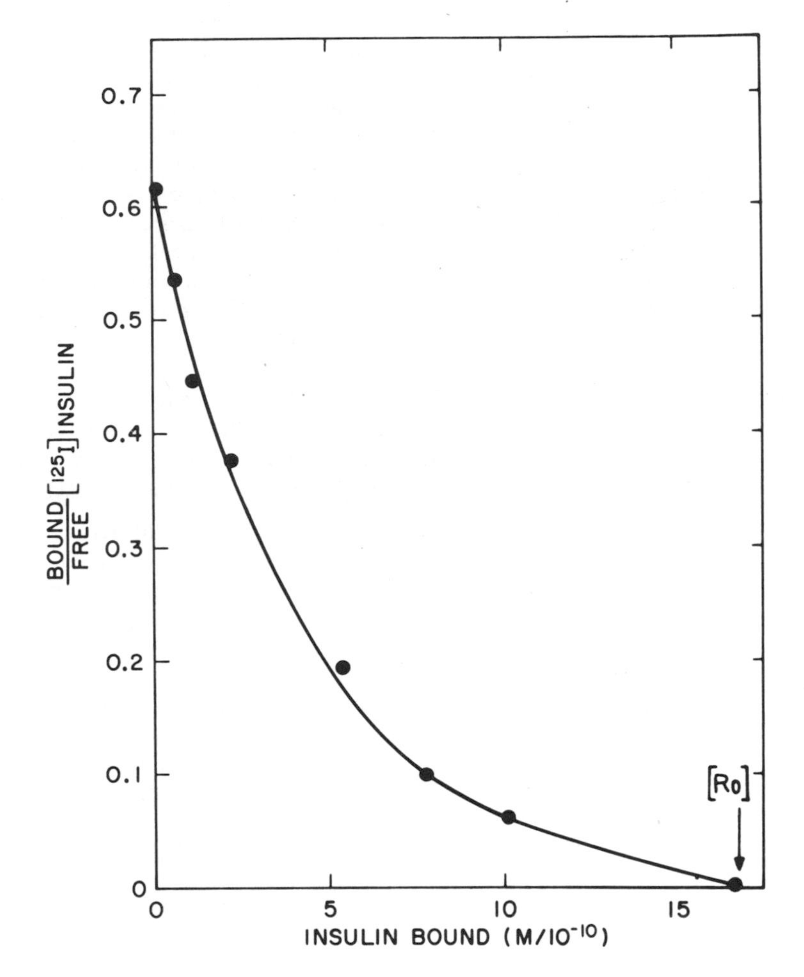

FIG. 10. Curvilinear Scatchard plot for insulin binding to cultured human lymphocytes (IM-9). [^{125}I] insulin (7×10^{-12} M) was incubated with 10^7 cells at 15°C for 90 min with a range of concentrations of unlabeled insulin. The bound/free ratio of [^{125}I] insulin is plotted as a function of the insulin bound. Nonspecific binding has been subtracted.

site heterogeneity has not been verified and, as already stressed, ignoring the latter part of the curve yields biased estimation of the "high-affinity constant", giving spuriously elevated values (see Note number 3).

Curvilinearity in Scatchard plots can arise from a variety of causes, including technical artifacts (errors in separation of bound and free hormone [127] and differences in affinity of labeled and unlabeled hormones

[128a]). An important alternative to consider is that negative cooperativity is present. In that case, the receptor sites do not have a fixed affinity, but the overall affinity of the sites decrease as occupancy of receptors with hormone increases, due to hormone-induced site-site interactions of a destabilizing type (see Sec. V). Since experimental evidence for this mechanism exists for insulin receptors, the degree of curvature in the Scatchard plot reflects at least partially the extent of site-site interactions; hence the discrete K_a values from Scatchard analysis do not have the same precise physico-chemical meaning. Therefore, arguing whether the "high affinity" K_a is 10^{10} M^{-1} or 10^9 M^{-1}, or what part of the curve is "specific," becomes obsolete. The degree of site-site interactions is accessible to experimental modulation (pH, temperature, ions, membrane fluidity, etc.), and the curvature is expected to vary to some extent according to the incubation conditions. A parameter which remains valid is the final abcissa intercept, which in all cases still measures the total concentration of binding sites $[R_o]$.

Analysis of steady-state data alone does not allow us to differentiate whether the hormone-receptor interaction fits a "multiple sites" model or a "cooperative" model [140, 143]. Some methods to analyze steady-state data are available which are valid in both cases. One of them is the computer "stepwise equilibrium analysis" of Fletcher et al. [140], which yields apparent binding constants to fit the curvature of the Scatchard model; no assumption is made as to the physical meaning of these constants, which might reflect the progressive alteration in affinity when site-site interactions increase, as well as the affinity of discrete subpopulations of sites.

Another method of analysis that is becoming increasingly popular in this field is the Hill plot.

6. The Hill Plot

In 1910, Hill [144] showed that the sigmoidal dissociation curve of oxyhemoglobin could be described by the equation:

$$y = 100 \frac{Kx^n}{1 + Kx^n} \tag{17}$$

where y is the percentage saturation of the hemoglobin with O_2, x is the tension of O_2 in the solution, K is the equilibrium constant, and n is a whole number > 1.

This equation derived from the application of the law of mass action to the reaction scheme

$$Hb_n + nO_2 \rightleftharpoons Hb_n (O_2)_n$$

expressing the hypothesis that the hemoglobin aggregated from a monomeric form into larger molecules, which then combined with oxygen. The number n, since called "Hill coefficient," would have equaled the number of binding sites on a hemoglobin "aggregate." It has since been shown that the Hill model was inaccurate, but the equation still fits the middle portion of the curve reasonably, with a value of $n = 2.3$ (smaller than the expected $n = 4$).

The Hill equation remains a satisfactory empirical equation for describing the main portion of a binding curve, and when applied to hormone-receptors systems, takes the form

$$\overline{Y} = \frac{K_a [H]^n}{1 + K_a [H]^n} \tag{18}$$

When $n = 1$, this equation reduces to the Michaelis-Menten equation, as previously shown, and describes a noncooperative system.

A system exhibiting positive cooperativity (such as O_2 binding to hemoglobin) (see Sec V) can be described by Eq. (18) with $n > 1$; in this case, K_a denotes an "average" affinity constant. The plot of $\overline{Y}$ versus [H] is sigmoidal (S-shaped).

A n value < 1 is consistent with either heterogeneity of binding sites (the equation is equivalent to the Sips equation describing the distribution of the adsorption energies of the sites of a catalyst surface [145], later widely applied to the reaction of antigens with heterogeneous antibodies) or with negative cooperativity. The plot of $\overline{Y}$ or [HR] versus [H] looks in this case hyperbolic, like in the Michaelis-Menten plot, but is flatter and, in extreme cases, may be biphasic with an intermediary plateau. K_a again denotes an average affinity constant.

It must be pointed out that the Hill equation describes accurately only the middle portion of the binding curve when $n \neq 1$; the nonfitting to the Hill model becomes extremely marked when $\overline{Y}$ approaches zero and when $\overline{Y}$ approaches 1.

An interesting property of the Hill equation is the following: from Eq. (18), it derives

$$1 - \overline{Y} = \frac{1 + K_a [H]^n - K_a [H]^n}{1 + K_a [H]^n} = \frac{1}{1 + K_a [H]^n} \tag{19}$$

therefore, dividing Eq. (18) by Eq. (19)

$$\frac{\bar{Y}}{1 - \bar{Y}} = K_a [H]^n$$

In logarithmic form

$$\log \frac{\bar{Y}}{1 - \bar{Y}} = n \log [H] + \log K_a$$

A plot of log $\bar{Y}/1 - \bar{Y}$ versus log [H] is then a straight line with a slope of n and an abcissa intercept of $-\log K_a/n$, and is called a "Hill plot" (Fig. 11) (see Notes 4 and 5).

Notice that

$$\log \frac{\bar{Y}}{1 - \bar{Y}} = \log \frac{[HR]/[R_o]}{1 - ([HR]/[R_o])} = \log \frac{[HR]}{[R_o] - [HR]}$$

(or $\log [B/(B_{max} - B)]$, which is the familiar form of the ordinate of the Hill plot). Notice also that in the usual representation of the Hill plot, the axes are arbitrarily set and do not correspond to the actual x and y axes; the x axis in fact is the horizontal line at $\log [\bar{Y}/(1 - \bar{Y})] = 0$ or $\bar{Y} = 0.50$, which corresponds to half-saturation of the binding sites or $[HR] = [R_o]/2$; the y axis is never on the graph in hormone-receptor systems, since $\log [H] = 0$ when hormone concentration is 1 M. The slope of the Hill plot gives a direct measure of the Hill coefficient.

If n = 1, there is no cooperativity. In this case, the intercept with $\log [\bar{Y}/(1 - \bar{Y})] = 0$ is $-\log K_a = \log K_d$, from which K_d can be derived (Fig. 11):

$n > 1$

indicates positive cooperativity; abcissa intercept $= -\log K_a/n$

$n < 1$

indicates negative cooperativity or sites heterogeneity; abcissa intercept $= -\log K_a/n$

When $n \neq 1$, the plot is usually linear over only a variable portion extending symmetrically around the half-saturation horizontal axis; departure from linearity becomes more important at distance from this axis, and the extremities finally asymptotically approach slopes of 1.0 when $\bar{Y}$ approaches zero and when $\bar{Y}$ approaches 1. (With "perfect data," the plot is, in fact, truly sigmoidal.)

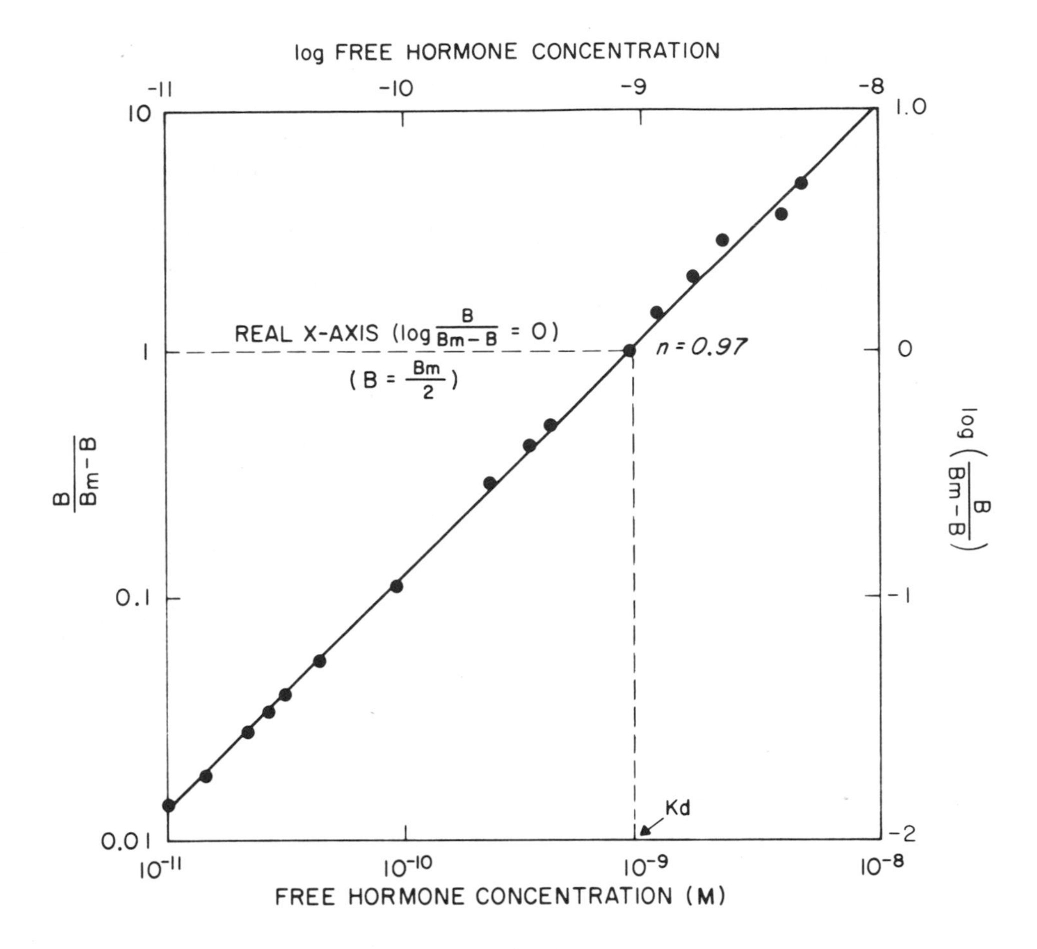

FIG. 11. Hill plot for human growth hormone binding to cultured human lymphocytes (IM-9). Data identical to Fig. 9. The Hill coefficient was 0.97, not significantly different from 1.0.

Some hormone-receptor systems may have an "asymmetrical" distribution of the energy of interaction among the binding sites, and the significant part of the slope is sometimes displaced under the half-saturation axis, at low fractional saturations. It is important therefore, in the hormone-receptor systems, to concentrate the data points under and around half-saturation of the sites to yield a valid estimate of n. This caveat is reinforced by the fact that, in the hormone-receptor systems, the subtraction of nonspecific binding artificially makes [HR] equal to $[R_o]$ (or $\overline{Y}$ to 1.0) for a finite value of [H]; whereas in theory, [HR] =

$[R_o]$ only when [H] approaches infinity. Consequently, log $[\overline{Y}/(1 - \overline{Y})]$, which theoretically should approach infinity with a slope of 1.0 only when [H] approaches infinity, approaches, in fact, an artifactual vertical asymptote at a finite value of log [H]. In effect, if $\overline{Y} = 1.0$, $1 - \overline{Y} = 0$, and $[\overline{Y}/(1 - \overline{Y})] = \infty$. Thus, the n value in this portion of the Hill plot will start to increase considerably and rapidly approach infinity. If one is not conscious of this, any weight in the curve-fitting procedure (or in eyeball drawing) given to points in this area may yield a plot with a spuriously elevated value of n, falsely interpreted as indicating positive cooperativity. Another way to get spuriously positive n values is to plot log $[H_o]$ instead of log [H] [160]. It is important to point out that [H] and $[H_o]$ are never interchangeable, except as an approximation when the percentage bound is small (< 25% [160]).

Hill plots of insulin binding to human cultured lymphocyte and peripheral monocytes yield Hill coefficients of 0.50 to 0.70 (Fig. 12), in contrast with the slope of 1.0 obtained for growth hormone (Fig. 11).

It should be pointed out that the various plots described, although equivalent in theory, may not have equivalent advantages in practice for statistical reasons, since they give weight to different parts of the binding isotherm and accentuate selectively the scatter of some regions of the

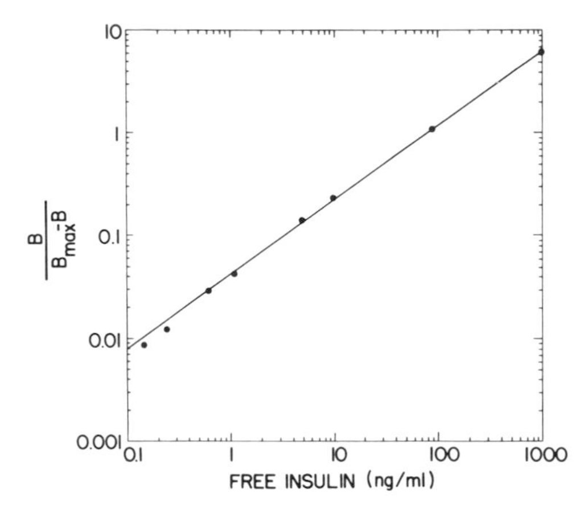

FIG. 12. Hill plot for insulin binding to human blood peripheral mononuclear cells. 5.6×10^7 per ml mononuclear leucocytes, of which 1.06×10^7/ml were monocytes, were incubated with 0.16 ng/ml (2.7×10^{-11} M) [^{125}I]insulin in the absence or in the presence of increasing concentrations of unlabeled insulin. Data were corrected for nonspecific binding. The B_{max} (1 ng/ml) was derived from a Scatchard plot. From Bianco, A. R. (unpublished data).

curve; the graphical representation of the data should be chosen to give maximal weight to the best determined data points [127].

In conclusion, the analysis of steady-state data in the case of growth hormone receptors in human cultured lymphocytes was consistent with a homogeneous class of receptor sites; whereas in the case of insulin receptors in cultured lymphocytes and monocytes, steady-state data were consistent with either multiple classes of sites or site-site interactions of the type known as negative cooperativity. Methods to detect site-site interactions based on kinetic analysis are described in the following section.

D. Quantitative Analysis of the Kinetics of Hormone-receptor Binding

Kinetics of hormone-receptor binding has been far less studied than steady-state binding. Values for association and dissociation rate constants have been derived from kinetic analysis based on the "$H + R \rightleftharpoons HR$" model in a number of cases; the results and methods have been recently reviewed [4] and will be evoked here only briefly. By the study of the concentration dependence of the initial association rates, it is theoretically possible to determine the order of the reaction and validate the "bimolecular, second-order" assumption. Unfortunately, the association is so rapid for insulin and growth hormone that the multiple determinations necessary for accurate measurements of initial association rates are likely to be perturbed by even the best methods to separate bound and free hormone. In hormone-receptor systems for which this has been attempted, the reaction often did not behave ideally [4]. Dissociation rates can be studied by either "infinite" dilutions of the hormone-receptor complex or by adding an excess of unlabeled hormone to the incubation medium after labeled hormone and receptors have reached a steady-state. In the latter case, unlabeled hormone will (1) saturate the empty receptor sites, (2) replace the labeled hormone which dissociates from the receptors, and (3) compete with the labeled hormone for reassociation so that no measurable reassociation of label will occur.

The "bimolecular second-order $\rightleftharpoons$ first-order" model predicts that the dissociation rates will be first-order (a plot of log [HR] against time of dissociation will be linear); moreover, the dissociation rates should be identical whether studied by dilution or by addition of unlabeled hormone.

In many systems, including growth hormone and insulin receptors in human cultured lymphocytes, the dissociation is not first-order. When both methods for studying dissociation are compared, dissociation by dilution and dissociation by an excess unlabeled hormone are often found discrepant, the latter being much more rapid. The dissociation curves have been subjected to curve peeling and slow and fast linear rate components derived, which were attributed to dissociation from the high and low affinity sites respectively [112, 8]; however, careful analysis showed that these data did not correlate with the proportion of high- and low-affinity sites derived from Scatchard analysis [112]. Only in the case of gonadotropin receptors, where all the kinetics are much slower, has it been possible to convincingly show by a sophisticated simultaneous computer curve fitting of association and dissociation data that the $H + R \rightleftharpoons HR$ model was valid [161]. Most of the discrepancies found in other systems are explainable by a model which assumes the existence of negative cooperativity among the receptor sites.

V. STUDY OF THE COOPERATIVE INTERACTIONS AMONG RECEPTOR SITES

A. Definitions

In the simple bimolecular model usually used for equilibrium and kinetic analysis, it is assumed that the receptor is univalent and that there are no cooperative interactions between the receptor sites.

An alternative assumption is that the binding of the hormone is cooperative. This means that binding of a hormone molecule to the receptor affects the binding of other hormone molecules, i.e., it may facilitate further binding (positive cooperativity) or make further binding more difficult (negative cooperativity). Consequently, the apparent affinity increases with fractional occupancy in the positive case, decreases in the negative case. Reciprocally, the binding of hormone molecules to the receptor may stabilize (positive cooperativity) or destabilize (negative cooperativity) the binding of already bound hormone molecules, with observable effects on the dissociation of the bound labeled molecules. Such cooperative effects are most commonly due to site-site interactions linked to conformational changes in oligomeric proteins, but some alternative possibilities should be carefully examined before concluding that site-site interactions are present [177]. Such conformational transitions are classically explained by two kinds of models: (1) "concerted" conformational changes where the ligand modifies the pre-existing equilibrium between two symmetrical conformational states in the binding macromolecule [151, 152] (known as the allosteric model); and (2) ligand-induced "sequential" conformational changes [162]. Whereas both models can explain positive cooperativity [162], only the latter can produce negative cooperativity in binding [162, 163]. A more general model has recently been proposed [164]. Negative cooperativity is a frequent occurrence in macromolecules [165-169] and all hormone-receptor systems for which a curvilinear Scatchard plot has been demonstrated are potential candidates for this

mechanism. To avoid semantic confusion, we want to make precise that we, as well as most authors [162], use the term "negative cooperativity" in a purely descriptive sense for interactions resulting in a decrease in the apparent affinity of receptors for the ligand when fractional saturation of the receptors increases. This is equivalent to saying that the shape of the binding isotherm is shallower than the curve for "statistical" binding, or that the Hill coefficient is less than 1.0. In no way do we imply that the "energy of interaction" is negative; that assumption would require some knowledge of the "unperturbed" state of the receptor [170, 171]. Nor do we make any assumption as to the molecular nature of the "conformational change" and of the "site-site interactions," which might include mechanisms as varied as intramolecular changes in tertiary or quaternary structure of the receptor, association or dissociation of receptor molecules [172-174], clustering of receptors in the membrane, or phase transitions in the membrane itself [175]. We have recently reported the existence of site-site interactions of a type consistent with negative cooperativity when insulin binds to its receptor in various tissues [176, 177]. We have developed methods, described in Sec. V. B to study the cooperativity and the influence on site-site interactions of various factors such as temperature, pH, ionic strength [177], structural modifications of insulin [178], and Concanavalin A [179]. We have, finally, considered alternative new models of hormone binding to cell membrane receptors [177, 180, 181].

B. Kinetic Method to Assess Cooperative Interactions in Hormone-receptor Binding

The basic experimental design assumes that in the case of negative cooperativity, the decreased affinity will result at least partially from an increased dissociation rate (see Note number 6). When this is the case, it is possible to detect the cooperative interactions by studying the dissociation of labeled hormone from the receptors in two conditions: in an "infinite" dilution of the hormone-receptor complex, and in an "infinite" dilution of the complex in the same medium containing an excess of unlabeled insulin. Slight differences in methodology are required for whole cells and membrane preparations.

Cells or membranes at high concentration in a single batch are reacted with the labeled hormone at low concentration such that only a minority of the receptor sites are occupied by bound tracer. Association is monitored by centrifugation of aliquots of the incubation mixture in a Beckman microfuge (1 min for cells, 3 min for membranes). When a steady-state level of occupancy is achieved, the cells or membranes are centrifuged at 4°C, the supernatant containing the unbound hormone is discarded and the pellet is resuspended immediately up to the initial volume with ice-cold buffer. At this point, only receptors filled with labeled hormone and free

receptors (the large majority), but no free hormone, are present. Aliquots (100 μl) are immediately distributed in two sets of tubes; half contain 10 ml of hormone-free buffer ("dilution only"), half contain 10 ml of buffer to which unlabeled hormone has been added ("dilution plus cold hormone"). To monitor the dissociation, duplicate tubes from each set are centrifuged at 4°C, 700 g for 2 min at regular intervals; the supernatants are discarded and the pellets counted. When membranes instead of whole cells are used, the content of the tubes is filtered through Millipore filters (0.45 μ); the membranes which are retained on the filter are washed once with ice-cold buffer and counted. The dissociation rate by "dilution only" and by "dilution plus cold hormone" are then compared.

The principles on which the interpretation of the results is based are the following (see also studies on dissociation of O_2 from hemoglobin [185]):

1. Dilution Only

In the "dilution only" situation, molecules of labeled hormone, once dissociated, will show negligible reassociation if the dilution factor is sufficient. If the reaction of hormone with receptor is a simple, reversible, bimolecular one which obeys the law of mass action, then

$$[H] + [R] \rightleftharpoons [HR]$$

where [H] is the concentration of free hormone, [R] is the concentration of free receptor, and [HR] is the concentration of bound hormone (or receptor).

And

$$K_a = \frac{[HR]}{[H]\ [R]}$$

where K_a is the affinity constant or

$$\frac{[HR]}{[H]} = K_a\ [R]$$

If only a small minority of the receptor sites are occupied, the concentration of free receptors $[R] = [R_o] - [HR]$ is essentially equal to the total concentration of receptors $[R_o]$ (see Note number 7).

Hence

$$\frac{[HR]}{[H]}\ \left(\text{or}\ \frac{B}{F}\right) \sim K_a\ [R_o]$$

Thus, at the end of the association period of our experiment

$$\frac{[HR_1]}{[H_1]} = \frac{B_1}{F_1} = K_a [R_o]$$

When aliquots are then diluted 100-fold, the reaction tends toward a new equilibrium, where

$$\frac{[HR_2]}{[H_2]} = \frac{B_2}{F_2} = K_a \frac{[R_o]}{100} = \frac{[HR_1]}{[H_1]} \cdot \frac{1}{100}$$

In other words, the B/F ratio after dissociation will be 1% of the B/F ratio at the end of association, which is trivial.

The experimental procedure then actually allows the dissociation rate to be monitored in the absence of measurable association. In any case, we prefer to control the absence of significant reassociation experimentally (see Sec. V. C).

2. Dilution Plus Cold Hormone

In the "dilution plus cold hormone" situation, the dilution factor of the complex being the same, the labeled hormone-receptor complex will also dissociate completely. The unlabeled hormone will bind to free receptor sites, which will not affect the rat of dissociation of tracer from filled sites unless the filling of the empty sites by unlabeled hormones induces site-site interactions. Unlabeled hormone will also replace the labeled hormone once the latter has left the receptor sites to which it was bound, but this should not affect the dissociation rate of the replaced labeled molecule either, since the rate-limiting step in the exchange is the dissociation rate of the label (see Note number 8), and since rebinding of the label is prevented by the dilution factor.

Thus, in the absence of site-site interactions, the dissociation of labeled hormone from the receptors in "dilution only" and in "dilution plus cold hormone" should be identical. This was the case for growth hormone and its receptors in human cultured lymphocytes [176], showing that the receptor sites were independent; this was consistent with the linear Scatchard plot of steady-state binding. By contrast, insulin receptors exhibit the behavior expected from a negatively cooperative system, as illustrated in Figure 13. The presence of a saturating concentration of insulin markedly speeds up the dissociation of labeled insulin, demonstrating the existence of site-site interactions between sites filled with unlabeled insulin and sites filled with labeled insulin.

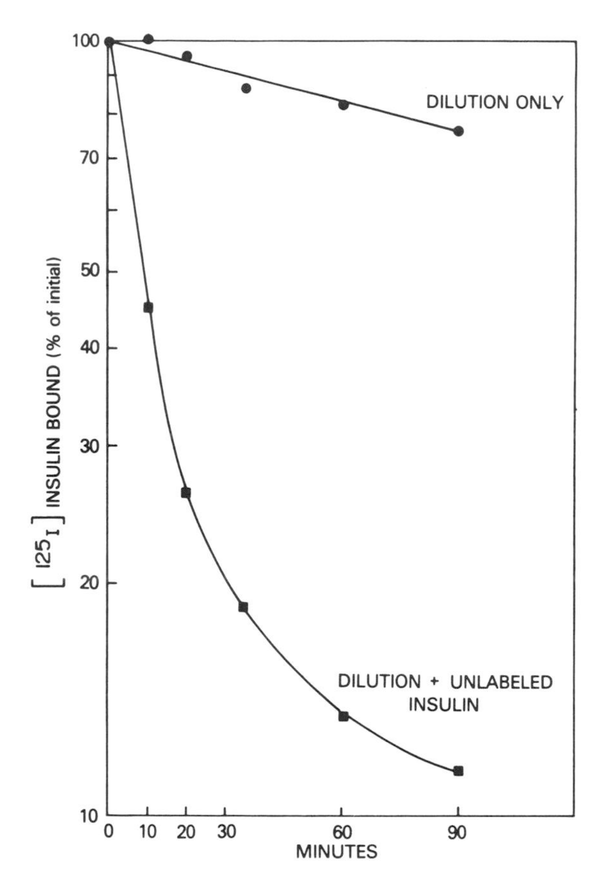

FIG. 13. Dissociation of [^{125}I]insulin from cultured human lymphocytes in a 100-fold dilution in the absence (dilution only) and in the presence (dilution plus unlabeled insulin) of 1.7×10^{-7} M unlabeled insulin. [^{125}I]-Insulin (1.7×10^{-11} M) was incubated for 30 min at 15°C with 5×10^{7} cells per ml, after which the cells were diluted according to the procedure described in Sec. V. B. When the fraction of sites occupied by the [^{125}I]insulin is small ($\sim < 5\%$) dissociation by dilution only appears to be first-order.

C. Experimental Controls of the Absence of "Rebinding"

If the dilution factor is insufficient, some labeled hormone will tend to rebind in the "dilution only" experimental situation. In the "dilution plus cold hormone" situation, this rebinding would be prevented by the presence of unlabeled hormone replacing the labeled on the receptor sites and by the "isotopic dilution" of the labeled hormone by the free unlabeled hormone, leading to an apparent increase in the dissociation rate. Two methods have been used as controls for the absence of rebinding [176]: (1) Dilution alone is compared with dilution plus unlabeled hormone over a wide range of dilutions: above a 50-fold dilution, the difference in dissociation between "dilution only" and "dilution plus cold hormone" is independent of the dilution factor; therefore, at this dilution, none of the difference is accounted for by prevention of tracer rebinding by unlabeled hormone. This difference thus purely reflects cooperative interactions [176]. (2) The two sets of cells ("dilution only" and "dilution plus cold hormone") are duplicated; in one of each, before distributing the aliquots for dissociation, an additional equal aliquot of fresh cells that have never been exposed to hormone is added. If rebinding occurs during dissociation, it will occur to a more-than-doubled population of free receptors and should then be magnified. This did not happen in our experimental conditions [176].

D. Effect of the Nonspecific Binding

Another important methodological control is to assess the participation of the "nonspecific" compartment in the observed dissociation rates. This was done in the following manner: a tracer size of labeled insulin is associated in the presence of a large excess of unlabeled insulin (10 μg/ml or 1.7×10^{-6} M); in this case, most of the binding is to the nonspecific compartment. The cells or membranes are then submitted to the dissociation experiment in the absence and in the presence of cold hormone.

For insulin binding to receptors in human cultured lymphocytes, the nonspecifically bound radioactivity, representing only 2.5% of the total tracer, and about 4% of the specific binding, dissociated almost instantaneously up to a trivial residual (and irreversible) binding of 0.3%; this was not affected by the presence of unlabeled hormone in the dilution medium. In these experiments, the dissociation rates observed are thus negligibly affected by the nonspecific binding. However, in systems where nonspecific binding represents an appreciable part of the total binding, it is necessary to include this control in each experiment and correct the dissociation curves.

E. Dose-Response Curve

The decrease (due to unlabeled insulin) in the amount of radioactivity remaining bound after a given period of dissociation has been used as a basic parameter to estimate the extent of site-site interactions as a function of the concentration of unlabeled insulin in the dilution medium: the dose-response curve is extremely sensitive (Fig. 14), with 20% of the total increase in dissociation observed with as low as $2\text{-}5 \times 10^{-10}$

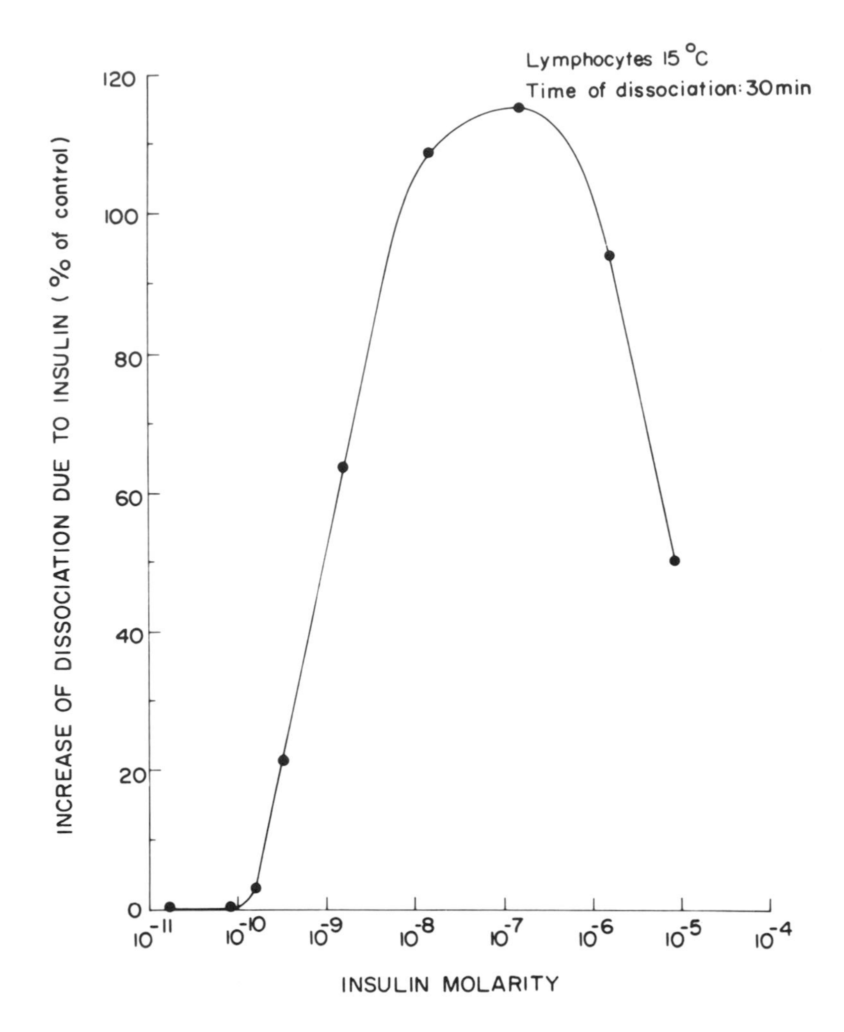

FIG. 14. Effect of insulin concentration on dissociation of $[^{125}I]$insulin. $[^{125}I]$Insulin (5×10^{-11} M) was incubated with 2.5×10^7 cells per ml for 30 min at 15°C and then diluted 100-fold according to the procedure described in Sec. V. B, in the absence and in the presence of the indicated concentrations of unlabeled insulin (10^{-11} to 10^{-5} M). After 30 min at 15°C, the cells were sedimented and the radioactivity in the cell pellet was counted. The increase of dissociation due to unlabeled insulin, reflecting the cooperative site-site interactions, is expressed as

$$\frac{[^{125}I]\text{ insulin dissociated with dilution + cold insulin}}{[^{125}I]\text{ insulin dissociated with dilution only}} \times 100$$

and plotted as a function of the concentration of unlabeled insulin. Notice the decrease in cooperativity at insulin concentrations above 10^{-7} M, which corresponds to the range of concentrations where significant dimerization of insulin occurs. From De Meyts et al. [176].

M insulin, and a maximum effect between 10^{-8} - 10^{-7} M. Further increase in insulin concentration results in a decrease in the cooperative effect, probably due to insulin dimerization [176-178, and manuscript in preparation]. Nondimerizing insulin species, such as tetranitro-insulin and guinea pig insulin, produced the cooperative effect expected from their relative ability to occupy the receptor sites, but showed no decrease in this effect at high concentration [178].

F. Influence of Fractional Saturation of Receptors

The sensitivity of the dose-response curve for increased dissociation suggested [177] that significant cooperativity may already be induced by a small fractional saturation of the receptors (1-5%). If cooperative interactions are already significant in the binding of the labeled hormone, the dissociation rate by dilution could already be accelerated and not truly "basal," with a consequent reduction of the difference in the rates observed by "dilution" and "dilution plus cold hormone."

The effect of the fractional saturation of the receptors on the dissociation rate of the complex was tested experimentally by incubating mixtures of labeled and unlabeled insulin at increasing concentrations and submitting the resulting complex to dissociation by "dilution" and "dilution plus cold hormone." It was shown that the larger the tracer (the higher the occupancy of receptors), the faster is the dissociation by dilution, until no difference is observed between dilution and "dilution plus cold hormone" (around 10^{-6} M: at this point, the specific sites are saturated and most of the tracer is bound nonspecifically). In other words, the dissociation rate by "dilution only" of a tracer which occupies an appreciable fraction of the sites may already be a "cooperatively" increased rate; it is then important—if one wants to observe a more "basal" rate by dilution only and optimize the experiment—to reduce the size of the tracer as much as allowed by counting accuracy (see Note number 9) [177].

If one considers the aforementioned principles, the experimental conditions (tracer concentration, receptors concentration, and dilution factor) can easily be adapted to a particular hormone-receptor system given its proper binding capacity and apparent affinity.

Various physico-chemical characteristics of the negative cooperativity of insulin receptors (see Fig. 15) have been reported elsewhere [176-181] and will not be detailed here.

G. Methodological Implications of Negative Cooperativity

When studying the effects on the hormone-receptor interaction of a given substance or of a given experimental manipulation, one has to explore its effects not only on the binding of a tracer size of labeled hormone, but

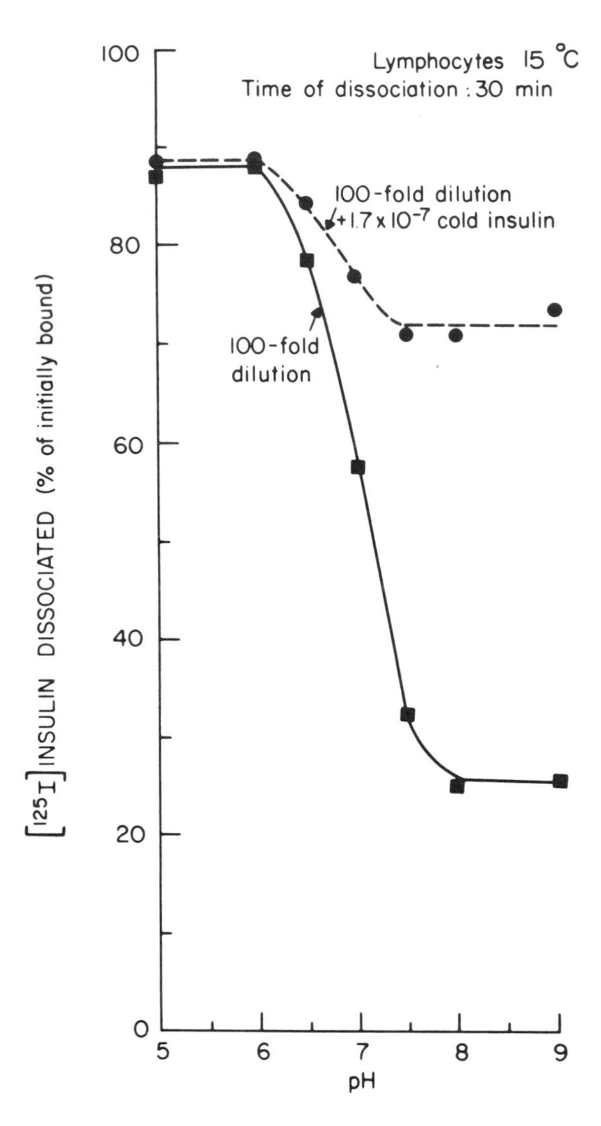

FIG. 15. Effect of pH on the dissociation of $[^{125}I]$ insulin from cultured human lymphocytes (IM-9). $[^{125}I]$ Insulin (5×10^{-11} M) was incubated with 2.5×10^{7} cells per ml for 30 min at 15°C and then diluted 100-fold in assay buffer at pH varying from 5 to 9, in the absence and in the presence of 1.7×10^{-7} M unlabeled insulin. After 30 min at 15°C, the cells were sedimented and the radioactivity in the cell pellet was counted. The $[^{125}I]$-insulin dissociated at t = 30 min, as percent of the initially bound, is plotted as a function of pH.

also on the whole binding curve. When a system is not cooperative, one may assume that alterations (or absence of alteration) in the binding of a tracer size of hormone is a reflection of what happens over the whole range of fractional saturation. In contrast, substances or experimental modifications which affect the site-site interactions might not modify the binding of a tracer size of hormone too small to induce site-site interactions, but might modify the binding at greater hormone concentration (and thus alter the shape of the binding curve). Many data concerning the characteristics of the insulin receptor have been gathered, often using the "tracer insulin binding" approach. Some of these studies might have to be reexamined for possible overlooked effects on cooperative interactions. For example, it was found [179; De Meyts et al., manuscript in preparation] that Concanavalin A (Con A), at 15°C and at concentrations which do not agglutinate the cells ($\leq 20\ \mu g/ml$) does not modify the association rate or the steady-state level of binding when a tracer size of insulin was incubated with cultured lymphocytes, indicating that Con A does not compete with insulin for the receptor. However, the steady-state level of binding of insulin at increasing insulin concentrations (when site-site interactions become significant) was increased, still without modifications in the association rate. Con A, in fact, linearized most of the Scatchard plot and raised the Hill coefficient from 0.53 to 0.85. This suggested that Con A acts in inhibiting the accelerating effect of site-site interactions on the dissociation rate, which was verified directly in the dissociation experiment by "dilution only" versus "dilution plus cold insulin." These results differ from reported effects of Concanavalin A on fat cell insulin receptors [186, 187].

VI. CONCLUSION: SOME PERSPECTIVES IN RECEPTOR RESEARCH

Since this series is mostly concerned with detailed review of methods, we have systematically avoided the description of results which we did not feel helpful in clarifying the methods. In conclusion, however, we would like to show that the study of hormonal receptors on blood cells does not constitute just a peripheral domain of interest, but that it has already proved to be a major tool in understanding the delicate mechanisms by which sensitivity to hormones is regulated in humans. We will briefly evoke some of the major findings and new perspectives which stemmed from the study of lymphocytes and monocytes.

A. Radioreceptor Assay of Hormones

Radioimmunoassay has been a major breakthrough in understanding physiological regulation of hormonal secretion and its disturbances. The limit of this technique is that it is based on recognition of the immunogenic characteristics of the hormone molecule, which is not systematically related to its biological properties. Although it is not likely to replace the convenient radioimmunoassay in routine assay of hormones, radioreceptor assay is irreplaceable in studying abnormal circulating hormones, heterogeneity in bioactive circulating forms, and nonimmunoassayable hormones such as NSILA-s [188]. The cultured lymphocyte constitutes a most practical and handy tool for assaying insulin, growth hormone [13-16], and calcitonin [11].

B. New Insights into the Molecular Topochemistry of Hormonal Bioactivity

The study of a large variety of insulin analogs for binding and induction of negative cooperativity in cultured lymphocytes showed that there is a site at the surface of the insulin monomer, distinct from the "bioactive" site, which is responsible for inducing the site-site interactions among the receptors [178; De Meyts et al., manuscript in preparation]. Identical structural specificity was found in cultured placental cells [88].

C. Search for New Biological Actions of Hormones

We have already mentioned the upsurge of interest in the role of hormones in mitogenesis, cell transformation, and malignancies, as well as in the hormonal regulation of the immune system in general [54, 55, 57]. Cultured lymphocytes constitute a convenient in vitro model, and circulating cells its counterpart in vivo.

D. Regulation of Receptor Affinity and Concentration

Through the study of lymphocytes and monocytes, it has become clear that cells do not react passively to acute and chronic fluctuations in hormone levels by just binding more or less hormone to a fixed number of receptors $[R_o]$ with a fixed affinity $[K_a]$. In fact, both $[R_o]$ and $[K_a]$ are subject to regulation [189]. The insulin receptor appears to be buffered against acute elevations of insulin levels through a mechanism by which $[K_a]$ decreases as [H] increases (negative cooperativity), while remaining exquisitely sensitive to low hormone concentrations.

Chronic elevation of hormone levels (insulin and hGH) appears to be associated with a down regulation of the receptor concentration on the cells; this is true in vitro, as demonstrated with cultured lymphocytes, as well as in vivo, possibly a major pathogenic factor in the insulin resistance associated with hyperinsulinemia (see Sec. VI. F). A similar regulatory effect was recently demonstrated for the β-adrenergic receptor (Lefkowitz et al., submitted for publication).

E. New Models of Hormone Action

The binding of hormone to receptors appears to be a complex mechanism. Insulin induces site-site interactions among the receptor sites, which result in a decreased affinity when fractional saturation of the receptors increases (negative cooperativity).

This model implies that the receptor undergoes ligand-induced changes in conformation, but does not allow one, in the absence of independent information, to make any assumption as to the precise nature of the conformational changes. The bulk of available data is consistent with a model in which the insulin-receptor sites are switched from a state in which insulin dissociates slowly, to a fast-dissociating state when occupancy by insulin increases; a purely "phenomenologic" illustration of such a model is presented in Figure 16, as the mirror image of the positive cooperativity in hemoglobin. Several possible mechanisms are currently being investigated [177], including conformational changes in the tertiary or quaternary structure of oligomeric receptors (the most extreme being reversible association $\rightleftharpoons$ dissociation), "clustering" of receptors through translational movements in the fluid membrane [190, 175] (which might explain the inhibition of site-site interactions of insulin receptors by Concanavalin A), or both. Recent electron microscopic studies of tagged insulin have directly demonstrated the occurrence of both dispersed and clustered distribution of insulin receptors on the membrane of fat cells [191] and liver [192]; if it is confirmed to be "physiologic," this might be a major breakthrough in understanding the mechanism of the negative cooperativity. Conformational changes in receptors amplified through a clustering mechanism, as recently emphasized by Levitzki (Fig. 17), could be the cause of the negative cooperativity in binding while "triggering" the membrane and inducing the biological effects. Such a mechanism constitutes thus the potential basis for a coherent theory of hormone action, including a role for excess or "spare" receptors [175, 193].

The negative cooperativity, first demonstrated for insulin receptors in cultured lymphocytes [176], was subsequently confirmed for insulin receptors in peripheral monocytes [6], purified liver membranes [177, 194], cultured placental cells [88], and turkey red blood cells [89]. By application of the method described in Sec. V. B, it was also discovered in receptors for the nerve growth factor in peripheral ganglia [182] and embryonic heart and brain [183], and in TSH receptors [184]. It appears thus likely to be a widespread phenomenon.

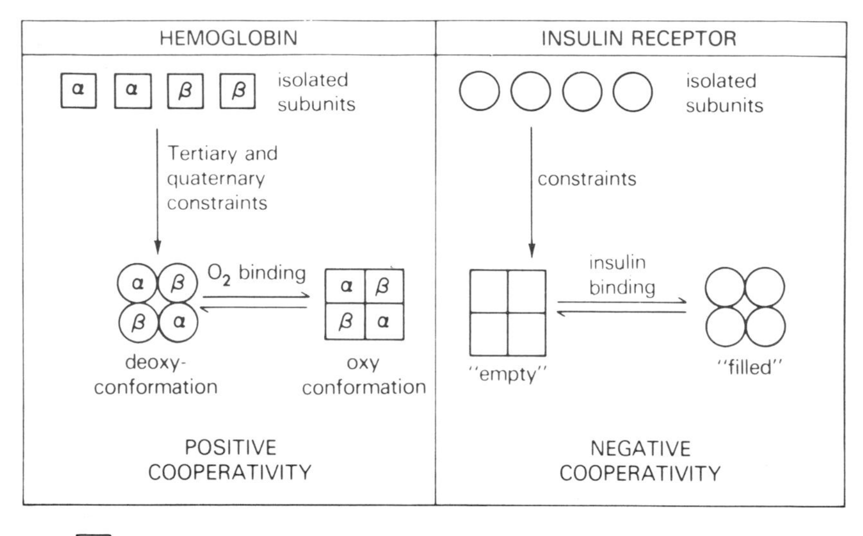

□ = subunit in a high affinity (slow dissociating) state

○ = subunit in a low affinity (fast dissociating) state

FIG. 16. A plausible model for the negative cooperativity in insulin receptors. This model was designed by analogy with the structure and function of hemoglobin. In hemoglobin, isolated chains (α and β) exhibit Michaelian saturation curves and high affinity for oxygen; the constitution of tetramer involves constraints which bring the subunits to a state of lower affinity for oxygen. The binding of oxygen is concomitant with a release of the constraints, which improves the affinity of neighboring subunits for oxygen (positive cooperativity). The rationale for this mechanism, in which affinity for oxygen is low at low partial pressure of oxygen, is that hemoglobin is functionally a carrier molecule and must release oxygen more easily when its tissue concentration is low. For a receptor, which must bind the hormone at low concentrations, high affinity at low concentrations is favorable, which is the case in a negative cooperative binding. The negative cooperativity buffers the system against high hormone concentrations. The model on the right, built as the "mirror image" of hemoglobin, is one among many plausible ones. It is not implied that all "subunits" change shape simultaneously: Possible intermediate steps (sequential model) have not been illustrated. From De Meyts et al. [177].

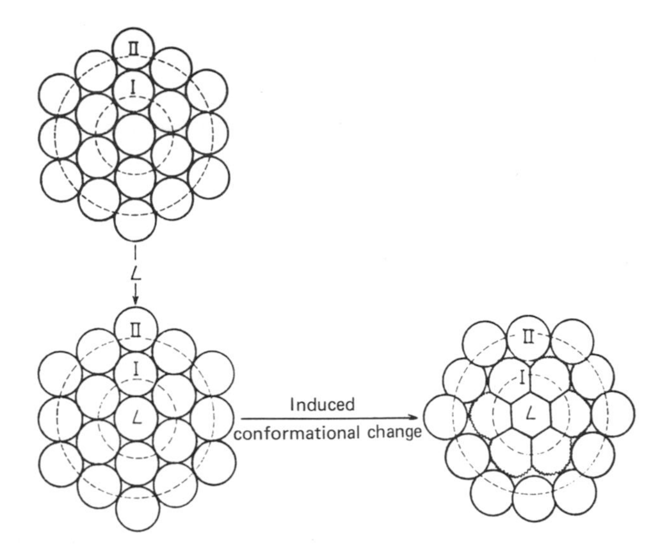

FIG. 17. A model for the negative cooperativity associated with receptor clustering. When one ligand molecule binds to one of the 19 available sites shown in the drawing, the receptor undergoes a structural transition. This transition propagates outwards towards layer I (6 receptor molecules). In this case, the amount of receptor molecules undergoing conformational change without binding of ligand is 86%. A modified model can also be offered where the receptor molecules on the membrane are not clustered a priori, and can move freely in the fluid membrane. Upon ligand binding, the receptor molecule undergoes a conformational change enabling it to interact with unbound receptor molecules and form a cluster; the unbound receptor molecules then undergo a conformational transition converting them into a nonreceptive state. This conformational change could propagate further than shown. Both models bring about negative cooperativity among receptors and explain "spare receptors." From Levitzki [175], with permission of Academic Press.

F. Molecular Pathology of Insulin-resistant States in Humans

It has been demonstrated that some insulin resistant states in humans are associated with a decreased binding of insulin to blood mononuclear cells [195, 196]. This paralleled results reported in the genetically obese (ob/ob) syndrome in mice [197, 198] and in vitro studies with cultured lymphocytes [199]. Careful quantitation of the monocyte subpopulation in peripheral mononuclear cells confirms this general pattern [7, 195; Bar et al., J. Clin. Invest., in press (1976).

The relevance of the study of receptors for insulin, NSILA-s, and growth hormone in diseased states in man has been recently reviewed in detail [200].

On the basis of the regulatory mechanisms described, one would predict that certain alterations in insulin sensitivity in vivo could be associated with some of the following defects at the receptor level: (1) down regulation by endogenous hyperinsulinemia of receptor concentration/cell; (2) primary defect in receptor concentration/cell; (3) primary or acquired defects in the molecular structures which determine receptor affinity and cooperativity (like the multiple abnormalities in affinity for oxygen and cooperativity found in hemoglobinopathies); and (4) perturbation of yet unknown metabolic or endogenous factors regulating the affinity or the site-site interactions.

ACKNOWLEDGMENTS

This review represents the work of a group, all of my friends and coworkers in the Diabetes Branch. Jesse Roth has been an invaluable sponsor in my confrontation with the receptor problem; working with him has been (and still is) an extraordinary experience. I am greatly indebted to James R. Gavin, III, Donald N. Buell, and Maxine A. Lesniak, who, besides providing precious advice and guidance, allowed extensive use of their published and unpublished data for this review. Phillip Gorden, David M. Neville, Jr., and C. Ronald Kahn continually shared their wisdom and extended their benevolent support. Angelo R. Bianco and Robert S. Bar have made extensive contribution to the methods and data concerning the monocytes. David Rodbard and Alex Levitzki have provided unpublished manuscripts and stimulating discussion. The generous gifts of insulin derivatives by Frederic H. Carpenter, Ronald E. Chance, Leslie F. Smith, Dietrich Brandenburg, and Tom L. Blundell have been essential to the work of this group. Dorothy E. Beall provided excellent secretarial assistance.

This work was supported by a PHS International Postdoctoral Fellowship FO5 TW 1918 and by the Fonds National de la Recherche Scientifique Belge.

NOTES

1. As noted by Bourne et al. [55], nonrandom distribution of receptors (as observed in cultured lymphocytes) may be an important argument in suggesting that these receptors may perform a physiological function (see Sec. II. C. 1). "Otherwise, receptors could be present on lymphocytes simply because they are protein coded in the genome of all cells; the

receptor could be expressed only because their presence confers no selective disadvantage to the organism. This would be the case if the lymphocytes never 'saw' the hormone in high enough concentration to affect its function."

2. Actually, if one increases the concentration of unlabeled hormone sufficiently, and uses a large amount of radioactivity for accurate counting, one can demonstrate a slope for the nonspecific component of the binding curve, corresponding to a very low affinity constant or K_a (10^6 - 10^5 M^{-1} for insulin binding to cultured lymphocytes) (De Meyts, unpublished data). The nature of the nonspecific binding sites has not been investigated.

It should not be assumed that, because nonspecific (i.e., "nonsaturable") binding is not competed for by unlabeled insulin in a steady-state study, it is irreversible: This is a separate kinetic concept which has to be tested experimentally for each particular case. For insulin binding to cultured lymphocytes, for example, 90% of the nonspecific binding dissociates readily upon dilution (as expected from a "very low affinity" binding), the remaining 10% (0.1-0.3% of the total radioactivity) being apparently irreversibly bound [177].

3. In the presence of a large receptor excess, the Scatchard plot becomes horizontal in the low range of fractional saturation, which corresponds to the initial horizontal part of the sigmoidal competition curve where $[HR]/[H_o]$ or $[HR]/[H]$ for $[^{125}I]$hormone is plotted versus log $[H_o]$. Since the abcissa of the Scatchard plot is not logarithmic, this usually does not markedly affect the graph and, in particular, its initial slope. This deviation at very low hormone/receptors ratios is easily explained if one considers the Scatchard equation formulated as $[HR]/[H] = K_a (R_o - [HR])$; it is clear that the larger R_o, the larger $[HR]$ needs to be to allow measurement of a significant decrease in $[HR]/[H]$. As $[H_o]$ becomes much smaller than $[R_o]$, $[HR]$ becomes much smaller than $[R_o]$ and $[HR]/[H]$ approaches $K_a [R_o]$.

4. In fact, this helpful derivation of the Hill equation as well as the plot itself, are not described in the often-quoted 1910 paper, [144], but can be found in a later paper, (Brown and Hill [146]) which would constitute the appropriate reference. Wyman later popularized this plot for describing cooperative binding [147-150] and proposed to call it a "Hill plot" [148]; he used it extensively to illustrate the statistical thermodynamics of linked functions [148, 149], which eventually led to the concept of "allosteric linkage" [149, 150] and to the famous allosteric model of Monod, Wyman, and Changeux [151, 152].

5. The fact that the Hill plot is linear can be more rigorously explained mathematically. Any binding curve, cooperative or not, when plotted as Y against ln [H], is sigmoidal and may be described with reasonable accuracy by a logistic equation [153] of the form:

$$P = \frac{1}{1 + e^{-(\alpha + \beta x)}}$$

(In this case, $P = \overline{Y}$; $\alpha = \ln K_a$; $\beta = n$; and $x = \ln [H]$.)

The most interesting mathematical property of the logistic equation is that $\ln [P/(1 - P)]$ (here $\ln [\overline{Y}/(1 - \overline{Y})]$, which is the logit of P (logit is "logistic unit"), equals $\alpha + \beta x$ (or, in our case, $\ln K_a + n \ln [H]$). This is equivalent to $\log [\overline{Y}/(1 - \overline{Y})] = \log K_a + n \log [H]$. So, a plot of logit $P = \ln [P/(1 - P)]$ against x is a straight line with a slope β. In fact, Brown and Hill [146] plotted $\overline{Y}$ versus O_2 pressure on a logit-log paper, which is the easiest way to make Hill plots. The physical interpretation of the logistic equation is beyond our purpose and the interested reader can find the details and references reviewed elsewhere [154, 155].

It is interesting that the logistic equation can be applied to a variety of biological and physico-chemical phenomena (from growth curves to autocatalysis), with the benefit of the linearizing properties of the logit-log transformation. Rodbard popularized its application in radioimmunoassay [156], and recently proposed interesting new methods for direct computer fitting to a logistic equation for data as varied as polyacrylamide gel electrophoresis [157], immunoradiometric assay [158], and detection of apparent cooperative effects in hormone-stimulated cAMP production [159, 160].

The Hill plot is thus only one among multiple applications of this mathematical wonder.

6. The cooperativity of oxygen and nitric oxide binding to hemoglobin, for example, appears to be due mostly to alteration of the dissociation rate. (E. G. Moore and Q. H. Gibson, J. Biol. Chem., 251:2788, 1976).

7. This approximation is made for the sake of simplicity. The demonstration stands true if the complete expression

$$[HR] = \frac{[H_o] + [R_o] + 1/K_a - \{([H_o] + [R_o] + 1/K_a)^2 - 4 [H_o] [R_o]\}^{1/2}}{2}$$

is used (De Meyts, unpublished).

8. This is true unless cold hormone can dimerize or polymerize with the label (ligand-ligand negative cooperativity), the polymer having a different affinity, which can be reasonably excluded in this case by the

high sensitivity of the cooperative effect, observed at insulin concentrations several orders of magnitude lower than the K_d for dimerization, by the fact that nondimerizing insulin species have the same effect, and by a variety of physico-chemical characteristics like inhibition by Concanavalin A [177]; or unless the dissociation rate of the labeled species can be affected by "mass isotopic effects" like in some cases of isotopic exchange: However, in that case, it is the heavier isotope which speeds up the dissociation of the light isotope; moreover, these kinds of interactions have been reported to be significant only at the atomic level. In particular cases, unstirred layers have to be excluded [176, 177]. These and some other alternative hypotheses are discussed elsewhere [177].

9. This might explain why dissociation of growth hormone from cultured lymphocytes is not first-order, although the Scatchard plot is essentially linear. It was, in fact, noted [10] that a tiny subpopulation of sites with higher affinity might be present at very low fractional saturation; if this, in fact, reflects site-site interactions, considerable nonlinearity might be present in the dissociation of a tracer size of growth hormone; however, since the fractional saturation by the tracer itself already induces all the site-site interactions possible in this receptor population, (the Scatchard is then linear over most of the curve), no increase in the dissociation will be found in the presence of unlabeled insulin, consistent with the Scatchard plot data. The growth hormone receptor system has then to be reexamined with lower fractional saturation in the "dilution only" situation to verify this hypothesis.

REFERENCES*

1. R. J. Lefkowitz, I. Pastan, and J. Roth, in NIH Fogarty International Center Proceedings (J. E. Rall, M. Rodbell, and P. Condliffe, eds.), No. 4, National Institutes of Health, Bethesda, Md., 1969, p. 88.

2. T. Goodfriend and S. Y. Lin, Clin. Res., 17:243 (1969).

3. J. Roth, Metabolism, 22:1059 (1973).

4. C. R. Kahn, in Methods in Membrane Biology (E. D. Korn, ed.), Vol. 3, Plenum, New York, 1975, p. 81.

5. J. R. Gavin, III, J. Roth, P. Jen, and P. Freychet, Proc. Natl. Acad. Sci. U.S.A., 69:747 (1972).

6. A. R. Bianco, R. H. Schwartz, and B. S. Handwerger, Diabetologia, 10:359 (1974).

7. R. H. Schwartz, A. R. Bianco, C. R. Kahn, and B. S. Handwerger, Proc. Natl. Acad. Sci. U.S.A., 72:474 (1975).

*Note added in proof: This reference list covers the literature available to us by December 1974.

8. J. R. Gavin, III, P. Gorden, J. Roth, J. A. Archer, and D. N. Buell, J. Biol. Chem., 248:2202 (1973).

9. M. A. Lesniak, J. Roth, P. Gorden, and J. R. Gavin, III, Nature, New Biol., 241:20 (1973).

10. M. A. Lesniak, J. Roth, P. Gorden, and J. R. Gavin, III, J. Biol. Chem., 249:1661 (1974).

11. S. J. Marx, G. D. Aurbach, J. R. Gavin, III, and D. N. Buell, J. Biol. Chem., 249:6812 (1974).

12. R. J. Lefkowitz, J. Roth, and I. Pastan, in Methods in Investigative and Diagnostic Endocrinology (S. A. Berson and R. S. Yalow, eds.), Vol. 2, Pt. II, North Holland, Amsterdam, 1973, p. 376.

13. P. Gorden, M. A. Lesniak, C. M. Hendricks, and J. Roth, Science, 182:829 (1973).

14. P. Gorden, J. R. Gavein, III, C. R. Kahn, J. A. Archer, M. A. Lesniak, C. Hendricks, D. M. Neville, Jr., and J. Roth, Pharmacol. Rev., 25:179 (1973).

15. J. Roth, in Methods in Enzymology (S. P. Colowick and N. O. Kaplan, eds.), Vol. 37: Peptide hormones (B. W. O'Malley and J. G. Hardmann, eds.), Academic Press, New York, 1975, p. 66.

16. J. R. Gavin, III, C. R. Kahn, P. Gorden, J. Roth, and D. M. Neville, Jr., J. Clin. Endocrinol. Metab., 41:438 (1975).

17. G. E. Moore, R. E. Gerner, and H. A. Franklin, JAMA, 199:519 (1967).

18. P. Gerber and J. H. Monroe, J. Natl. Cancer Inst., 40:855 (1968).

19. P. R. Glade, I. M. Paltrowitz, and K. Hirschhorn, Bull. N.Y. Acad. Med., 45:647 (1969).

20. B. Clarkson, A. Strife, and E. de Harven, Cancer, 20:926 (1967).

21. G. E. Moore, E. Ito, K. Ulrich, and A. A. Sandberg, Cancer, 19:713 (1966).

22. M. A. Epstein and Y. M. Barr, Lancet, i:252 (1964).

23. P. R. Glade and K. Hirschhorn, Am. J. Pathol., 60:483 (1970).

24. G. Moore and J. Minowada, New Engl. J. Med., 288:106 (1973).

25. J. R. Gavin, III, in Immunopharmacology (J. W. Hadden, F. Fereafico, and F. Garattini, eds.), Comprehensive Immunobiology Series, in press (1977).

26. M. Benyesh-Melnick and J. S. Butel, in The Molecular Biology of Cancer (H. Busch, ed.), Academic Press, New York, 1974, p. 403.

27. J. F. Enders, in The Harvey Lectures, Series 59, Academic Press, New York, 1963, p. 113.

28. S. D. Douglas, H. H. Fudenberg, P. R. Glade, L. N. Chessin, and H. L. Moses, Blood, 34:42 (1969).

29. G. E. Moore, H. Kitamura, and S. Toshima, Cancer, 22:245 (1968).

30. J. L. Fahey, D. N. Buell, and H. C. Sox, Ann. N.Y. Acad. Sci., 190:221 (1971).

31. G. E. Moore, in "Long-term lymphocyte cultures in human genetics," Birth Defects, Original Article Series, Vol. IX, No. 1:31 (1973).

32. P. S. Moorhead, In vitro, 10:143 (1974).

33. Proposed usage of animal tissue culture terms. Proceedings of Annual Meeting of Tissue Culture Association, June 3, 1966 (Quoted in Ref. 29).

34. I. Finegold, J. L. Fahey, and H. Granger, J. Immunol., 99:839 (1967).

35. G. E. Moore and R. E. Gerner, Ann. Surg., 172:733 (1970).

36. R. A. Adams, G. E. Foley, S. Farber, E. E. Hellerstein, and L. Pothier, in "Long-term lymphocyte cultures in human genetics," Birth Defects, Original Article Series, Vol. IX, No. 1:200 (1973).

37. J. A. Schneider, in "Long-term lymphocyte cultures in human genetics," Birth Defects, Original Article Series, Vol. IX, No. 1: 212 (1973).

38. H. Zur Hausen, Int. Rev. Exptl. Pathol., 11:233 (1972).

39. M. Nonoyama and J. Pagano, Nature, New Biol., 233:103 (1971).

40. G. Miller, Yale J. Biol. Med., 43:358 (1971).

41. G. Klein, in The Herpes-viruses (A. S. Kaplan, ed.), Academic Press, New York, 1973, p. 521.

42. W. Henle and G. Henle, Cancer, 34:1360 (1974).

43. M. A. Epstein, B. C. Achong, and Y. M. Barr, Lancet, i:702 (1964).

44. G. Klein, T. Lindahl, M. Jondal, W. Leibold, J. Menézes, K. Nilsson, and C. Sundström, Proc. Natl. Acad. Sci., U.S.A., 71: 3283 (1974).

45. M. Jondal and G. Klein, J. Exptl. Med., 138:1365 (1973).

46. P. J. Fialkow, G. Klein, E. R. Giblett, B. Gothosker, and P. Clifford, Lancet, i:384 (1971).

47. Y. Matsuoka, G. E. Moore, Y. Yagi, and D. Pressman, Proc. Soc. Exptl. Biol. Med., 125:1246 (1967).

48. J. Minowada and G. Moore, Proc. VI International Symposium on Comparitive Leukemia Research, University Park Press, Tokyo, New York, in press.

49. A. Hellman, M. N. Oxman, and R. Pollack, Biohazards in Biological Reasearch. Cold Spring Harbor Laboratory, 1973.

50. P. R. Glade and S. W. Broder, in In Vitro Methods in Cell-mediated Immunity (B. R. Bloom and P. R. Glade, eds.), Academic Press, New York, 1971, p. 561.

51. G. E. Moore, in "Long-term lymphocyte cultures in human genetics," Birth Defects, Original Article Series, Vol. IX, No. 1:31 (1973).

52. R. W. Smith, W. D. Terry, D. N. Buell, and K. W. Sell, J. Immunol. 110:884 (1973).

53. M. F. Greaves, J. J. T. Owenn, and M. C. Raff, T and B lymphocytes, Origins, properties and roles in immune responses, Excerpta Medica-American Elsevier, 1974.

54. C. W. Parker, T. J. Sullivan, and H. J. Wedner, in Advances in Cyclic Nucleotide Research (P. Greengard and G. A. Robison, eds.), Vol. 4, Raven Press, New York, 1974, p. 1.

55. H. R. Bourne, L. M. Lichtenstein, K. L. Melmon, C. S. Henney, Y. Weinstein, and G. M. Shearer, Science, 184:19 (1974).

56. A. Boyüm, Scand. J. Clin. Invest., 21, Suppl. 97:77 (1968).

57. U. Krug, F. Krug, and P. Cuatrecasas, Proc. Natl. Acad. Sci. U.S.A., 69:2604 (1972).

58. S. A. Eisen, H. J. Wedner, and C. W. Parker, Immunol. Commun., 1:571 (1972).

59. M. F. Greaves and G. Brown, J. Immunol., 112:420 (1974).

60. M. H. Julius, E. Simpson, and L. A. Herzenberg, Eur. J. Immunol., 3:645 (1973).

61. J. Olefsky and G. M. Reaven, J. Clin. Endocrinol. Metab., 38:554 (1974).

62. D. Zucker-Franklin, J. Immunol., 112:234 (1974).

63. J. J. Oppenheim, B. G. Leventhal, and E. M. Hersch, J. Immunol., 101:262 (1968).

64. M. M. Wintrobe, in Clinical Hematology, 6th ed., Lea and Febiger, Philadelphia, 1967, p. 241.

65. M. J. Cline and R. I. Lehrer, Blood, 32:423 (1968).

66. L. T. Yam, C. Y. Li, and W. H. Crosby, Amer. J. Clin. Path., 55:283 (1971).

67. O. Hechter and A. Calek, Jr., Acta Endocrinol., 77, Suppl. 191:39 (1974).

68. L. Birnbaumer, S. L. Pohl, and A. J. Kaumann, in Advances in Cyclic Nucleotide Research (P. Greengard and G. A. Robison, eds.), Vol. 4, Raven Press, New York, 1974, p. 239.

69. S. P. Martin, G. R. McKinney, R. Green, and C. Becker, J. Clin. Invest., 32:1171 (1953).

70. M. E. Dumm, Proc. Soc. Exptl. Biol. Med., 95:571 (1957).

71. A. N. Weinberg and J. Field, Clin. Res., 7:247 (1959).

72. V. Esmann, Diabetes, 12:545 (1963).

73. N. Kalant and R. Schucher, Can. J. Biochem. Physiol., 40:899 (1962).

74. J. A. Antonioli, J. P. Felber, and A. Vannotti, Acta Haematol., 37:161 (1967).

75. T. P. Stossel, R. Murad, R. J. Mason, and M. Vaughan, J. Biol. Chem., 245:6228 (1970).

76. T. P. Stossel, R. J. Mason, J. Hartwig, and M. Vaughan, J. Clin. Invest., 51:615 (1972).

77. (a) J. W. Hadden, E. M. Hadden, E. E. Wilson, and R. A. Good, Nature, New Biol., 235:174 (1972); (b) C. H. Li, in Advances in human growth hormone research (S. Raiti, ed.), DHEW publication No. (NIH) 74-612 (1973).

78. I. D. Goldfine, J. D. Gardner, and D. M. Neville, Jr., J. Biol. Chem., 247:6919 (1972).

79. J. D. Boyett and J. F. Hoffert, Hormone Metab. Res., 4:163 (1972).

80. P. M. Lundin and L. Angerwall, Path. Eur., 5:273 (1970).

81. G. Astaldi, G. R. Bargia, A. Astaldi, Jr., B. Yolcin, G. Meardi, and G. Gotti, Lancet, 2:709 (1972).

82. D. Maor, T. Englander, E. Eylan, and P. Alexander, Acta Endocrinol., 75:205 (1974).

83. V. W. Adamkiewicz, Can. Med. Assoc. J., 88:806 (1963).

84. D. A. Long, A. A. Miles, and W. L. M. Perry, Lancet, ii:902 (1951).

85. G. E. Thompson and R. J. De Falco, Cornell Vet., 55:66 (1965).

86. G. E. Thompson, Nature, 215:748 (1967).

87. M. Koltai, A. Ottlecz, E. Minker, and G. Blaszó, Int. Arch. Allergy, 46:261 (1974).

88. J. M. Podskalny, J. Y. Chou, and M. M. Rechler, Arch. Biochem. Biophys., 170:504 (1975).

89. B. H. Ginsberg, C. R. Kahn, and J. Roth, submitted for publication.

90. P. Freychet, J. Roth, and D. M. Neville, Jr., Proc. Natl. Acad. Sci. U.S.A., 68:1833 (1971).

91. P. Freychet, C. R. Kahn, J. Roth, and D. M. Neville, Jr., J. Biol. Chem., 247(12):3953 (1972).

92. M. D. Hollenberg and P. Cuatrecasas, in Control of proliferation in animal cells (B. Clarkson and R. Baserga, eds.), Cold Spring Harbor Conferences on cell proliferation Vol. 1, 1974, p. 423.

93. B. A. Cunningham, J. L. Wang, G. R. Gunther, G. N. Reeke, Jr., and J. W. Becker, in Cellular Selection and Regulation in the Immune Response (G. M. Edelman, ed.), Raven Press, New York, 1974, p. 177.

94. A. E. Powell and M. A. Leon, Exptl. Cell. Res., 62:315 (1970).

95. D. Rabinowitz, B. J. Fowlkes, and D. N. Buell, Ninth Leucocyte Culture Conference, Abstract Volume, 174, p. 62 (1974).

96. J. Roth, in Methods in Enzymology (S. P. Colowick and N. O. Kaplan, eds.), Vol. 37: Peptide Hormones (B. W. O'Malley and J. G. Hardmann, eds.), Academic Press, New York, 1975, p. 223.

97. W. M. Hunter and F. C. Greenwood, Nature, 194:495 (1962).

98. P. Freychet, J. Roth, and D. M. Neville, Jr., Biochem. Biophys. Res. Commun., 43:400 (1971).

99. J. H. Marchalonis, Biochem. J., 113:299 (1969).

100. J. I. Thorell and B. G. Johansson, Biochim. Biophys. Acta, 251: 363 (1971).

101. J. L. Hamlin and E. R. Arquilla, J. Biol. Chem., 249:21 (1974).

102. J. C. Sodoyez, F. Sodoyez-Goffaux, and E. R. Arquilla, Diabetologia, 10:387 (1974).

103. J. C. Sodoyez, F. Sodoyez-Goffaux, M. M. Goff, A. E. Zimmerman, and E. R. Arquilla, J. Biol. Chem., 250:4268 (1975).

104. S. Linde and B. Hansen, Int. J. Peptide Protein Res., 6:157 (1974).

105. S. A. Berson and R. S. Yalow, in Methods in Investigative and Diagnostic Endocrinology (S. A. Berson and R. S. Yalow, eds.), Vol. 2A: Peptide Hormones, American Elsevier, New York, 1973, p. 84.

106. Principles of Competitive Protein-binding Assays (W. D. Odell and W. H. Daughaday, eds.), J. B. Lippincot Co., 1971.

107. N. E. Good, G. D. Winget, W. Winter, T. N. Connolly, S. Izawa, and M. M. S. Raizada, Biochemistry, 5:467 (1966).

108. M. Rodbell, M. J. Krans, S. L. Pohl, and L. Birnbaumer, J. Biol. Chem., 246:1861 (1971).

109. D. J. Merchant, R. H. Kahn, and W. H. Murphy, Handbook of Cell and Organ Culture, 2nd ed., Burgess, 1964.

110. H. J. Phillips and J. E. Terryberry, Exptl. Cell. Res., 13:341 (1957).

111. D. M. Neville, Jr., C. R. Kahn, A. Soll, and J. Roth, in Protides of the Biological Fluids, 21st Colloq. (H. Peeters, ed.), Pergamon Press, Oxford, 1973, p. 269.

112. C. R. Kahn, P. Freychet, J. Roth, and D. M. Neville, Jr., J. Biol. Chem., 249:2249 (1974).

113. P. Cuatrecasas, J. Biol. Chem., 246:6522 (1971).

114. I. A. Mirsky, Recent Progr. Hormone Res., 13:429 (1957).

115. H. H. Tomizawa, M. L. Nutley, T. Hiromichi, H. T. Narahara, and R. H. Williams, J. Biol. Chem., 214:285 (1955).

116. H. H. Tomizawa, J. Biol. Chem., 237:428 (1962).

117. H. M. Katzen and F. Tietze, J. Biol. Chem., 241:3561 (1966).

118. M. L. Chander and P. T. Varandani, Diabetes, 23:232 (1974).

119. G. Rosselin, R. Assan, R. S. Yalow, and S. A. Berson, Nature, 212:355 (1966).

120. J. W. Uhr, E. S. Vitetta, and U. K. Melcher, in Cellular Selection and Regulation in the Immune Response (G. M. Edelman, ed.), Raven Press, New York, 1974, p. 133.

121. J. R. Gavin, III, D. N. Buell, and J. Roth, Science, 178:168 (1972).

122. J. R. Gavin, III, D. L. Mann, D. N. Buell, and J. Roth, Biochem. Biophys. Res. Commun., 49:870 (1972).

123. G. Scatchard, Ann. N.Y. Acad. Sci., 51:660 (1949).

124. S. A. Berson and R. S. Yalow, J. Clin. Invest., 38:1996 (1959).

125. I. M. Klotz and D. L. Hunston, Biochemistry, 10:3065 (1971).

126. E. E. Baulieu and J. P. Raynaud, Eur. J. Biochem., 13:293 (1970).

127. D. Rodbard, in Receptors for Reproductive Hormones (B. W. O'Malley and A. R. Means, eds.) Plenum, New York, 1973, p. 289.

128. (a) D. Rodbard, in Receptors for Reproductive Hormones (B. W. O'Malley and A. R. Means, eds.), Plenum, New York, 1973, p. 327; (b) A. R. Midgley, Jr., A. H. Zeleznik, H. J. Rajaniemi, J. S. Richards, and L. E. Reichert, Jr., in Gonadotropin and Gonadal Function (N. R. Mougdal, ed.), Academic Press, New York, 1974, p. 416.

129. J. Roth, in Insulin Action (I. Fritz, ed.), Academic Press, New York, 1972, p. 183.

130. A. Katchalsky and P. F. Curran, Non-equilibrium Thermodynamics, Harvard Univ. Press, Cambridge, Mass., 1965.

131. I. Prigogine, Introduction to the Thermodynamics of Irreversible Processes, Charles C. Thomas, Springfield, Ill., 1955.

132. C. M. Guldberg and P. Waage, Avhandl. Norske Videnskaps Akad. Oslo, I. Mat. Naturv. Kl. 1864; Reproduced in The Law of Mass Action. A centenary volume: 1864-1964, Det Norske Videnskaps, Akademi I Oslo, Universitetsforlaget, Oslo, 1964, p. 7.

133. L. Michaelis and M. L. Menten, Biochem. Z., 49:333 (1913).

134. I. Langmuir, J. Amer. Chem. Soc., 40:1361 (1918).

135. H. Lineweaver and D. J. Burk, J. Amer. Chem. Soc., 56:658 (1934).

136. G. Scatchard, J. S. Coleman, and A. L. Shen, J. Amer. Chem. Soc., 79:12 (1957).

137. H. A. Feldman, Anal. Biochem., 48:317 (1972).

138. H. G. Weber, J. Schildknecht, R. A. Lutz, and P. Kesselring, Eur. J. Biochem., 42:475 (1974).

139. D. Rodbard and G. R. Frazier, Radioimmunoassay Data Processing, 2nd ed., National Technical Information Service, Springfield, Va., 1973, p. 224.

140. J. E. Fletcher, A. A. Spector, and J. D. Ashbrook, Biochemistry, 9:4580 (1970).

141. K. J. Catt, T. Tsuruhara, and M. L. Dufau, Biochim. Biophys. Acta, 279:194 (1972).

142. P. Cuatrecasas, Ann. Rev. Biochem., 43:169 (1974).

143. E. T. Harper, J. Theoret. Biol., 39:91 (1973).

144. A. V. Hill, J. Physiol., 40:iv (1910).

145. R. Sips, J. Chem. Phys., 16:490 (1948).

146. W. E. L. Brown and A. V. Hill, Proc. Roy. Soc., B94:297 (1923).

147. J. Wyman, Advan. Protein Chem., 4:407 (1948).

148. J. Wyman, Advan. Protein Chem., 19:223 (1964).

149. J. Wyman, J. Mol. Biol., 11:631 (1965).

150. J. Wyman, J. Amer. Chem. Soc., 89:2202 (1967).

151. J. Monod, J. Wyman, and J. P. Changeux, J. Mol. Biol., 12:88 (1965).

152. M. M. Rubin and J. P. Changeux, J. Mol. Biol., 21:265 (1966).

153. J. Berkson, J. Amer. Statist. Assoc., 39:357 (1944).

154. W. D. Ashton, "The logit transformation, with special reference to its use in bioassay," Griffin's Statistical Monographs and Courses, No. 32 (A. Stuart, ed.), Hafner, New York, 1972.

155. L. J. Reed and J. Berkson, J. Phys. Chem., 33:760 (1929).

156. D. Rodbard, P. L. Rayford, J. Cooper, and G. T. Ross, J. Clin. Endocrinol., 28:1412 (1968).

157. R. N. Frank and D. Rodbard, Arch. Biochem. Biophys., 171:1 (1975).

158. D. Rodbard and D. M. Hutt, in Proceedings, Symposium on radioimmunoassay and related procedures in clinical medicine, Int. Atomic Energy Agency, Vienna, Austria (Unipub, New York), 1974, p. 165.

159. D. Rodbard, Endocrinology, 94:1427 (1974).

160. D. Rodbard, W. Moyle, and J. Ramachandran, in Hormone Binding and Target Cell Activation in the Testis (M. L. Dufau and A. J. Means, eds.), Plenum, New York, 1974, p. 79.

161. J. M. Ketelslegers, G. D. Knott, and K. J. Catt, Biochemistry, 14:3075 (1975).

162. D. E. Koshland, Jr., G. Némethy, and D. Filmer, Biochemistry, 5:365 (1966).

163. A. Goldbeter, J. Mol. Biol., 90:185 (1974).

164. J. Hertzfeld and H. E. Stanley, J. Mol. Biol., 82:231 (1974).

165. A. Conway and D. E. Koshland, Jr., Biochemistry, 7:4011 (1968).

166. A. Levitzki and D. E. Koshland, Jr., Proc. Natl. Acad. Sci. U.S.A., 62:1121 (1969).

167. B. W. Matthews and S. A. Bernhardt, Ann. Rev. Biophys. Bioeng., 2:257 (1973).

168. M. Lazdunski, Progr. Bioorg. Chem., 3:81 (1974).

169. R. N. Ferguson, H. Edelhoch, H. A. Saroff, and J. Robbins, Biochemistry, 14:282 (1975).

170. H. A. Saroff, Science, 175:1253 (1972).

171. H. A. Saroff, Biopolymers, 12:599 (1973).

172. C. Frieden, in The Regulation of Enzyme Activity and Allosteric Interactions, Universitets Forlaget, Oslo; and Academic Press, New York, 1968, p. 59.

173. A. Levitzki and J. Schlessinger, Biochemistry, 13:5214 (1974).

174. K. C. Ingham, H. A. Saroff, and H. Edelhoch, Biochemistry, 14:4745 (1975).

175. A. Levitzki, J. Theoret. Biol., 44:367 (1974).

176. P. De Meyts, J. Roth, D. M. Neville, Jr., J. R. Gavin, III, and M. A. Lesniak, Biochem. Biophys. Res. Commun., 55:154 (1973).

177. P. De Meyts, A. R. Bianco, and J. Roth, J. Biol. Chem., 251:1877 (1976).

178. P. De Meyts, J. Roth, D. M. Neville, Jr., and P. Freychet, Endocrinology, 94:Suppl. 267 (1974).

179. P. De Meyts, J. R. Gavin, III, J. Roth, and D. M. Neville, Jr., Diabetes, 23, Suppl. 1:355 (1974).

180. P. De Meyts, in Cell surface receptors, abstracts volume, ICN-UCLA Winter Conferences on Cellular and Molecular Biology, March 2-7, 1975.

181. P. De Meyts, J. Supramol. Struct., 4:241 (1976).

182. W. A. Frazier, L. F. Boyd, and R. A. Bradshaw, J. Biol. Chem., 249:5513 (1974).

183. W. A. Frazier, L. F. Boyd, M. W. Pulliam, A. Szutowicz, and R. A. Bradshaw, J. Biol. Chem., 249:5918 (1974).

184. L. Kohn and R. Winand, in Molecular Aspects of Membranes Phenomena (H. R. Kaback, G. Radda, and R. Schwyzer, eds.), Springer-Verlag, Heidelberg, in press 1975.

185. Q. H. Gibson and F. J. W. Roughton, Proc. Roy. Soc., B143:310 (1955).

186. P. Cuatrecasas, J. Biol. Chem., 248:3528 (1973).

187. P. Cuatrecasas and G. P. E. Tell, Proc. Natl. Acad. Sci. U.S.A., 70:485 (1973).

188. K. Megyesi, C. R. Kahn, J. Roth, and P. Gorden, J. Clin. Endocrinol. Metab., 38:931 (1974).

189. J. Roth, Diabetes, 23, Suppl. 1:353 (1974).

190. S. J. Singer and G. L. Nicolson, Science, 175:720 (1972).

191. L. Jarett and R. M. Smith, J. Biol. Chem., 249:7024 (1974).

192. L. Orci, C. Rufener, F. Malaisse-Lagae, B. Blondel, M. Amherdt, D. Bataille, P. Freychet, and A. Perrelet, Israel J. Med. Sci., 11:639 (1975).

193. A. Levitzki, L. A. Segel, and M. L. Steer, J. Mol. Biol., 91:125 (1975).

194. A. S. Soll, C. R. Kahn, and D. M. Neville, Jr., J. Biol. Chem., 250:4702 (1975).

195. J. A. Archer, P. Gorden, J. R. Gavin, III, M. A. Lesniak, and J. Roth, J. Clin. Endocrinol. Metab., 36:627 (1973).

196. J. A. Archer, P. Gorden, and J. Roth, J. Clin. Invest., 55:166 (1975).

197. P. Freychet, M. H. Laudat, P. Laudat, G. Rosselin, C. R. Kahn, P. Gorden, and J. Roth, FEBS Lett., 25:339 (1972).

198. C. R. Kahn, D. M. Neville, Jr., and J. Roth, J. Biol. Chem., 248:244 (1973).

199. J. R. Gavin, III, J. Roth, D. M. Neville, Jr., P. De Meyts, and D. N. Buell, Proc. Natl. Acad. Sci. U.S.A., 71:84 (1974).

200. J. Roth, C. R. Kahn, M. A. Lesniak, P. Gorden, P. De Meyts, K. Megyesi, D. M. Neville, Jr., J. R. Gavin, III, A. H. Soll, P. Freychet, I. D. Goldfine, R. S. Bar, and J. A. Archer, Recent Progr. Hormone Res., Vol. 31, "Proceedings of the 1974 Laurentian Hormone Conference," 1975, p. 95.

201. P. Freychet, D. Brandenburg, and A. Wollmer, Diabetologia, 10: 1 (1974).

202. J. Gliemann and S. Gammeltoft, Diabetologia, 10:105 (1974).